Clash of Spirits

THE HISTORY OF POWER AND
SUGAR PLANTER HEGEMONY
ON A VISAYAN ISLAND

Filomeno V. Aguilar, Jr.

University of Hawai'i Press
Honolulu

© 1998 University of Hawai'i Press
All rights reserved
Printed in the United States of America

03 02 01 00 99 98   5 4 3 2 1

Library of Congress Cataloging-in-Publication Data

Aguilar, Filomeno V.
    Clash of spirits : the history of power and sugar planter hegemony on a Visayan island / Filomeno V. Aguilar.
      p.   cm.
    Includes bibliographical references (p.   ) and index.
    ISBN 0-8248-1992-6 (cloth : alk. paper). — ISBN 0-8248-2082-7 (pbk. : alk. paper)
    1. Negros Island (Philippines)—Civilization. 2. Negros Island—(Philippines)—Social conditions. 3. Negros Island (Philippines)—Economic conditions.   4. Sugar trade—Philippines—Negros Island.   5. Plantations—Philippines—Negros Island.   I. Title.
DS688.N5A34   1998
959.9′5—DC21   98-16629
                                      CIP

University of Hawai'i Press books are printed on acid-free paper and meet the guidelines for permanence and durability of the Council on Library Resources.

Designed by Omega Clay

May kaibahan na pagguirumdom asin pagmabalos ki Papa na yaon na sa Kagurangnan

Contents

Preface ix *Introduction* 1

PART I · Colonial Enchantments, Indigenous Contests

1 A Clash of Spirits: Friar Power and Masonic Capitalism 15
2 Cockfights and *Engkantos:* Gambling on Submission and Resistance 32
3 Elusive Peasant, Weak State: Sharecropping and the Changing Meaning of Debt 63

PART II · The World of Negros Sugar after 1855

4 The Formation of a Landed *Hacendero* Class in Negros 97
5 "Capitalists Begging for Laborers": *Hacienda* Relations in Spanish Colonial Negros 126
6 Toward Mestizo Power: Masonic Might and the Wagering of Political Destinies 156
7 The American Colonial State: Pampering Sugar into an Agricultural Revolution 189

Abbreviations 229 *Notes* 231

References 275 *Index* 305

Preface

With the deepening rapprochement between history and the social sciences, the past is no longer what it used to be. The past has ceased to be viewed as merely another country: the past lives on as an arena of debate and a flourishing field of interpretations. In the particular case of societies that were under the sway of imperial colonial powers, local scholars often write history with the clear intent of meeting the emotional demands of the nation. It must be a usable past, as Renato Constantino has said.

Usually having received a Western-style education, the historian of and from a former colony confronts a host of challenges. History is written to edify the modern nation and speak to a culture whose traditional forms and aims of retelling the past differ from those of the West, while at the same time abiding by certain canons of Western historiography. These aims may be difficult to reconcile, although Soedjatmako's argument about the polyinterpretability of history may be comforting. In writing this book, however, I found that history's polyinterpretability can be located not only in a range of authors who represent different perspectives but even within myself as crafter of the historical narrative.

In the course of my research but especially during the period of actual writing, I found that moments of uncertainty, speculation, and imagination were interspersed with moments of objectivity, empirical certainty, and factual rectitude. Moments of theory and logic welded with moments of sentiment and invention. Moments of ironical detachment joined moments when I felt the great need to thresh out misconceptions. Moments when I sprang to the defense of my natal land mingled with moments when I consciously fought the inordinate celebration of the indigenous. Moments when I wanted to glare back at the colonial gaze intersected with moments when I sought to overcome the traps of colonialist discourse. Moments when I was convinced that, despite the hazards of translation, local and national histories must be understood on their own terms rather than through Western lenses combined with those when I was also convinced such histories would be the poorer if they remained parochial and insular. Moments when I believed the past to be recoverable competed with those when I felt history is all fiction. These various moments converge in the text which, in its entirety, is of course meaningful to me. Despite its many failings, which are wholly my own, and notwithstanding errors I did not detect, which should not be held against the people whose names appear in the following paragraphs, I hope that, if not in toto, various sections and aspects of the text will be found meaningful by others as well.

Luang Wichit Watthakan suggested that the historian is more than God because, with the might of a pen or a computer keyboard, he or she can change the past. Despite my inadvertent mistakes and lapses, my avowed goal could not be farther from the Thai intellectual's suggestion of playing God. Rather, as someone struggling for existential consistency as a Christian, I find one biblical incident during the claiming of the Promised Land (Josh. 5:13–14) highly instructive. After crossing the Jordan and just before leading a bloody conquest, Joshua sees "a man standing in front of him with a sword drawn in his hand." During this vision, he inquires of the man, "Are you for us or for our enemies?", to which the response is "Neither." In the case of this book's narrative, the standpoint I take is encapsulated in that response. Not that such matters as identities and allegiances are unimportant, for they definitely are. But, despite the incomprehensibilities of the past, it is my personal belief that the purposes of the Lord of history, whose thoughts are higher than my thoughts, are being accomplished in rather unfathomable ways. This belief I find to be liberating because within its bounded limits, I gratefully take the liberty of exploring and reinterpreting the past.

The making of this work, which was originally presented as a doctoral thesis at Cornell University, has its own little history with a number of unexpected twists and turns. My first three weeks of archival diggings, which happened to be in London, had to be cut short by the death of my beloved father in Naga, in the Philippines. After a period of mourning, I returned to Europe and was surprised to receive the generosity of Rene Salvania, who selflessly shared with me a handwritten catalog of Philip-

Preface

pine documents he had compiled at the Archivo Historico Nacional. Sadly, this catalog is now irretrievably lost, and the only other person to have used it is Xavier Huetz de Lemps when he was in Madrid. I caught up with Xavier at the Philippine National Archives where we pored over documents amid the crippling brown-outs that then gripped Manila; to him I owe special thanks for sharing with me a bundle of documents unenticingly classified as "Cargadores" that turned out to contain crucial materials on Negros.

At various stages of research, writing, the search for a publisher, and beyond, I was the recipient of various acts of kindness, intellectual stimulation, and warm encouragement from colleagues, teachers, and friends to whom I am very grateful. They include Benedict Anderson; Belinda Aquino; the late Milton Barnett; Coeli Barry; Geoffrey Benjamin; Dominique Caouette; Jose Cruz, S.J.; Maribeth Erb; Shelley Feldman; Hing Ai Yun; Napoleon Juanillo, Jr.; James Jesudason; Anita Kendrick; Benedict Kerkvliet; Alicia Magos; Philip McMichael; Lily Mendoza; Leoncio and Virginia Miralao; the late Rene Ocampo, S.J.; Mina Roces; James Rush; Clifford and Jane Scherer; Takashi Shiraishi; William Sunderlin; Marjatta Tolvanen; Benito Vergara, Jr.; Edgar Wickberg; Paul Yamauchi; Liren Zheng; and Julie Zimmerman. Takashi, Shelley, and Philip served on my graduate committee, Takashi giving me unforgettably bold instructions, and Shelley and Philip tolerating my sociological idiosyncrasies and foibles.

The many informants in Negros who shared with me their stories and points of view cannot be identified, but this study would not have taken its present contours had they not been generous with their trust, which I have kept to the best of my ability. Virgilio Aguilar, Rowena Bañes, Lyndon Caña, Jhoana Cañada, Mario and Violeta Gonzaga, Ramon Lachica, Mercy Lapuz, Nene and Susan Losande, Maribel Lozada, Manette Mapa, Cecille Nava, and Raymundo Pandan, Jr., were invaluable in facilitating and enlivening my research with their readiness to help and their good humor and warm friendship. In welcoming me to her home in Bacolod, Lourdes Dionio Montelibano treated me like a dear nephew. To these wonderful informants and friends in Negros, *daghan salamat gid*. In Europe, overseas Filipino workers took pity upon a miserable graduate student, and I am truly indebted to Nita and Mary Morta and Susan Pacleb in London, and to Noli and Ellen Ramones in Madrid, for succoring me: *Dios iti agngina*.

My teachers at the Ateneo and in Swansea have left an indelible imprint in the ways I perceive and analyze the world. How could one not be influenced by the late Horacio de la Costa, S.J., and by Gavin Kitching, Edna Manlapaz, Mary Racelis, and Ramon Reyes? They made learning exciting and challenging, and their teaching skills are among the best I have experienced. I hope this work does not terribly disappoint them. My earliest teacher was my late father who never even stepped into secondary school; he taught me to listen and to observe, and he taught me for life. Affectionately calling me "Dr. Laway," Mama has been supportive, patient, and loving; Manoy Pabling's interest in my stories from the field prompted him to say I should write two books instead of one; Manay Ciony and Kuya Eddie never tire of warmly welcoming me to Marikina in my countless comings and goings; Manay Nini proudly displays my graduation photo; Manoy Santy and Manoy Boy in their own ways take pride in what I do. Jun-Jun makes me feel like the coolest uncle in the world. Even though unnamed for lack of space, numerous friends in and out of academia have been unswerving in their moral support.

Research for this study was supported by an International Doctoral Fellowship grant from the Social Science Research Council and the American Council of Learned Societies, with funds provided by the William and Flora Hewlett Foundation. With their generous support, I had a smashing time conducting research in four countries. In addition to its exciting intellectual environment and supportive staff, Cornell's Southeast Asia Program provided financial assistance for a year of writing, and even lavished upon me a Lauriston Sharp Prize. In the course of my teaching sojourn at the Department of Sociology, National University of Singapore, I had the opportunity to ventilate some of the ideas contained here. The manuscript received its final polish in the congenial atmosphere of the School of History and Politics, James Cook University.

At the University of Hawai'i Press, Pamela Kelley patiently nurtured this project even as I moved from Singapore to north Queensland. To the anonymous reviewers at this and another press, many thanks for their constructive comments and helpful suggestions. Because a number of individuals are mentioned by name, particularly in Part II, I should like to say that no offense is meant against any of them or their relations nor is this work intended to cast aspersions upon them.

Although no single chapter of this book as presently constituted has been published previously, various sections have earlier appeared in print. I have been richly rewarded by the kind invitation from James Eder and Robert Youngblood to present a paper at a conference in Tempe, which served as the first forum for presenting some of the ideas in this study. That paper appears as a chapter in the book they coedited, titled *Patterns of Power and Politics in the Philippines: Implications for Development* (1994). My thanks to the Program for Southeast Asian Studies of the Arizona State University for permission to use materials from that chapter. Similarly, parts of chapters 5 and 7 have previously been used in an article that appeared in the October 1994 issue of the *Journal of Peasant Studies;* I am grateful to Frank Cass and Company for permission to reprint those materials here. During the 1996 International Conference on the Centennial of the Philippine Revolution held in Manila, the warm reception of my views on what transpired in Negros was heartening. That paper has appeared in the September 1997 issue of the *Journal of Southeast Asian Studies;* I thank the publisher, the National University of Singapore, for permission to reuse materials now contained in chapters 1 and 6. The cover illustration is taken from volume 6 of *Filipino Heritage: The Making of a Nation* (Lahing Pilipino Publishing, Inc., 1978), and I am grateful to its editor-in-chief, Alfredo Roces, for his kind permission to use it here. Unless otherwise stated, all translations into English are my own.

Introduction

In the late 1960s a steam-filled tank at the Victorias centrifugal sugar mill on the island of Negros exploded, killing about sixteen unsuspecting workers. More than two decades later, I was told the mishap occurred because of the then new management's refusal to conduct the *daga* rite of "baptizing" new machinery with the ritual blood of chicken; the factory spirits were upset that they had not been propitiated. After this episode, the management was said to have relented and has since sponsored the periodic performance of *daga* and other rites of appeasement. Workers attest that no devastating mechanical failures or accidents have since occurred. It is essential to pay respect to the spirits, they assert, which is why the *daga* is performed in all sugar mills, and analogous offerings are made on the farms. They go on to explain that, at Victorias, the mill's engineer from its inception had been Sinyor Norberto, but—alas—he passed away in 1952. Norberto had friendly relations with spirits, which allowed the factory to gain stature and amass wealth. There had been a plan to construct a waterway made of steel and concrete that would have straightened and totally altered a small crooked river within the sugar estate, but Norberto objected because that river was the passageway of a golden vessel—whose whistle blows could be heard but was itself invisible to the ordinary eye—which shipped sugar and brought riches to the mill. Because he respected the spirit-world, Sinyor Norberto was honored with magical prowess, and therefore wealth. One of his abilities was knowing exactly when a particular piece of equipment would malfunction, an insight he could obtain by simply gazing at the smoke billowing out of the sugar factory's chimney. Once Norberto was gone, practices that offended the spirit-world were introduced by the new modernizing management. It took the loss of many lives, according to my informants, before management accepted the wisdom of appeasing the spirits and conforming with tradition.

This and countless other stories were narrated to me in the course of my research in Negros, one of the Visayan islands in the central Philippine archipelago (see Figure 1). They were told by ordinary people, by students and professionals, by local healers, by farm and factory workers, as well as by those sympathetic to or even affiliated with the Communist Party's New People's Army. These stories, which are part of the locality's vibrant oral tradition, convey a profound understanding that integrates the realities of capitalism with those of the spirit-world. In this view, there is no contradiction or dissonance in resisting, if not fighting, the established social order and simultaneously holding the spirit-world in awe and fear. A comparable bridging of the twin dimensions of the palpable and the unseen is absent, however, in the academic and popular literature on Negros.

Since the 1960s, news reports, radical studies, and scholarly writing on this predominantly sugar-producing island have tended to focus on the poverty of its sugar workers, the inequitable labor relations, the debilitating dependency of workers on employers, and the injustice suffered by the masses at the hands of the wealthy and powerful. Collectively they express moral indignation at the iniquities of an extremely lopsided social system. They also communicate a genuine search for alternatives to a setup dominated by "sugar barons" who had gained immense wealth from exports to the U.S. market since early this century and wielded formidable political influence, which was shattered only by the Marcos regime's imposition of a government sugar-trading monopoly in the late 1970s. Despite its merits, this literature (except for highlighting the ostentatiousness of consumption behaviors by elites) has been virtually silent about such cultural elements as the belief in spirits, the love of cockfighting and gambling, the workers' odd reference to their planters and employers as *"toto"* (brother) and *"inday"* (sister), the acceptability of wealth obtained by whatever means, and the constant sizing up of others to pigeonhole them in a mental hierarchy, among others. Whenever noticed, such cultural features have been brushed aside, usually as the stifling effects of feudalism or the remnants of a passing tradition. The enigmatic combination of such cultural practices with capitalist profit making, international trade, wage labor, and state power—features associated with a different sort of rationality—has scarcely received atten-

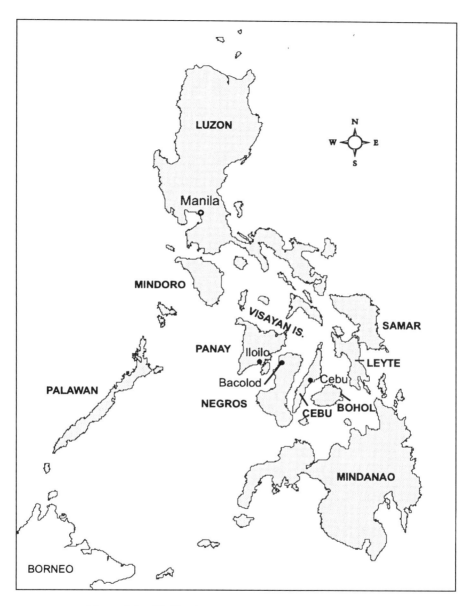

FIGURE 1. Map of the Philippines

tion. The hybrid configuration that Negros represents is what this study seeks to understand. It weaves together the various strands of politics, economics, and culture even as it aims to comprehend the seamless complexity of social life. Without privileging either cultural or structural analysis, this work seeks to fuse the two in the form of an analytic historical narrative.

Fluidity and the Dialectics of Culture and Structure

Undoubtedly, the island of Negros presents an apparent paradox. Underpopulated and virtually unexploited until the opening on adjacent Panay Island of the port of Iloilo to world trade in 1855, Negros ceased to be a terra incognita and became the primary sugar producer and exporter in the colonial Philippines. Its sugar farms, commonly known as *haciendas,* were worked by migrants from adjacent islands, who labored for a similarly migrant group of planters known as *hacenderos.* Emerging during the era of a rapidly expanding global capitalism, the sugar *haciendas* adopted a mode of production both advanced and archaic. However, this perceived anomaly is largely an artifact of earlier studies.

Among the first to analyze the Negros phenomenon was Alfred McCoy (1982a, 323–326), who has claimed that "By the century's turn the Negros plantation economy had evolved into a social system without parallel elsewhere in the Philippines or Southeast Asia," the *hacenderos* relying upon "capitalist instincts" by employing "supervised work-gangs paid a nominal daily wage" while concomitantly using "debt bondage" and "corporal punishment" to control labor. This view has been echoed by several other writers (Gonzaga 1987, 16, 1991, 13; Hawes 1987, 86; Rivera 1982, 2–4).[1] Recently, John Larkin (1993, 60, 81) has advanced a similar proposition: in that era of "unfettered capitalism," labor was reduced to the wage nexus and posed nothing more than "incidental difficulties" to the Negros planter who, as "the unchallenged master," was said to have "ruled absolutely."

The conventional understanding of Negros rests upon the assumption that the planters enjoyed solid and unequivocal dominance as soon as the *haciendas* were formed in the late nineteenth century. Consequently, Negros society has been seen as unchanging, an ossified view that closely approximates the more general perception of the Philippine class structure as dominated by an immovable, permanent, even eternal, landed elite whose rule is sanctioned by tradition. Continuities overshadow change, as in one book title's succinct description, borrowing novelist F. Sionil Jose's phrase, of the Philippines as *A Changeless Land* (Timberman 1991). There is a "timelessness," David Timberman laments, in the personalistic politics, and a "sad constancy to the poverty, inequity, and injustice that characterize Philippine society" (xii). To the same pattern but from a more sanguine outlook, Frank Lynch (1979, 44) has written that the Big People and the Small People, "the wealthy and the poor, the patrons and the clients of the Philippines, have lived in symbiotic union through the centuries." Elements of that relationship are postulated as givens: debt, for example. The cornerstone of the tenant–landlord relationship that has glued society together, debt, Larkin argues, "is a phenomenon woven into the fabric of Philippine society and is prehispanic in origin" (Larkin 1993, 110). Molded in antiquity, elements of society are seen as essentially unchanging. Thus, historical studies unwittingly end up having a static and ahistorical quality about them.

In this study's analytic narrative, however, the dominance of Negros planters—and, by extension, the ruling classes of the Philippines—is problematized rather than assumed. Contrary to its image of a granite-like monolith, the social structure is portrayed as porous, fluid, and in dynamic movement. The complex set of social relations that make up society, rather than fixed, are pictured as flowing in time and space, and continually subject to contestation by diverse actors. The purpose is to see the social structure in motion and thereby to understand the continuous processes of the formation of classes and systems of stratification. The apparent continuity of given social phenomena is questioned in order that, apart from its visible edifice, one may also observe the hitherto unnoticed fluidity of social life.

In addition to the problem of social fixity, studies of Negros have also failed to give its peoples sufficient voice in their own history; social actors are presented as consigned to their designated role as pawns of unstoppable economic forces. With reference to the first British vice-consul who played a strategic role in the emergence of Negros' sugar industry, McCoy, for instance, declares that "If we are to simplify a complex causality, then Nicholas Loney was not the architect of Iloilo's prosperity, he was its assassin" (McCoy 1982a, 298).[2] The proposition that one man can cause everlasting havoc (or, for that matter, progress) is an example of an extreme determinism that, like a bulldozer, flattens all historical beings in its path. Characteristic of other works on the Philippines, this paradigmatic approach reduces human beings to blind surrogates of powerful external economic forces; local actors bear no responsibility even as they are denied human agency. Seeking a break from determinism, this study strives to depict local peoples, planters and workers alike, as actively forging *hacienda* life amid the sway of macro-historical forces.

In the Tagalog area, the voice of the ordinary peasant has found eloquent expression in Reynaldo Ileto's *Pasyon and Revolution* (1979). In other parts of the Philippines, however, no comparable study has been written. In the literature on Negros, the one historical voice of local peoples that stands out is the so-called *babaylan* movement, which, under the leadership of native shamans, took center stage at the turn of the century. But the magical worldview that animated this movement is presented in two questionable ways: first, as forming a great divide; magic and its perceived disorders belong to shamans and farm workers, whereas rationality and legal order belong to the westernized Roman Catholic elite planters (McCoy 1982b, 173, 1992, 108) [3] (in effect, two distinct cultures are manifest in predictable political alignments—a dualism that parallels the oversimplified two-class image of Philippine society); second, as springing into action only in the form of millenarian movements that, in turn, are considered as issuing from the inability of peasant culture to catch up with the demands of modernity, a case of consciousness lagging behind structural change, resulting in stress and strain (Cullamar 1986, 73, 77; cf. Ileto 1992, 234; McCoy 1982b, 166).

On the whole, studies of Negros represent the culture by which the peoples of Negros have viewed life and participated in history as splintered and incidental. Not accorded a primacy of its own, but rather considered the by-product of the economic base, culture is seen in such studies as merely responding to economic triggers that allow it to break through the surface during certain, otherwise transitory, interludes. Thus consigned to the margins of history, culture is treated as having no significant bearing upon the formation, reproduction, and transformation of social relations.

Accounting for the role of culture has been problematic not just in studies of Negros but in other attempts to understand the Philippines, a condition signaled by the ease with which James Fallows' (1987) vacuous phrase "damaged culture" has gained wide acceptance. Renato Constantino (1985) proffers the term "synthetic culture" to suggest its character as a mediating channel that distorts reality for "miseducated Filipinos," the hapless victims of American cultural imperialism. These views denigrate culture for being either the culprit responsible for national problems or for somehow being unreal or wanting in authenticity, leading to oppression. Both views deny the role culture plays in history and, given the predilections of nationalist ideologies, fuel the romanticized search for cultural purity.

In this book, the historic role of culture is based on the fundamental view in the field of anthropology that culture provides tools for appropriating the world, for interpreting reality and temporal existence, as well as for organizing, legitimating, and at other times undermining, the relations human beings sustain among themselves and with the world.

Economic and other social activities would not be possible without culture. Moreover, rather than treating culture as peripheral, this study starts from the premise that culture is dialectically intertwined with political economy. It seeks to show that the processes of culture formation are inseparable from those of class structuring and state formation. Furthermore, precisely because of such inseparability, culture is both unified and differentiated. Because culture is acted out and acted upon by social actors who may belong to the same society but are located in different structural positions, history yields a cultural tree with many branches, natural and grafted, that nonetheless possess internal articulation. By emphasizing the linkage of culture with politics and economics, this study also constructs a narrative that, unlike the current vogue in cultural studies, presents culture as far from disembodied from the materiality of society and history.

Given this preferred view of class formation, culture, and political economy, I eschew the economistic definition of the concept of relations of production.[4] Instead, I follow Derek Sayer, who takes this fundamental analytic category as including "any and all social relations which are demonstrably entailed in a given mode of production" (Sayer 1987, 75), hence any and all material and ideational activities implicated in a particular mode of producing and reproducing social existence. From this perspective, production relations encompass the totality of social, cultural, political, and ecological relationships that provide the conditions for the production of subsistence, a surplus and a way of life. They encompass the conditions by which surpluses, physical and symbolic, are claimed and appropriated by certain groups and classes from others in society, and the conditions by which differential control and privilege are maintained and prolonged, or challenged and transformed.

In showing culture as integral to the structuring of production relations and the fashioning of the social world in Negros, this book endeavors to give voice to different categories of social actors as they have played their parts in the social game and sought to alter the rules by which such games are played in history. It thus reconstructs indigenous systems of knowledge and thought by which local peoples have perceived, resisted, accommodated, and maneuvered through the power relationships and the world-historical forces that defined the social matrix in which they had to lead their lives. In large part, the study attempts to reenter the minds of past generations by analyzing myths and folk concepts, ideational factors intimately related to the modes of producing and reproducing social life.

In recreating the voices of the past in the context of formidable global structures, this narrative grapples with the problematic of agency and structure, what Philip Abrams (1982, xiv) calls the "unbudgeable fulcrum of social analysis."[5] In the enduring legacy of Karl Marx (1963 [1852], 15), the unforgettable lines of *The Eighteenth Brumaire of Louis Bonaparte* remind us that "Men [and women] make their own history, but they do not make it just as they please, they do not make it under circumstances chosen by themselves, but under circumstances directly encountered, given and transmitted from the past." Affirming that human beings make history within definite bounds of historical possibilities, this study shows generations of social actors who, though facing profound constraints, chose to act under the given circumstances. The narrative hopes to elucidate the paradox that individuals historically constitute society even as society historically constitutes its individual members, that individuals shape and create societal structures even as society imposes its structures upon individuals who are thereby shaped by it. Following Abrams, I take the process of structuring as the key to the patterning of society, a process in which agents and structures become mutually determining.

As people engage with the present, they invent and reinvent social practices that are forged and eventually bequeathed to future generations. In an unintended consequence, the products of human activity become, in the illuminating exposition of Gavin Kitching (1988, 46), the wellspring of alienation for later generations for whom, after some indeterminate period, the earlier actions, innovations, and creative solutions appear as autonomous, given phenomena that constrain, coerce, or oppress those who inherit them. But social action does not cease. A later sociological generation, whose identity is assembled on the basis of a different system of meanings and possibilities (Abrams 1982, 256), will endeavor to modify, even overhaul, the very structures meaningfully erected by an earlier generation. Seeking change, people act in ways that eventuate in pivotal conjunctures, which set the context for another generation to arise.

The reconstruction of such pivotal conjunctures is a major task of this study. Formed by an event or interrelated set of events, each conjuncture leads to a reshaping of lived experiences and social structures as reflected in changes in modes of thinking, in ways of behaving, and in systems of producing. Such conjunctural transitions are marked by both progressive and negative aspects: change is both liberating and alienating, a proposition that may not be inconsistent with indigenous thought that, as discussed in the text, takes reality as suffused by complementary oppositions. Every conjuncture then contains its own inherent contributions and contradictions, which may remain entrenched until another pivotal conjuncture occurs that various groups of another sociological generation will attempt to seize to their advantage. Moving from one conjuncture to the next, this analytic narrative strives to illumine the structure–agency dynamic, the inseparability of culture and political economy, and the fluidity of social structures.

Archives, Myths, and Folk-Historic Categories

In writing this story, I pieced together information, insight, and inferences derived from archival documents, published materials, key-informant interviews, and field observations. I consulted archives in Bacolod, Chicago, London, Madrid, Manila, Marcilla, and Washington, D.C., as well as libraries in various locations ranging from Iloilo to Ithaca, from Sto. Tomas in Manila to St. La Salle in Bacolod and Silliman in Dumaguete, from Madrid's Biblioteca Nacional to Manila's National Library. In Negros Occidental Province, I observed rituals and listened to fragments, sometimes whole chunks, of folklore during fieldwork in 1990. Some fifty-seven respondents and informants were formally interviewed in at least nine towns of Negros Occidental. I have also learned much from talking to friends and strangers alike in the course of my repeated visits to Negros since the early 1980s. While in the Visayas, I began to formulate reasoned conjectures and plausible hypotheses, for what I saw and heard propelled me to reimagine the collective thoughts that informed past social relations and ways of life in this and other parts of the Philippines. In so doing, I reviewed the broad patterns of Philippine history from a non-Tagalog position, from a periphery of the periphery.

The outcome of that revision is Part I of this study, which offers a reinterpretation of the broad currents of history that flowed through the islands colonized by Spain from the late sixteenth century. The limited historical sources for the propositions I make, particularly as these relate to Negros Island prior to the advent of its sugar economy in the second half of the nineteenth century, have led me to sketch a general portrait, an ideal type of events that I believe encompassed the islands to which Negros was definitely linked. But the Visayan Islands, including Negros, did have their own stories to tell. Whenever possible, documentary materials pertaining to the Visayas and to Negros were utilized for the reconstructions in Part I. With the establishment of the sugar economy after 1855, as presented in Part II, Negros would appear to have its own autonomous history. The more abundant source materials certainly made it easier to discuss Negros separately from other parts of the Philippines. During this period, however, its history would be even more incomprehensible without some appreciation of its deepening linkages with neighboring islands, not least of all Panay, whence Negros sugar was exported through the Iloilo port. On the whole, therefore, this study treats Negros as refracting certain pervasive themes in Philippine history, even as its sugar-growing areas telescoped the processes of change arising from closer integration with global capitalism.

Moreover, the Negros I encountered in the early 1990s proved to be a most stimulating platform for an examination of its past, a past that I realized I shared despite my having originated from a different ethnolinguistic group in Luzon. Consequently, Negros offered me a sense of detachment as well as intimacy, especially as I began to listen to myths and legends that, although not part of my original research plan, were too compelling to be ignored. Stories about sugar planters with horns, who flogged Christian crosses, who flew on magical horses, alerted me to rethink the spirit-world, the realm of the *taong-lipod* with which I had grown up in Bikol but which, when I became a Christian, I learned not to fear but also to ignore. By taking the spirit-world as an analytical field, however, I revised my understanding of Negros history and various aspects of broader Philippine history. The resulting narrative of this study thus blends more conventional written sources with transcriptions of tales recounted from informants' memories, as well as my own interpretations.

The methodological import of myths and legends is underscored by Terence Turner (1988), who argues in the South American context that cultural forms in their mythical genre provide indigenous peoples with the ingredients to form historical awareness and explain their own participation, as human agency, in complex patterns of action that render both submission and resistance possible. Edward Thompson (1977) has also advanced the case for the critical use of folklore in social history to recover past states of consciousness and the texturing of social beings. Rather than being denigrated as false consciousness, myths and legends can be seen as dynamic components of social structuring. As I hope to show, they have been extremely salient in shaping social relations in Negros. And since apparently no power system is completely hegemonic or totally alienating, certain elements of culture can lead to the uncanny disclosure of deeper layers of meaning in society.

Admittedly, social memories are problematic, and reminiscences are circumscribed by the social and political circumstances of the past as well as of the present. Moreover, for my own part, I am cognizant that I have reimagined the past from the perspective of, and the need to understand, the present. Unavoidably, it is the present that organizes my argument and provides analytical devices to reconceive the past, combine muted voices in the archives with the remembrances of folklore, and find a pattern in the myriad of spoken and written words. Notwithstanding my temporal and personal limitations and the historicities of my sources, I hope I will be excused for peering into the past. In this undertaking, I have utilized popular ideas, or folk-historic categories of lived experience, which I consider integral to the analytic reconstruction of pivotal conjunctures. Evinced by everyday words, sometimes by recorded history, these folk-historic categories of thought include: the concepts of luck and gambling; of spirits and power encounters; of debts, gifts, and social rank; and the idea of capital as evil or, as we shall see, Masonic. Providing glimpses into indigenously rooted notions of reality and the world, these folk-historic concepts are interrelated and change in ways that shed light upon the ebb and flow in the economic, cultural, and power relations between groups and classes in society.

Of the various folk-historic concepts, capital's filiation with evil and Masonry fulfills in many ways the function of a guiding thread through the narrative of social structuring in Negros. The local discourse revolving around what I call Masonic Capitalism situates Negros within the universal discourse of capital-as-evil. This construct has taken disparate forms: the belief in *Supay*, who was transmogrified into the devil and propitiated by miners in Bolivia, analyzed by June Nash (1979); the devil belief among proletarianized black peasants of Colombia's Cauca Valley, essayed by Michael Taussig (1980); the *hantu* (evil spirits) that attack factory women in Malaysia and Indonesia, discussed by Aihwa Ong (1987), Diane Wolf (1992), and others (Ackerman and Lee 1981; Lim 1978, 32–33); the proposition by an Islamic Communist in 1920s Java that capitalism is Satan's work, elucidated by Takashi Shiraishi (1990); the reaction by farmers in the late-nineteenth-century United States to futures trading in commodities as gambling hells, narrated by Ann Fabian (1990); and the literary works of Wordsworth, Dickens, T. S. Eliot, and others portraying the city in the Industrial Revolution as dark, monstrous, and satanic, expounded upon by Raymond Williams (1973).

From poets to radicals, from rural cultivators to factory workers, the imagery of the devil has been widely used critically to dissect capitalism, which, as the range of capital-as-evil constructs indicates, has been experienced as more than a material reality. But the evident similarities, born of the common heritage of the capitalist epoch, need not conceal historical specificities. Closer analysis would show that the various beliefs and metaphors refer to nuanced dissimilarities in the devils and evils with which people in different parts of the world have had to come to terms. Furthermore, popular notions of evil/the devil need not be outrightly dismissed by brandishing reificatory tags like "populism," "reverse romanticism," or "idealized negation of capitalism," although these elements could well be present in the various capital-as-evil constructs. By analyzing rather than dismissing, or labeling as mere superstition, the version of the devil construct in Negros, I have been provoked to rethink the historical sociology of the sugar plantations. As a critical folk-historic concept, the unraveling of Masonic Capitalism has guided my reconstruction of the history of charisma and magical power, the entry of Western merchant capitalists, and the consolidation of sugar-planter class hegemony in Negros. In the narrative that follows, the changing contents of Masonic Capitalism signal the movement of the social structure in different historical periods and the concomitant social contests over status, cultural capital, and the economic surplus.

Taken together, the three chapters of Part I provide a reinterpretation of the broad currents of Philippine history without which the social structuring of Negros' settler society would be unintelligible. These chapters reconstruct the preconquest cosmological perspectives and social arrangements of the islanders before they fell under Spanish imperial dominion in the late sixteenth century, which reduced them to the category of the colonial subject, *indio*. These chapters also present the political, economic, and cultural changes crafted by both colonizer and colonized prior to the nineteenth-century development of Masonic Capitalism. Part II begins with chapters 4 and 5, both of which discuss various aspects of the processes of erecting sugar *haciendas* on Negros Island after the 1855 opening of Iloilo port to world trade: the origins of *hacenderos* and farm workers, their strategies of adaptation and resistance, and the games of chance they cunningly pursued as one tried to impose its will on the other. The turn of the century depicted in Chapter 6 presents the decisive moment in which native shamans and sugar planters struggled to edge each other out with the common object of dismantling Spanish colonialism, a contest whose outcome was the conjuring up of an atavistic charisma for the planters. In Chapter 7, the various actors of the sugar industry are shown maneuvering through the altogether different context of the American colonial state in a way that led to the simultaneous consolidation and fracturing of sugar capitalist interests, while the industry itself underwent a profound phase of transformation. The narrative ends by unmasking labor's commodified condition as revealed by the spirits of popular culture. In the way the story is presented, the creative tensions binding continuity to change, linearity to circularity, liberation to alienation, fluidity to closure, resistance to submission, sameness to differentiation, the material to the transcendental, and local contingencies to global structures will, I trust, gradually emerge as evident to the reader.

Part I

Colonial Enchantments, Indigenous Contests

A Clash of Spirits:
Friar Power and Masonic Capitalism

The Historical Construction of Capital-as-Evil

Occurring as part of an Asiatic pandemic, the outbreak of a cholera epidemic in October 1820 claimed thousands of lives in Manila and the nearby towns following a devastating typhoon that ravaged the colonial capital.[1] Because the epidemic was particularly fatal in the villages along the Pasig, the Spanish authorities decided to prohibit the use of river water. They also mounted a relief operation to which the medical personnel of the non-Spanish ships anchored at Manila Bay volunteered their services. In the midst of these extremely unsettled conditions, an invidious rumor began to circulate that the *extrangeros* or "foreigners" (also "strangers") intended to annihilate the *indios*. They allegedly had perpetrated the cholera by poisoning the water and air of the capital, a scheme they were said to have furthered when their medical personnel administered poison, instead of medicine, to the victims of the epidemic. The evidence adduced for this supposed plot were the specimens of insects, reptiles, and other creatures found in the collection of a visiting team of French naturalists.

The *indios* sought to avenge their grief. As the Russian consul at Manila, who happened to have been in Macao at that time, later reported: "On the 9th of October about 10 or 11 in the morning they collected, to the number of about 3,000 Men armed with pikes knives and bludgeons and proceeded coolly and deliberately to plunder and Massacre all the

Strangers on whom they could lay their hands" (Dobell 1907, 41). Two detachments of troops failed to quell the tumult, which left twenty-eight persons dead: twelve French, two Dutch, and fourteen English and American merchants and seamen—the only pogrom of Caucasians in recorded Philippine history. The following day about eighty-five Chinese were also killed, having been accused of aiding the foreigners in spreading the poison.[2]

On 20 October 1820, the Spanish captain-general issued a decree to the natives of Tondo Province. In it he rebuked them for their credulity and blamed "certain malicious persons" in authority for having disseminated erroneous ideas that provoked the natives to rise against the foreigners.[3] Though only obliquely blamed in the decree, members of the friar community, according to later accounts, were deemed responsible for the macabre episode (Pardo de Tavera 1905, 345; Regidor 1982, 5). That friars were the instigators of the massacre had a strong logical basis, however.

Although an isolated incident, the 1820 carnage encapsulated the primary tensions of colonial society in the early nineteenth century, created by the long-standing dominance of the religious orders and their opposition to the entry of foreign merchant capitalists, who had been quietly admitted to the colony by liberal-minded governors/captains-general from around the late 1780s. The admission of foreigners was a drastic reversal of Spain's ancient policy of sealing off colonial possessions from rival Western powers, particularly in the islands Spain had called *Las Islas Filipinas,* where for centuries international commercial intercourse had been restricted to the galleon trade between Manila and Acapulco.[4] The strain in the relations between ecclesiastical and certain types of civilian authorities in the colony reflected the mounting challenge that confronted Catholicism in Spain itself. As their statement against the liberal drift of colonial administration, the clergy apparently exploited the 1820 cholera epidemic and succeeded in doing so owing to the profound influence they exerted upon the *indio.*

In contrast to the civilian authorities, the friars, as Captain-General Rafael Aguilar wrote in 1804, possessed *"el arte de dominar el espiritu del Yndio."*[5] In an Englishman's concurring opinion of 1820, "The degree of respect in which 'the Padre' is held by the [native], is truly astonishing. It approaches to adoration, and must be seen to be credited" (Anon. 1907, 113). Given their unique "art of dominating the *indio* spirit," the friars would appear to have been the most plausible instigators of the massacre. In Cebu, a rebellion by natives in 1814 "may have been staged" to thwart wealthy Chinese mestizos from penetrating areas claimed by the Augustinians; if anything, the incident proved the friars' capacity, if they wanted to, to contain *indio* wrath and persuade them to lay down their arms (Cullinane 1982, 259, 262). During the two-day rampage in 1820, the friars were nowhere visible as a voice of temperance. On the contrary, on several earlier occasions prior to the massacre, the clergy had made public denunciations of foreign merchants as "Protestants" and "Masons," sometimes as "Jews," at all events as enemies of Catholic Spain (cf. Myrick 1969, 123; Regidor 1982, 4–5). Conjoined with the negative image of the Jew in the *pasyon* and related Holy Week dramatizations of the crucifixion, the Protestant label would have created an overall anti-Christ imagery. These labels and imageries were potent tools that prepared the natives for mobilization by the friars.

However, the friars' name-calling stratagem, which relied upon medieval inquisitorial labels, was rather anachronistic for the 1820s. Not only was the record of the Holy Office of the Inquisition against Heretical Perversity and Apostasy largely uneventful in the Spanish Philippines, but as early as the opening of the eighteenth century the threat of Protestant ideas had diminished and political coexistence with Protestant states had become the norm in Europe (Greenleaf 1966; Lea 1908, 299–317). In Mexico, whence Filipinas was ruled by the metropole, the Inquisition had been moribund. By the eighteenth century, Mexico's Holy Office had become lenient, eventually losing interest in sexual magic, witchcraft, and pacts with the devil, "base superstitions" the Inquisitors by this time had fully equated with the lower classes (Behar 1987). The earlier racial and caste categories (mestizos, mulattoes, blacks, *pardos*, etc.) that guided the Inquisition—from whose strictures the natives were exempted—had given way to sociocultural differentiation along class lines. Contrary to earlier ideas about race, the dregs among the Spaniards in Mexico could already be lumped together with the *gente vil*.

The resurgence of a new type of inquisitorial mentality, however, was triggered by the rise of Freemasonry in Europe, a movement whose rites were cloaked in secrecy. From the 1720s the Spanish Inquisition began to sense the potential dangers of Masonry, which had spread to Mexico where the movement recruited members of the nobility and the professional and commercial classes (Fisher 1939; Greenleaf 1969). In Filipinas, the inquisitorial mood was heightened by the British occupation of Manila in 1762–1763 in conjunction with the Seven Years' War. As attested to by Mexican soldiers stationed in the colony, lodges belonging to the Scottish Rite were formed in Manila, which was not surprising as British Masons of this era zealously vied among themselves in creating lodges in whatever part of the world they found themselves (Fay 1935, 120–122, 216). In all likelihood, the natives were excluded from those lodges, but this did not stop the friars from publicly denouncing the British as Jews and heretics (Greenleaf 1969, 99; Myrick 1969, 123). By the 1780s, secret lodges affiliated with French Masonry, the dominant strain in the Peninsula, had become active in the colony, coincident with the incipient liberalized entry of foreigners.

That there was an inextricable association of Catholic Spain's enemies with merchant capitalism and political agitation was evident from the moment the metropole took formal notice of Masonry's presence within the empire. In 1754 the Spanish Inquisition warned Catholic bishops to guard against Freemasons who might be in their sees "for reasons of commerce" (Greenleaf 1969, 94). Foreign merchants were envisaged as engaged not simply in trading goods but in peddling dangerous ideas as well. All over Spanish America, Enlightenment ideas that fostered nationalist sentiments drew further inspiration from the American War of Independence and the French Revolution, two historic social movements with links to Masonry (Fay 1935; J. Lynch 1973, 24–36). In condemning a book published in Philadelphia, the Mexican Inquisition in 1794 derisively referred to the writer as "a bankrupt merchant" who traded in "sublime goods" that consisted of "impiety and insolence" toward both "royal authority" and "divine will" (Greenleaf 1991, 258). The resurgent inquisitorial mood engendered the blurring of conceptual boundaries: foreign merchants, religious heretics, and treasonous agitators became intertwined and indistinguishable.

With disloyalty to the Crown becoming a new canon of the General Edicts of Faith in 1752, the Holy Office in the viceroyalty of Mexico was increasingly used for political repression. Over a decade later, there was a marked increase in politically motivated inquisitorial charges falling under such rubrics as criticism of the king, materialism, republicanism, Freemasonry, and heretical propositions against the state. As the eighteenth century drew to a close, numerous Frenchmen were so arraigned (Tambs 1965). But the Inquisitors knew their battlefield, for Masonic influence did have an intimate link to the revolutionary ferment in Spanish America. From the Hidalgo Revolt onward, Masonic organizations provided inspiration and support to the wars of independence in Mexico and played a crucial role in the immediate postindependence period (Fisher 1939, 198–214), presaging the Masonic influence in the late-nineteenth-century nationalist campaign by native elites in the Spanish Philippines.

The turbulent viceroyalty's Holy Office, however, could not enunciate a policy toward Freemasonry for fear of treading upon the toes of powerful colonial officials and creole priests, many of whom had Masonic ties. Thus, despite the royal edicts of 1812 and 1814 proscribing Masonry in the Spanish realms (particularly in light of Masonic collaboration in Napoleon's 1808 invasion of Spain), the Mexican Inquisition treated offenders, especially insurgent clergymen, with leniency. Moreover, the Holy Tribunal restored by Ferdinand VII had become emasculated following its earlier abolition by the Cortes in 1813 (Greenleaf 1969). In all, Mexican Masonry was not stemmed.

Precipitated by such factors as the disintegrating empire in the Americas, liberals allied with Masonic lodges were suddenly, in 1820, returned to power in the Peninsula, dashing hopes of clerical revival nurtured by the monarchy's restoration in 1814 (Callahan 1984, 110–118). The disappointment over this turn of events, I would think, was widely shared by the colonial Philippine church where, unlike in the Americas, the peninsular-dominated ecclesiastical hierarchy remained staunchly opposed to Masonry and liberalism. Friar emotions leading to the October massacre were also conceivably intense, as that year marked the beginning of the final phase in Mexico's war of independence, the consummation of the creole-led dream of an autonomous Spanish America. It must have been gravely painful for the churchmen in Filipinas to countenance the loss of Mexico, the last blow in a series of revolutions that gradually shattered the empire.

Producing two-thirds of the world's silver by the end of the eighteenth century, Mexico, through the Bourbon policy of *comercio libre,* generated revenues that amounted to a staggering fourteen million pesos a year, four million of which were used to subsidize other colonies, while six million represented pure profit for Madrid's royal exchequer (J. Lynch 1973, 301). As for the church, which controlled the real-estate market and interest rates, its vast financial reserves had formed "the principal lubricant of the Mexican economy" in the wake of currency deficits and excess of imports over exports (Farriss 1968, 164–172; J. Lynch 1973, 302). Despite the belated cessation of two centuries of Manila's dependence on Mexican subsidy, made possible by the profitable tobacco monopoly, the separation of Mexico, the empire's most valuable possession, was a loss of gigantic economic and psychological proportions.

In seeking to understand the empire's fragmentation, Spaniards in Filipinas came to the conclusion that the process of generating wealth, which inevitably pulled countries into the international arena, then rife with libertarian ideas, was responsible for the ultimate debacle that was Spanish America. Even in the early 1800s the friars, along with some civilian officials, equated the presence of a foreign commercial body with political unrest and revolution.[6] Acting on that belief, the religious in Filipinas, whose brethren had led the crusadelike resistance to Napoleon's invasion of the Peninsula and the earlier British occupation of Manila

(Callahan 1984, 85–91; Myrick 1969), militantly opposed the accelerating pace of the colony's economic liberalization.

The growing presence of English, American, French, and other European merchants in Manila began to be perceived by Spaniards in markedly negative terms. An English visitor in the 1820s described conditions at the time as follows:

> Foreigners have been . . . gradually admitted since 1800; and they have supplied the wants of the country by introducing European articles, and carrying off the surplus produce, when a sufficient quantity could be procured to employ their capital, which rarely happens without much delay. So rapid has been the augmentation of this trade, that though in 1813 only 15,000 pekuls of sugar were exported, it had increased in 1818 to 200,000, at from 6-1/2 to 9 dollars per pekul. (Anon. 1907, 152–153)

Spanish reaction to this changing economic landscape was one of resentment, founded on the axiom that "a dollar gained by foreigners was one taken from the pocket of a Spaniard" (145), a sentiment that deepened the already existing animosity toward foreign merchants.

Finally, on 17 September 1820, the French commercial ship *Orion* arrived in Manila Bay. On board was the newly designated intendant general, who brought a copy of the liberal constitution to which Spaniards, including all religious, were required to swear allegiance.[7] While the issue of oath to the constitution was a mere "irritant" to the peninsular church (Callahan 1984, 119), the situation in Filipinas was far different. The document that demanded loyalty extended broad privileges to foreigners who wished to settle for an extended period of time in the colony and conduct business not only in the colonial capital but in provincial areas as well.[8] News of more liberal reforms triggered what the Russian consul described as the "envious disposition, on the part of Spaniards, [which] increased daily, against the Strangers, until an opportunity presented itself of gratifying their malignant hatred . . . and without themselves appearing to have any thing to do in the business" (Dobell 1907, 41). Seeking to preserve Filipinas for Throne and Altar, the friars who were in the best position to ignite the events of October 1820 must have felt completely justified in launching a crusade against a policy of economic change they believed was fraught with devastating political consequences.

In waging their resistance against the foreigners, the friars succeeded in disseminating a system of cultural categories in which the capitalist-cum-heretical-cum-political enemies of the Spanish empire were viewed as the evil opponents of the Catholic Church. The terms of opprobrium "Protestant," "Mason," and even "Jew," became synonymous with the heretical and the diabolical (but also monied and seditious), in contradistinction to the "Catholic" (and supposedly loyal, though not as wealthy) Spaniards. Diverging from the pattern in Mexico, the new theocratic terms in Filipinas became surrogate concepts of class. Although ethnicity and race had reconfigured the colony's sociolegal classification scheme by the early 1800s (Wickberg 1965, chap. 1), informal religious categories emerged to delineate an inchoate capitalist stratum. As a result, foreign traders who set foot in Filipinas (and the sugar-planter class that would later arise in Negros) were apprehended from within the inquisitorial matrix.

As late as the 1860s, "economic reasoning" was condemned at a Catholic university in Manila as a "science of the devil" (LeRoy 1968, 116). As enemies of the friars, the merchant capitalists were made to personify an evil to be shunned: "it is necessary to keep the people away from every point of contact with foreigners," as even a peninsular official counseled in 1827 (Bernaldez 1907, 208). Ostensibly, the friars' labeling technique was aimed at preserving religious purity. However, there was also an important pragmatic consideration: to drive an economic wedge between natives and foreign merchants. The export business was dependent upon native middlemen, primarily Chinese mestizos, for the distribution of imported goods to and the collection of exportable produce from the countryside. From 1760 to 1850, the Chinese mestizos dominated the local trading networks after the ethnic Chinese were expelled from the colony in view of their collaboration with the British occupation of Manila. The native mestizos were therefore crucial to the nascent foreign commerce. What better way of disengaging the local population from external trade links than by dramatizing the maleficence and treachery of foreign merchants as purveyors of poison who wanted to gobble up the islands and seize all of their riches?

Death, therefore, would seem to be the just reward for the foreigners' heinous intent. Death was also the occasion for the final sociospatial segregation dictated by the logic of the colonial society's classificatory system. Because heretics were seen as polluting the hallowed perimeters of a Catholic cemetery, it was the church's policy to deny non-Catholics the rites and locus of a religious burial. The fate of natives who resisted Spanish rule, a non-Catholic burial was accorded to the foreigners who fell during the October 1820 massacre. Although most of the dead were Roman Catholics, as the Russian consul averred, their corpses were simply dumped into a mass grave. They were, in the consul's words, "thrown into a hole together without the shadow of a ceremony or a stone to mark their graves" (Dobell 1907, 42).

As suggested by the 1820 cholera episode, the negative coloration of capital was forged in the context of the colonial church's political economy that crystallized in response to the shattering experience of a chaotic, crumbling Spanish empire. Drawing upon inquisitorial categories framed in relation to the social revolutionary ideas of the time, the ecclesiastical battle against saboteurs of what in the friars' minds ought to be an indissoluble unity of church and state, itself fiercely disputed in the Peninsula, led to the equation of merchant capitalism with heresy and sedition. The purported truth of the inquisitorial labels, in turn, acquired an immediacy to the natives through palpable events that converged in the 1820 cholera epidemic.

The mental construct of capital-as-evil was thus fashioned in the dialectics of a specific historical and structural conjuncture in which merchant capitalism was intertwined with the activities of multiple actors across the globe engaging in rumors, rituals, and revolutions the consequences of which were mediated in the colony by the Catholic establishment. The association of capital with evil was not intrinsic to the commodity form, as Michael Taussig (1980) has argued in the Latin American case, as though commodity relations could be abstracted from their historical embeddedness. Indeed, having been woven in history, the meaning of capital-as-evil was also transformed as Philippine history changed in ways that made possible the eventual triumph of capitalist trade. The intractable economic misfortunes of Spain led to a modification of colonial practices, at the same time that aggravation of the political turmoil in the Peninsula led the Catholic orders to redefine their relationship to the state and their role in the colony.

Capitalist Penetration and Its Mythic Ascendancy

Recurrent economic troubles forced Madrid to pursue a pragmatic continuity in policy within a colonial framework replete with contradictions. The events of October 1820 made the Peninsula realize the urgent need to pursue two goals simultaneously: (1) preserve Filipinas within the Spanish realm, and (2) further promote its economic development.[9] Of all the measures considered—notwithstanding apparent prevarications and a long history of conflict between state functionaries and ecclesiastics—the deployment of more priests and missionaries to the colony was identified by both Manila and Madrid as the key to the attainment of both ends.

The dictum that the ability of one Spanish friar to control the *indio* was unmatched by even a hundred soldiers was felt to be more pertinent than at any other time since the conquest of the islands in the 1560s. The whole world, it was said, was under agitation. Hence, the friar's art of domination had to be instrumentalized even more. But only friars from the Peninsula were to be involved in the new imperial thrust, a lesson the metropole decided it would learn from the involvement of creole and mestizo priests in the Spanish-American liberation movements. Designed to keep in check the native clergy's influence, the fielding of more peninsular priests was seen as strategic to achieving Spain's twin goals.[10]

Concomitantly, however, the definitive triumph of liberalism in Spain after 1833 saw the dismantling of the absolutist ancien régime church, leaving its remnants the difficult task of adapting to new political realities (Callahan 1984). Among the state-directed measures to restructure the church was the *exclaustración* of 1837, which led to the closure of seminaries for the regular clergy. This move, however, would have thwarted the strategy of sending more priests to the colony for an explicitly political goal. The solution was found in the retention of three schools, which were placed under the jurisdiction of the *Ministerio de Ultramar* (Ministry of Colonies), exclusively to train missionaries "for Asia."[11] Recruitment to the mission schools at Valladolid, Monteagudo, and Villa de Ocaña was stepped up, with quotas fixed by the *Ministerio de Gracia y Justicia* (Ministry of Grace and Justice). Imbued with patriotic zeal for one of the very few mission fields available—indeed, virtually the only religious (and monetarily rewarding) activity open to the Spanish, mainly rural, youth at the time—many enrollees from the countryside eagerly joined the Ultramar's mission schools.

In addition to theology and philosophy, the curriculum of the mission schools was modified to impart skills in agriculture, engineering, and the sciences in order to fulfill the economic dimension of the new colonial strategy. The religious who would henceforth embark for Filipinas were to pursue the multiple tasks of winning souls, pacifying the *indio* spirit, controlling the native clergy, and enlivening the treasury through their participation in the colony's economic progress. Although the antagonism directed at Protestants, Masons, and other enemies of Spain's Catholic empire remained, economic liberalization itself was no longer to be dreaded. Friar opposition to foreign capital was tamed: at least, the heinous scheming behind the October 1820 catastrophe would not be repeated.[12]

Running parallel to the reconsideration of the Catholic Church's strategic role in the colony was a more candid reappraisal of Spain's colonial economics. Indicative of Madrid's desperate financial strait was the creation in November 1832 of a separate, elaborately named *Ministerio de Fomento General del Reino* (Ministry for the General Promotion of the Kingdom), the imperial arm dedicated to the "general development" of what remained of "the Kingdom."[13] Two years later, the application filed by Jonathan Peele, an American businessman, for a residence permit in Manila impelled the Section on the Indies of the *Consejo Real* (Royal Council) to issue a memorandum that served, in effect, as a lengthy form of self-criticism.[14]

The memorandum stated that the defunct Council of the Indies had misinterpreted past prohibitions concerning the business and residence of foreigners in Filipinas, as the restrictions contained in the voluminous Law of the Indies were meant to be applied only to the Portuguese. Even more significant was the admission of error in the official view (informed by a mineral-extraction concept of wealth) that foreign trade siphoned off the colony's riches to the detriment of Spain. Facing their ignorance of capitalism, the memorandum admitted that Spanish officials "had no exact idea about the nature of capital, neither about the role of commerce in the formation of wealth." Foreign commerce, it was argued, had to be seen in a new light. Far from being piratical, it was generative of new wealth in the country where foreign capital was invested, to the mutual advantage of the foreign capitalist and the host country. The deepening financial crisis pressed home the conclusion that only foreign trade could ensure the colony's "salvation."

Quite typically, the reappraisal of Spain's long-held international trade doctrine met with stiff resistance in official circles, and a protracted period elapsed before policy reforms could be introduced. Notwithstanding official impediments, however, the new economic perspective that had begun to sway Madrid resulted in progressively more liberal, if vexed, colonial policies and practices. For example, a tariff system was imposed to stimulate the economy and raise customs revenue, but the schedule of duties was lopsidedly protectionist, as it exempted Spanish ships but raised the fees on foreign vessels.[15]

Despite such difficulties, Manila was opened to world trade in 1834,[16] an event that pulled the colony deeper into the global arena. The colony's assimilation to the international capitalist system was signaled a decade later by the synchronization of the Philippine calendar with that of the West through the suppression of the last day of 1844.[17] Ushering in business confidence, the opening up of Manila brought some of the biggest commercial houses to the colony. To overcome the unpredictable supply of local produce that characterized the Manila market, the merchant houses instituted a system of cash advances to middlemen and producers.[18] With the merchant capitalists acting as commission agents, bankers as well as commodity speculators, the import–export trade was raised to unprecedented levels. Sugar production, for example, though largely confined to parts of Luzon and to Cebu and Panay Islands, almost doubled, with export figures rising from 11,602 tons in 1835 to

21,529 tons in 1844, about 70 percent of which went to England and the British colonies.[19]

The foreign business community eventually became a force to reckon with, its relations with colonial society reaching a climactic turning point at mid-century. Although barred from owning land owing to a fusion of Spanish commercial and ecclesiastical interests (Fast and Richardson 1979, 14), "Protestant" merchant capitalists came in increasing numbers. In the late 1820s, at least seven British and American merchant houses were doing business in Manila (16). About two decades later, there were as many as thirty-nine merchant houses in Filipinas, at least a dozen of them decidedly foreign companies (Benitez 1954, 233; Cushner 1971, 198). By 1846 Manila had forty-six British residents.[20] Numerically, the corps of foreign merchants had become a more visible component of colonial society, they who had previously formed the despicable Protestant/Masonic fringe. Their numbers would grow and their activities pervade colonial society even more. But how was the prevailing negative attitude, not least by the native, toward foreigners-qua-Protestants decisively overcome? For an answer, we need to turn to a little-known episode in Philippine history: the marriage of the leading merchant, George Sturgis, of the renowned American commercial house, Russell, Sturgis, and Co., to a Catholic lady in the first Protestant ceremony ever solemnized in Filipinas.

Having been previously associated with the merchants at Canton, Russell and Sturgis began operations in Manila around 1828 and reportedly dominated the sugar trade by the 1830s (Legarda 1972, 30–34; Tarling 1963, 306–310). With Sturgis at the helm, the firm enjoyed the enviable position of representing Baring Brothers, one of the largest and most prestigious English banking houses then active in the Orient. Owing to the superior credit enjoyed by the firm, it essayed what became the customary practice of providing clients a 50 percent advance. To further outflank their competitors, Russell and Sturgis reveled in ostentatious advertising, hosting lavish dinner parties "almost nightly" so that their reputation "spread over the Archipelago" (Regidor and Mason 1925, 27).

Sturgis' marriage to a Catholic on 22 April 1849 aboard a British warship anchored in Manila Bay heightened the firm's reputation in a singularly peculiar manner (cf. Santayana 1944, 43). As recounted by a leading Spanish Mason, the event was

> at once novel, and yet typical of the repressive influences that had been holding in check, for two centuries, the development of the Philippines. He married at Manila, a young Filipino girl of Spanish descent, Miss Josefina Borras, according to Protestant rites. . . . The [native] Fili-

pinos, taught by the Friars of the awful penalties attached to any connection with Protestantism, expected to see this desecration of their creed meet with some immediate supernatural punishment. In their minds, the event assumed the importance of a trial of strength between the Friars and the firm of Russell, Sturgis, and on the day of the ceremony . . . they gathered about the landing place, anticipating that the bride would be snatched to the infernal regions. The absence of any unusual episode was attributed to the power and strength of the American firm, and attracted further business from the natives. (Regidor and Mason 1925, 27) [21]

This wedding ceremony represented a pivotal conjuncture in the capitalist penetration of the Spanish Philippines, as it ramified in increased native cooperation with the burgeoning export economy. To the segment of indigenous society that then directly dealt with the merchant capitalists, the Chinese mestizos "thro' whose hands pass almost the entire produce of the Country," this episode bestowed upon the firm of Russell and Sturgis what a British merchant was later to allude to as "an attitude of power" that produced "an overwhelming influence" (Tarling 1963, 302). Thus the aura of power and influence sought by foreign merchants was won unwittingly by Sturgis through a historic rite of passage that native observers interpreted as a sign of approval by the entire cosmic polity. In the indigenous worldview, the Protestant/Masonic merchant capitalists had earned spiritual favor through the sheer demonstration of superior strength in their contest with the friars.

The wedding episode served to elucidate "the evil" the foreign merchants represented. Merchants and friars were locked, not in a fight between the two absolutes of good and evil as understood in the Judeo-Christian sense, but rather in the amoral struggle between competing supernatural forces, a struggle decided solely by the superior power of the victor. The capitalist penetration of the colony was thus seen by the natives as a contest between contending cosmic forces, the resolution of which was dictated by sheer might. In many ways, the native's view of a clash of spirits that pitted foreign merchant capitalists with Friar Power was consistent with the political and economic rivalry taking place among Western powers. That the backward colony would resist merchant capitalism with a show of force was also congruent with the Marxist notion of an opposition between capitalist and precapitalist modes of producing and reproducing social life.

Cosmic Struggles, Power Encounters, and Social Stratification

The islanders conquered by Spain in the late sixteenth century saw the physical world as thoroughly suffused by a spiritual dimension. The people of the islands believed in, feared, and worshipped a host of highly differentiated deities and spirits, preternatural beings the early Spanish accounts referred to as *anito* among the Tagalog and *diwata* among the Visayan. These ancestral and environmental spirits were believed to inhabit their representations in wood and stone, explaining the native belief in the potency of icons. Spirits were not always benign, for many were in constant conflict with humans. Reflective of their relative strengths, the islanders' social world was locally circumscribed while the spirit-world was not.

Unless placated by priestly ritualists, the preconquest shamans, who were usually elderly women known as *babaylan* in the Visayas and *catalona* in Tagalog, or effeminate/transvestite men *(asog)*, the unseen beings would cause misfortune to befall such essential human ventures as warfare and the cultivation and harvesting of food crops, especially rice. The late-sixteenth-century Boxer Codex (Anon. 1979, 336) reported, for instance, a certain highly malevolent deity who had to be fearfully implored to spare people from harm as he ambulated about the fields and mountains. The female divinity that reigned on Negros Island's volcano, variously called Laon or Lalahon, also needed to be invoked during harvests, failing which she would send a horde of locusts to destroy the crops (de Loarca 1903, 135).[22]

Spirits were also believed to be responsible for causing illness. During the healing ritual, the shaman would offer prayers, food, drinks, and even gold and other valuable gifts to appease the offended unseen beings. She would go into a trance, possessed by spirits with whom she had established friendly relations. If health was not regained, other more powerful spirits were believed to have intervened and overwhelmed the shaman's spirit-guides (Anon. 1979, 344). As an ultimate symbol of the adversarial relationship between the spirit-world and the human realm, the early Visayans attributed death to the consuming envy of a deity called Macaptan (de Loarca 1903, 133, 135).

The islanders' mythical understanding of the world portrayed the inherent conflicts that pervaded the cosmos. In the Visayan origin myth, the creation of land was attributed to the strife precipitated by a scheming bird that succeeded in making the sky and the sea engage in battle, the earth being formed when the sky angrily threw islands at the sea (Anon. 1979, 315–316; Colin 1906, 73; de Loarca 1903, 125–127).[23] The antagonism later extended to the opposition between land and sea such that "Those who journeyed ashore could not mention anything of the sea; and those who voyaged on the sea could not take any land animal with them, *or even name it*" (Colin 1906, 78; my italics). In everyday life, the islanders were witnesses to a cosmic struggle in the ceaseless chase and marital conflict between the female Moon and the male Sun, which erupted when the latter touched and embraced their children, causing them to perish and rousing the Moon's maternal fury.[24] The mythically divine origin of warfare also conferred upon human conflict a natural quality and sense of inevitability (de Loarca 1903, 141).

In the indigenous worldview, reality was also permeated by continual conflict among humans, an interpersonal battle involving what still today is known in Ilonggo and Kinaray-a as one's *dungan*.[25] Literally meaning "together," the *dungan*—in Alicia Magos' pioneering exposition of this critical concept—is a life force, an energy, as well as an ethereal entity, a spirit with a will of its own that resides in the human body and provides the essence of life. Apart from denoting an alter ego and soul stuff, the *dungan* as presently understood refers to such personal attributes as willpower, knowledge, and intelligence, and even the ability to dominate and persuade others (Magos 1992, 47–50).

This multireferential concept is indispensable for our understanding of the primeval Visayan notion of what it meant to be human. Present-day shamans have retained the ancient notion that everyone has a *dungan* that begins to inhabit the body of its choosing at around the time of a person's birth. The *dungan* undergoes a process of adaptation and of knowing its world, and must feel comfortably lodged and domesticated in the host person's body; otherwise it could be coaxed by spirits to stray and even be held in captivity, the loss of soul resulting in prolonged illness or death of the host person. When an individual is startled, the *dungan* jumps out of the body, and the regaining of composure is linked to the *dungan*'s return and the forgetting of its fright. When the person is asleep, the *dungan* voluntarily leaves the body, allowing one to see one's other self in dreams. While away, the *dungan*'s wanderings must be safe so it can return at the appropriate time, for which reason someone asleep is never rudely awakened.[26]

Differentials in *dungan* strength—commonly referred to as "height"—signify the latent antagonism in human relationships; this conflict is settled by an act of power on the part of the stronger, metaphorically taller individual.[27] These power encounters start early in life, in the form of the still widely believed phenomenon of *usug,* which occurs when the weak and unknowing *dungan* of an infant is frightened at the sight of a stranger, who normally has a more forceful *dungan*. In its fright, the *dungan* causes the child to cry, vomit, and experience abdominal pains. To prevent *usug*—hence to subdue the weak *dungan*—the adult wets a finger with his or her saliva and applies this to the infant (Tan 1987, 21). A child prone to illness or crying means that its *dungan* has not fully known and felt confident in its corporeal dwelling, necessitating the performance of a series of rites to familiarize and nurture the *dungan* in its new habitation. Well-being is therefore synonymous with a *dungan* that spatially and socially "knows" its place and feels sufficiently domiciled in its new abode, the result of overpowering acts by the adult world.

The *dungan* is also the traditional idiom for the power encounter and recurrent negotiation over rank among adults of all social strata. An individual with a stronger *dungan* can overpower and wittingly cause illness in another with a weaker *dungan*. This can occur through forms of speech ("flattery") or gesture ("a malicious touch") that have the effect of sapping energy and stealing the desired trait of another. The contemporary notion of the *dungan* (although influenced by what would later be discussed as the gambling mentality and the realities of capitalist accumulation) nonetheless has strong affinity with the idea that the person with unsurpassed *dungan* exhibits acute intelligence, vast knowledge, indomitable willpower, and self-confidence; generates wealth and an awesome reputation; and exudes capacities to rule, dominate others, and subdue enemies. Feared and respected, individuals of such extraordinary qualities are known in Ilonggo as *dunganon*,[28] the equivalent of what in Bikol are referred to as "big people" (Lynch 1979). In all, the working out of *dungan* competition orders personal status, ranks individuals, and stratifies society, with the acceptance or toleration of the outcome easing social conflict and the fluidities of hierarchy.[29]

Based on the contemporary Visayan concept, I surmise that the apogee of *dungan* strength in the precolonial world was anchored in the person's ability to relate to the spirit-world. Gravitating around such persons would be a set of followers, warriors, and dependants, wanting to participate in the spiritual and material benefits and accomplishments that accrue to persons of strong and tall *dungan*. The precolonial *datu*— leader of each of the scattered bands and settlements over which no unifying superordinate political entity existed—most assuredly possessed the ancient vitality and intensity of a forceful *dungan*, demonstrated in a robust physique, sharp mind, masterful oratorical style, good fortune, bravery, and a loyal, servile, and dependable followership. The *datu* was a Big Man whose exceptional *dungan* was indicative of favor from and rapport with the spirit-world, a spiritual relationship that allowed him to fight other *datu*, attain success, and perform supernatural feats.[30]

Each *datu* was resolutely independent, especially among the Visayan *Pintados*, so-called because of the elaborate tattoos that covered the bodies of both men and women (de Loarca 1903, 175). Valorous accomplishments in life conferred upon them in death the self-consciously sought status of an ancestor known as *papu*. Precolonial *datu*ship as inextricably linked to *dungan* strength suggests that the islanders shared with their regional neighbors a cultural theme prevalent throughout early Southeast Asia: the dominance of what Oliver Wolters (1982) has called "men of prowess."

Because not everyone had the same ability to relate to the spirit-world, each band or settlement was stratified according to the distribution of supernatural prowess and the attendant economy of prestige. In an apparent gender division of power, the male *datu*'s elevated status was matched by the highly respected female shamans who were specialists in propitiating the spirits.[31] In rendering their priestly services, the shamans received a gift that signified a preemptive propitiatory act, a form of advanced appeasement to obviate punishment from the spirits who stood with the shaman in providing the service—just in case the recipient turned out to be "unworthy" of the service (Anon. 1979, 335, 343–344; Chirino 1969, 301; Rosell 1906, 217).[32] (Friars would later misinterpret this gift as a form of payment making for a lucrative trade.) As in the *datu*'s headship, wealth obtained through shamanship and displayed in dress and personal adornment articulated the differential status and rank among individuals as ordained by the spirit-world (Chirino 1969, 301; de Loarca 1903, 133).

Social existence in the precolonial world was thus characterized by an endless series of power encounters settled through various acts of potency, victory, and strength. Rooted in origin myths and in explanations of illness, in the antagonism of spirits to humans, in the contests of *dungan,* in the *datu*'s leadership, and in the *babaylan*'s priestly centrality, the view of reality possessed by the inhabitants of the islands Spain colonized was one pervaded by cosmic struggles. With neither side being inherently good or evil, conflicts were omnipresent in the entire cosmos. The world was experienced and life lived as a material-cum-spiritual arena for the contest of strength and favor, counterstrength and counterfavor.

It was, I believe, from such a cultural window that *indios* observed and participated in the power encounter between Spanish friars and foreign merchant capitalists. It was from the same vantage point that natives meaningfully interpreted the wedding of George Sturgis as endowing the merchants with mystical prowess capable of challenging Friar Power, the essence of Spanish colonial rule. By virtue of the feat that Protestant/Masonic capital symbolically demonstrated in that wedding, the natives, I surmise, began to ascribe to the merchants a different role in the cosmic order. As mercantile economics was crowned with mythical might, the trading sections of indigenous society found an unequivocal sign that they were transacting with the powerful. No longer were foreigners to be relegated to the social and spatial fringes of colonial society. No longer were they prey to the native's wrath, despite the episodic mortality from cholera epidemics that occurred throughout the nineteenth century. Consistent with the structural changes at the level of empire, the native interpretation of changing colonial realities led to intensified commercial relations and, through networks of intermediaries, encouraged further export-crop production in the countryside. Sturgis' wedding served as pivot and augury of things to come.

The basic cultural constructs the natives used to interpret the triumph of merchant capitalism were the same ones they relied upon to comprehend the earlier power encounter with Iberian conquerors. Despite phenomenal continuity, however, the islands' indigenous cultures and societies had already undergone a radical reconfiguration by the time of capitalist incorporation in the nineteenth century. The depth of the colonial transformation of the islands needs to be understood from the moment of conquest, the task set out for the next chapter.

2

Cockfights and *Engkantos:*
Gambling on Submission and Resistance

Spiritual Conquest and Colonial Enchantment

By imperial design Catholic priests were at the forefront of Spanish colonialism. For the first two centuries of colonial rule, natives had virtually no contact with Spaniards other than the friars.[1] Engaged in their solemn duty of fighting heathenism, the friars distributed themselves throughout the archipelago, which, for missionary purposes, was administratively subdivided and allocated to different religious orders. Initially considered an alien enemy, the friars eventually overwhelmed and overpowered the natives. With minimal military support, the friars gradually but decisively extended the area of Spanish control.

The friar was seen through indigenous cosmological lenses, and justifiably so, for the *indio* and the friar were one in their belief in a spiritual realm inhabited by preternatural beings. It was on that common ground that colonial domination was built and the colonial state's foundations were laid. The ubiquitous friar set in train the beginnings of a collective memory for the *indios* who, to whatever important population center they traveled, saw a friar who could speak the locality's lingua franca. Under the aegis of friar dominance, the internecine warfare that had characterized the preconquest epoch also largely ceased. Spanish imperial hegemony thus could not be understood apart from an explication of the friars' relationship with the native population. And this relationship—founded on the art of dominating the *indio* spirit—I propose to analyze using the splendid phrase the friars themselves used to describe their mission, which to them was a *conquista espiritual*.²

Despite the absence of conventional historical evidence, it can be argued that the circumstances of an imperial conquest led by a priestly caste impressed upon the natives a veritable "spiritual invasion," a massive intrusion of Hispanic spirit-beings into the islands. That this was the indigenous formulation of the Iberian conquest and the natives' way of coming to terms with the radical changes wrought by it can be inferred from Spanish words appropriated into various Philippine languages.³ Denoting preternatural entities of a distinctively Spanish origin, commonly used words in the contemporary Filipino spirit-world include *engkanto, engkantu* or *ingkanto*, referring to a generic spirit-being, a word derived from *encanto* (charm/enchantment/spell) or *encantado* (enchanted); *dwende* from *duende* (elf); *multo* or *murto* (meaning ghost) from *muerto* (dead); *maligno* (an evil spirit) from *maligno* (malicious/malignant); *kapre* (a dark, hairy, otherworldly giant) from *Cafre* (Kaffir);⁴ *santilmo* (a spirit or soul in the appearance of fire) from *fuego de Santelmo* (Saint Elmo's fire); *sirena* (sea nymph) from *sirena* (mermaid); *tag-lugar* (environmental spirit) in a *lugar* (place, spot, or site).⁵

The features and qualities ascribed to these imagined preternatural entities are particularly revealing, the first in the list being an exemplary case. In a pioneering paper on "The *Engkanto* Belief," the Jesuit Francisco Demetrio presents a portrait of the *engkanto* based on some eighty-seven folk narratives obtained from the Visayas and northern Mindanao (Demetrio 1968). The *engkantos* are described as being "of both sexes and varying ages" and "of fair complexion, golden haired, blue eyed; they have clean-cut features and perfectly chiselled faces" (137–138).⁶ Demetrio adds:

> Though beautiful and fairskinned, *engkantos* are said to be romantically attracted to a brown-skinned girl or boy. Although spirits, they are said to indulge in dalliance with mortal beings. Though known to dislike noises, they themselves sometimes indulge in raucous noises while feasting or punishing a mortal who has refused their love or abandoned them.
>
> They are whimsical and unpredictable; they play jokes on people; making them go astray in the forest at night, or transform themselves into the likenesses of mortal friends and relatives in order to dupe the objects of their desire. (Demetrio 1968, 137)

Interestingly, *engkantos* have purportedly been seen singly or as families, but hardly as communities, unlike indigenous spirits (Ramos 1971, 54).

It is my contention that the characteristics of the folkloric *engkantos* have been culled from the friars' idealized physiognomy and their historic sacerdotal misdemeanors. The *engkanto* belief mirrored those Caucasians dispersed throughout the islands who, because of their extremely small number, could hardly be considered as constituting a community in any given locality. In their imposing presence, the friars laid down new rules of the social game only they could break. They demanded silence in the rectory but broke it with their own noisy gatherings; their orders had to be obeyed lest the *indio* receive a severe beating; and their cravings for sexual gratification could not be spurned. Despite pretensions to clerical celibacy, those white men left Spanish mestizo offspring. As the friars were the first to infringe the rules they themselves laid down, colonial rule was founded upon their arbitrary word, a fact reflected in the *engkanto*'s "whimsical" and "unpredictable" character. Indeed, the *engkanto* figure constituted a telling critique of the colonizers who "duped," "led astray," and "made fun" of the natives.

But the *engkanto* belief had significance other than as a trope, for to the natives the *engkanto* represented a "real" entity in the spirit-world. That these alien preternatural beings had landed on their shores was a way of explaining the new sensations the natives had begun to experience with the Spanish conquest. For instance, the story was told of a Tagalog who "wandered off towards the mountains, as if in a daze, and roamed from hill to hill . . . impelled to wander away . . . against his will and control" (Chirino 1969, 378). A later generation of natives would have easily diagnosed the man's behavior as the work of an *engkanto*. The image of the dreaded *kapre* was evident in the experience of a Cebuano who "was afflicted with horrible visions" of "hideous black men" who "threatened him with death" (381). In Leyte, while some *indios* began to recite Catholic prayers and with no one else in sight, "stones began to fall on the house from outside, making a great noise and knocking down objects that they had left out in the open" (397). The novelty of such strange experiences impressed upon the natives the tangibility of the Spanish spirit-world, a force that had impinged upon the islands.

That Spanish preternatural beings existed would not have been at all odd to the islanders. For if the friars who had boldly set foot on their soil were authentically human, they too, like the natives, would have possessed *dungan*,[7] and they too would have come from a place filled with spirit-beings that mattered in the Spaniards' lives. What was even more certain was that those beings were no longer confined to wherever those white men originally came from, but were actual companions in the men's journeying. How else could they have subdued the best native warriors—and, by implication, the native deities—with a never-before experienced cosmic force? Confronted by a superior power, the islanders grasped the meaning of colonial conquest in terms of the Spaniards' alliance with their spirit-world. The main link to that newly present yet unseen realm was the priest, curiously a male, dressed in a drab, dark cassock.

As the foreign male ritualists began to live in newly founded settlements and commenced their evangelizing mission by mastering the local language and performing Catholic rites, the *indios* received confirmation concerning the activities of alien spirits. The friars, in turn, interpreted the bizarre reports of their fresh subjects as signs of the latters' diabolical ties. The *indios*' accounts of "extraordinary accidents" arising from the conquest were conceptually framed by friars specifically in terms of "enchantments," namely, *encanto* (see de San Antonio 1906, 345–346). In the complex process of finding a correspondence between the preternatural entities whose presence the natives had discerned in their midst and the foreign words they heard used by the friars to refer to the natives' strange experiences, Spanish nouns and adjectives were appropriated, jumbled, and converted into proper nouns. Those nouns became the words by which the natives learned to call the alien spirit-beings by name: *engkanto, dwende, murto, maligno, kapre, sirena, santilmo, tag-lugar*. And the spirits had to have names, for to the *indio* the act of naming constituted the formidable step of confronting and objectifying the altered realities triggered by the colonial conquest. Refusal to name would have been a sign of unquenchable fear, and inability to name an indication of the total absence of knowledge, consequently of absolute vulnerability.[8] Although the native's fear remained, the Hispanic spirits at least had become knowable, even familiar. Those imagined preternatural beings were, in a sense, the genuine conquerors of Filipinas: they had inundated the islands and could not be made to depart. Truly it was a *"conquista espiritual."*

The multiplicity of beings that inhabited what I believe the natives conceived of as the Spanish spirit-world, including the numerous *santos* and *santas* (saints), was not inconsistent with the structure of native cosmogony, which accommodated the alien spirits in their respective niches in the cosmic order. As Alicia Magos' (1992, 51–52) indispensable reconstruction of the contemporary shamanist worldview in Antique suggests, preconquest reality was hierarchically divided into seven strata, each of which had its distinctive territorial occupants, a belief system the islanders shared with other parts of the ancient "Hindu world." Based on the cosmological map redrawn by Magos, the various layers can be seen as having been infiltrated by a host of Hispanic spirit-beings. For instance, the fourth layer, the earth's surface, is said to be inhabited by invisible terrestrial beings in direct contact with human beings. Along with indigenous preternatural creatures such as the *aswang, tikbalang, kama-kama, sigben,* and so on, can be found the Spanish *engkanto, kapre, murto,* and so on. In the sixth layer located at the "top" of the earth dwell the natives' ancestors *(kapapuan),* as well as the Catholic saints and angels, an uncanny classification that combined "real" people who had achieved marvelous deeds in the Spaniards' distant past with the spiritually favored "real" people in the islanders' past.

The natives deftly imposed their logic in apprehending a spiritual reality to which the Spaniards adhered. And rightly so, for their interpretation was reinforced by several defining parameters of the historical situation. First of all, in the Europe of the conquest period, the dominant cosmology divided reality into three domains: the truly supernatural (God's unmediated actions), the natural (what happens always or most of the time), and the preternatural (what happens rarely, but nonetheless by the agency of created beings and spirits such as angels, demons, ghosts, and other terrestrial beings; cf. Daston 1991). Only in the late seventeenth century would the erasure of the preternatural domain commence and the concept of spiritual power as centralized in the Supreme Being be theologically thinkable. At the time of conquest, the *indio* and the Spaniard shared an intrinsically similar worldview founded upon a solid belief in a nonmaterial yet palpable reality, particularly in a decentralized preternatural domain populated by spirit-beings with power to affect and even determine worldly affairs. With that spiritual realm humans communicated through words and actions performed by individuals possessing specialized sacral knowledge, hence the mediating role of priests and shamans.

Moreover, the centuries-long *reconquista* of Spain from the Moors, the emergence of Protestantism, and the avenging Inquisition of the Middle Ages set the Spanish belief system in a thoroughly aggressive and bellicose mood—the so-called crusading spirit. Eager to subdue and exterminate their spiritual opponents, the Spaniards, particularly those numbered among the Catholic missions, were unquestionably enacting the *indio*'s language of cosmic struggles.

The "golden age" of the missionary enterprise in Filipinas also coincided with the "golden age" of Catholicism in Spain in the second half of the sixteenth century, when piety encompassed all of human existence.[9] Disruptions of everyday life immediately provoked a religious response: "floods or prolonged droughts, invasions of locusts, frosts, food shortages, epidemics, all evoked a cycle of processions and prayers, conjuratory or expiatory ceremonies which the end of the public calamity transformed into expressions of thanksgiving" (Bennassar 1979, 70). Religious devotion "assumed a propitiatory nature" (Defourneaux 1970,

118) closely resembling the *indio*'s religious practice. During this golden period the Marian cult began to flourish, a faith that made the Spanish belief system, with its emphasis on female power, more intimately proximate to the native worldview.

As a result of the Inquisition, little was required from the multitude but their attendance at ceremonies and the reproduction of officially sanctioned words and gestures. The prevailing orthodoxy encouraged the popularity of *ensalmadores* (casters of spells) and *saludadores* (healers) in the seventeenth and eighteenth centuries. They used magical objects and performed incantatory rituals whose formulas *(ensalmos)* were culled from the drama of the Crucifixion—practices treated indulgently by the Inquisition (Bennassar 1979, 87). Relying on "white magic," the Spanish folk practitioners evinced a striking similarity with the indigenous shamans of the islands colonized by Spain. Notwithstanding formal theology, unofficial but tolerated superstitions blended superbly with their official variants, producing the peculiar Catholicism the friar missionaries propagated in the Philippines.

The degenerate state of the Spanish church was aggravated by the process of recruitment to the Catholic priesthood. With the onset of economic decline in the Peninsula in the early seventeenth century, the priesthood became a safety valve for many Spanish youth who detested manual labor. In 1624 a bishop lamented: "Some say that religion has now become a way to gain a living, and many become religious just as they would enter any other occupation" (Defourneaux 1970, 107). The mediocrity of priestly morals was felt in the prevalence of curates living with concubines, called devil's mules, and having children by them, a breach also treated indulgently by the Inquisition. In the mission field, Spanish priests were quick to seize the monetary benefits that came with the occupation. For example, the Jesuits who were reputed to be the least prone to poor discipline early on in mission history paid disproportionate attention to the profitable beeswax trade in the Visayas (Phelan 1959, 37–38). Although it has been claimed that members of the regular clergy abided by "higher standards of discipline" and so were "better prepared for missionary work" than members of the secular clergy (31), the friars who went to the colony did not transcend the cultural norms and moral proclivities prevalent in the Peninsula.

Overall, the islands claimed by Spain became the meeting point for two religious systems that were fundamentally alike. Sharing with the natives the same universe of discourse, the European clergy could easily locate the native religion in direct cognitive opposition to their Catholicism. The Spaniards' sensitivity to religious difference made it possible for the inventory of native beliefs and practices to be recorded in colonial chronicles and for these to be branded not merely as superstition but as creditable works of the devil.

On the other hand, cosmological parallels allowed the natives to perceive the colonizers as similar to themselves despite overt signs of difference. The discernment of a basic alikeness made it possible for the natives to classify and localize the saints along with their ancestors, and the *engkantos* along with indigenous terrestrial beings. Just as the *Santo Niño* (Holy Child) image was initially known in Cebu as the *diwata* of the Spaniards (Chirino 1969, 235), so Catholic idols were seen as corresponding to, and hence were treated in the same way as, native icons, which until the nineteenth century were used in Bukidnon to touch "the ailing member, or the painful part, in order to find relief and even a total cure" (Clotet 1906, 296).

In the midst of such critical sameness, the colonizer triumphed. Backed by a host of Hispanic preternatural beings, the friars were seen by the colonized as founding their legitimacy and domination on the basis of their potency and superior cosmic strength. Viewed from the indigenous cultural framework, the friars were seen as alien shamans who engaged, both willfully and unintentionally, in innumerable acts of healing that the missionaries, for their part, interpreted as the windfalls of faith. As discussed in the next section, the friars' medical prowess was proven efficacious in subduing not only Hispanic spirit-beings but local entities as well. In the course of waging their spiritual battle, the friars overwhelmed the indigenous cosmology, shattered the precolonial meaning system, and altered the configuration of the islanders' social world.

Friar Power and the Submission of the *Indio*

As the friars went about their mission work, they projected the image of shamans whose magical ensemble included the Catholic sacraments, which served as powerful intercessory devices with the spiritual realm. In the late sixteenth century, countless missionaries extolled baptism as a most efficacious remedy for leprosy and other afflictions, as indicated by reports from Leyte, Samar, and Negros; baptism was also employed to revive a comatose man in Butuan (Chirino 1969, 367, 388–389, 396–397, 440, 487). Extreme unction cured an old woman and penance a sick man in Cebu, while confession stilled the "death rattle" of a woman in Negros (385, 440). The Jesuit Pedro Chirino made the happy report that "Many were cured of serious illnesses after receiving the Holy Sacraments, so that they all asked for them persistently and received them most devoutly" (355).

The friars' paraphernalia were transformed into inherently potent objects. Catholic icons, medallions, rosaries, scapulars, the cross, and water blessed by the priest became novel media for the transference of power from the spiritual to the physical domain. As in Spain, a reliquary crucifix warded off a swarm of locusts in Luzon (Chirino 1969, 346–347; Christian 1981, 184). Holy water became known as a medicine, and the rapid spread of its popularity overshadowed the native curative practice of drinking water from a place where an idol had been dipped (Clotet 1906, 296). In Bohol, those who drank holy water were spared from death in an epidemic that caused "pains in the head and stomach" (Chirino 1969, 333). From Negros, Cebu, and Bohol, reports suggested that holy water was also an effective remedy when sprinkled on or applied to a patient's body (333–334, 385–386, 439–440).

The resulting "general custom all over the islands" saw natives abiding by "this holy devotion" of drinking blessed water, even from the church stoup (Chirino 1969, 334). Undoubtedly, the wellspring of the natives' new religious practices was none other than the friars themselves, who were the first to believe in the efficacy of holy water and who taught the natives to drink it as medicine. An illustrative case is that of a friar who, while officiating at a healing ritual, began by asking the woman if she believed that holy water could cure the sick: "She answered in the affirmative, whereupon he gave her a little of it to drink" (333).

It was only logical that, as men with healing powers, hence with special acumen to negotiate with the spirit-world whence diseases were believed to originate, the friars would be perceived by the natives as possessing forceful *dungan*. The islanders must have drawn such an inference from the moment of their initial contact with the friars, who, in advancing through dangerous unfamiliar territories, exuded courage and fortitude. The friars did fit the role of the strong *dungan*, as they cast themselves in the role of crusaders, some being so consumed with self-confidence that they deemed their very presence medicinal (Chirino 1969, 376).

The friars' *dungan* astounded the natives. In 1720, in a controversial but acutely observant letter, the Augustinian Gaspar de San Agustin (1906, 265) noted that "one must not shout at them, for that is a matter that frightens and terrifies them greatly, as can be seen if one cries out at them when they are unaware—when the whole body trembles; and they say that a single cry of the Spaniard penetrates quite to their souls." In the social contest of strength, many a startled native found his "soul penetrated," his *dungan* jarred and unable to withstand the friar's overbearing speech and thunderous voice. Repeatedly, the islanders were shaken by the male shamans who had chosen to live in their midst.

The friars' *dungan* were tested and their strength confirmed through their proven ability to appease and subdue the Hispanic spirits that impinged upon the native imagination. The friars effected a cure for the wandering disease prompted by the *engkanto* by using Catholic magical objects. The Tagalog who roamed aimlessly "against his will" was told "to put his trust in the power of the holy *Agnus Dei*"[10] pendant, which a religious put "round the man's neck. From that moment on the man felt at peace" (Chirino 1969, 378–379). The Cebuano haunted by "hideous black men" called for a priest, who heard his confession, and "Thereupon the man felt very much relieved and recovered his peace of mind" (381). Life for the *indio* was proving to be inextricably bound up with the Iberian shamans, who had become indispensable to overcoming the physical and emotional ailments and cultural crises precipitated by the spiritual conquest.

The friars were further vital to native existence as their intervention became necessary in subduing, not only the foreign spirits such as the *engkanto,* but local spirit-beings as well, which previously had been amenable to appeasement by native shamans alone. For instance, a friar used an Agnus Dei medallion to counter a sorceress' spells on a woman who, as a result of a quarrel, experienced violent convulsions (Chirino 1969, 466). To this day, Demetrio reports that "The *agnus* medal is used by natives as amulets, together with the *carmen* and the cross, to protect one from all sorts of danger, accidents or the bad intentions of others" (Demetrio 1970, 136–137). As a result of the exercise of Friar Power, Catholic objects and rituals gradually replaced preconquest mechanisms for dealing with afflictions attributed to the spirit-world.[11] The friars, in demonstrating their potency and centrality as shamans, began to attract a following and core of adherents who, for their health and well-being, resolved to identify, at least overtly, with the dominant power and the colonizing culture.

Working under the instruction that they were to extirpate native beliefs and practices and not to rest until paganism was stamped out (Ortiz 1906, 105), the philistine friars sought to discover the images, implements, and meeting places the *indios* guarded with secrecy. Whenever indigenous icons were ferreted out, the friars celebrated their success by contemptuously desecrating them, to the bewilderment of the natives, who believed that anyone who committed such acts of sacrilege would perish. In Bohol a missionary awed villagers by touching their *anito* without dying; furthermore, he spat on the local idols, trampled them, and then had them burned and thrown into the river (L. de Jesus 1904, 384).

Unable to retaliate against the marauding Spanish shamans, local preternatural beings were evidently losing the battle instigated by the friars. In Zambales, "amid the great shouting and lamentations" of the natives, the fathers ordered a servant to fell a venerated bamboo thicket the natives thought could not be cut down lest they die (de San Nicolas 1904, 179). On another occasion, a missionary climbed a feared *paho* tree and gathered its fruits while reciting a Latin chant:

> The [natives] were very sorrowful because father Fray Rodrigo had decided to eat of the fruit, and they accordingly begged him earnestly and humbly not to do it. But the good religious, arming himself with prayer and with the sign of the cross, and repeating the antiphony, *Ecce crucem Domini: fugite partes adversae. Vicit leo de tribu Juda*, began to break the branches and to climb the tree, where he gathered a great quantity of the fruit. He ate not a little of it before them all. . . . The [natives] looked at his face, expecting every moment to see him a dead man. (de San Nicolas 1904, 145.; cf. de la Concepcion 1904, 276–277)

Thus, with the aid of Catholic magical words, relics, and gestures, the friars demonstrated superior cosmic strength by their ability to vanquish local spiritual strongholds and break age-old taboos. The display of incomparable *dungan* in the fearless and successful confrontation with the indigenous spirit-world was a historic achievement of the Spanish friars.

With the unremitting success of Friar Power, the *anito* and *diwata* began to lose their abilities to cause as well as heal illnesses and, in general, to affect the course of human existence. Symbolic of the destruction of the islanders' precolonial identity and world of meaning, the indigenous deities eventually vanished. In their place today are found Hispanic spirits who, interestingly, exhibit the same behavior as the preconquest gods. For instance, like a pagan spirit the *Santo Niño* (Holy Child), which contemporary Filipinos have localized, is said to possess shamans bodily during a trance.[12] Similarly, Negros' fabled volcano, Kanlaon, formerly the abode of the female Laon of native antiquity, has become the regal seat of the Hispanic-inspired and uncertainly gendered entity called Sota.

Friar conquest of the indigenous spirit-world also resulted in local spirits becoming benign, or even innocuous (the present-day translation of *diwata* being simply "fairy"), while the Hispanic spirits assumed the maleficent role of bearers of illness. Demetrio observes that in contemporary beliefs: "*Engkantos* are known to possess power to inflict diseases: fevers, boils and other skin diseases as result of their curse or *Buyag*. Without knowing it someone may brush against the invisible *engkanto*, and suddenly he is slapped in the face or his skull is cracked by a blow" (Demetrio 1968, 138). In agriculture, *engkantos* have replaced the ancient environmental spirits as those whose favors must be obtained before peasants can proceed to ready the land and harvest the crops; offerings must also be made to the *engkantos* before timber can be felled from the forest (Demetrio 1970, 364, 378).

By acting as dependable shamans deliberately performing the sacraments as healing rituals, the friars made themselves an integral part of native strategies for coping with life's vicissitudes. The legitimacy established through Friar Power became the anchor of the Spanish imperial presence. Friar authority also became the basis for the extraction of surplus from the native population. Contrary to a royal edict, the Spanish priests began to charge sacramental fees (Phelan 1959, 63). The fathers were emboldened to reap their material rewards, for they saw that the natives would "bestow esteem, confidence, adoration and anything they own" on "anyone who can furnish [medical remedy] or promise to do so" (Chirino 1969, 300). There was therefore a monetary angle to the friars' denunciation of the *babaylan* as frauds.

But although the natives succumbed to the cosmic power and strength of the Iberian colonizers, they were not reduced to total passivity in the face of cataclysmic change. On the contrary, the alien shamans were possessed of a prowess that so mesmerized the *indios* that, as John Phelan speculates, "parents may even have encouraged their daughters to make liaisons with the clergy," a comment that must be understood in the context of the elevated status of preconquest women and of an indigenous sexuality unencumbered by European precepts (Phelan 1959, 39).[13] Phelan's point is highly plausible and is buttressed by the observation that the native desire for amorous contact with the alien shamans has found a parallel expression in local folklore. For all the irascibility of the *engkantos,* Demetrio notes that twentieth-century Filipinos paradoxically entertain "a certain deep-seated attraction [to] or fascination for these creatures," prodding them to "secretly wish they enjoyed the special attention of these strange and dreadful but fascinating beings" (Demetrio 1968, 138).

We can take a further cue from the *engkanto* belief that the human victim is said to disappear into the nether world of the *engkantos,* there to taste its extraordinary pleasures. With the unfolding of Friar Power, the desire to commune with the friar-qua-shaman, or at least the predisposition to respond positively to the friars' advances, could be interpreted as the route chosen by natives to penetrate and know the colonizer's awesome power. But as a mark of separation from all previously meaningful realities the islanders had known, to enter the world of the alien shamans through carnal union meant to disappear and depart from indigenous society. Yet that disappearance, that departure, also signified that the friar's/*engkanto*'s victim would be transported into another realm of power, allowing the local maiden and her allies to enter into a special relationship with the regnant order. Her role as broker would be, in a sense, a mere variant of preconquest women's role as negotiators with the spiritual. The fair-skinned offspring of friar–native trysts, today said to be the fusion of *engkanto* and native,[14] also disappeared metaphorically, as the native's deep-brown skin pigmentation was diluted by Castilian "blood"—causing the Spanish mestizo complexion to be widely admired, even desired, by natives. Friar concubinage, it would appear, became a mechanism for resolving the question of power for an earlier generation of *indios*, but it bore unintended consequences for later generations, who had to wrestle with issues of cultural integrity and identity.

The alliance between some natives and the colonial power was echoed by entities in the indigenous spirit-world that, in an apparent switch of allegiance, started to behave in full accord with the friars. Called *nono* as one of the embodiments of ancestor spirits, the crocodile, by ancient tradition, was implored not to harm the islanders, who knelt and clasped their hands in supplicating the creature (de Morga 1904, 131–132). Under the regime of the friars, however, the *indios* were told to kneel before a different set of objects. The demand for the transfer of sacral gestures—hence of loyalty, emotion, and identification—became compelling when the crocodile itself began to be portrayed as favoring the colonizer's religion.

In San Juan del Monte, a man who allegedly mocked those who attended a Catholic prayer session and decided to remain in the river to bathe ended up being bitten by a crocodile, to the *indios'* "great horror and their renewed respect for the disciplines and the *Salve* [Litanies] of Our Lady" (Chirino 1969, 425). The reptile had ostensibly fully turned around to ally with Friar Power, as suggested by an incident in early-seventeenth-century Binalbagan, Negros: "a converted [native] woman, having been convicted of a grave sin, in order to deny it cursed, saying: May a crocodile eat me before I reach my house, if what I said was untrue. God punished her immediately, for while near her native place, called Passi, in the island of Panai, a crocodile attacked . . . and swallowed her" (L. de Jesus 1904, 244–245). Subjugated, disarmed, and finally converted, the self-aggrandizing crocodile *(buaya)* was emblematic of some natives' response to colonial rule that would provide institutional mechanisms for opportunistic alliances with the dominant power.

Not all islanders submitted to colonial rule, however. A number of chiefs and native shamans resisted it. But given the nature of Spanish hegemony, which reinforced, perpetuated, but also altered indigenous cultural constructs, their resistance was ineluctably articulated in religious terms. Despite its failure to overturn the conquest, native resistance persisted, and to a degree subverted, colonial authority and weakened its grip over colonial space. Because of the Iberian rulers' inability to eliminate this resistance, the natives who overtly submitted to colonial rule soon found themselves in the middle of a power struggle between two opposing spheres of power—in effect between two conflicting claims to loyalty and identity. As though inflicted by the *engkanto,* the collective native soul *(dungan)* suffered from the disease of not being comfortably lodged and domiciled in the colonial corpus. To the natives, colonialism was a sorcerous enchantment. As though collectively struck by a spell, native society was prompted to wander between two realms of power. The resulting *indio* politics emanated from this tension and ambivalence.

Cultural Entrapment and the Colonial Cockpit

The indigenous spirit-world was not entirely defeated by the tempest caused by Spain's spiritual conquest; some local entities did fight back in the cosmic struggle that had enveloped the islands. In 1885 a missionary reported from Mindanao, pointing to traditional female shamans, that "those women are the most difficult to attract to our holy faith, and even to enter the presence of the father missionary" (Rosell 1906, 217–223). But as early as 1599 a *catalona* in Manila had told the people that "at first the God of the Christians had prevailed over their Anitos, but . . . the latter were now returning in triumph and were punishing those who had abandoned them" (Chirino 1969, 373). In Butuan indigenous spirits were said to have appeared to the natives, "persuading them not to admit those fathers into their country, because of whom . . . dire calamities and troubles must happen to them" (L. de Jesus 1904, 221). The warnings of spiritual reprisal were not unfounded, as indeed the global expansion of Iberian colonialism did fuel the transoceanic spread of diseases and gave rise to other "dire calamities."[15] Believing in the battle waged by the indigenous spirit-world, native shamans were emboldened to challenge the colonial order.

Even within Spanish-controlled areas, the indigenous religion continued to be practiced clandestinely, with the cooperation of the old precolonial elite. In San Juan del Monte a silent procession was held "in the thick of night" to transfer an idol from the house where a deceased underground shaman had lived to that of her successor; although close to Manila, the forbidden religion was not discovered until it was disclosed two years later by a lower-stationed *indio* (Chirino 1969, 302–303). In Zambales the local elite, serving as the "principal priests," led their community's covert observance of animism, until children befriended by the Dominican friar divulged the secret (V. de Salazar 1906, 52). Local religion was betrayed by those who were either too young to have any affective attachments to the old practices or those who had little or no interest to protect, which derived from the ancient cosmology. The shamans and *datus,* in contrast, colluded to defend the indigenous belief system that provided legitimacy to their respective social positions, both of which the friar singularly usurped.

No doubt, many chiefs willfully converted to the friar's religion. These important personages, in an apparent quest to reestablish the ancient pattern of legitimacy, exhibited "great zeal for bringing pagans," presumably their followers, to be proselytized and baptized (Chirino 1969, 454). Desirous of retaining the central role they had once played in the preconquest social milieu, the converted chiefs in a part of Leyte, for instance, sought preeminence during Holy Week by guarding the Blessed Sacrament with their "customary arms" (403–404). However, other chiefs, including some the Spaniards referred to as the truly "big fish," obstinately resisted conversion (356, 359–61).

Buffeted by the rising tide of conversions, some natives expressed their resistance in a less passive manner. In Zambales those "respected and venerated as the greatest chiefs" killed the resident missionary and set fire to the church and convent before fleeing to the mountains (de la Concepcion 1904, 282–283; de San Nicolas 1904, 180–181). Others poisoned friars or stoned them to death (de la Concepcion 1904, 274; de Morga 1904, 100). With unswerving passion and obduracy, many chiefs defended the indigenous religion and sought to regain their authority in terms of the old cultural framework.

Other natives resisted Spanish rule by simply withdrawing to the wild interior beyond the reach of the conquerors. Sometimes, entire mission villages retreated along with their *babaylan,* whom the friars blamed for causing their converts to apostatize (Chirino 1969, 377, 458; V. de Salazar 1906, 56). In following the shaman, those who fled decided on a course of noncompromise with the alien power. The resort to flight expressed their unyielding faith in the indigenous religious system as the sole provider of meaning and the only balm for the travails of human existence. Such outright resistance most probably accounted for the relative failure, particularly in the Visayas, of the *reducción,* the imperial program that intended to bodily aggregate the natives into compact settlements as in Mexico (Phelan 1959, 44–49). A fully successful *reducción* might have been the equivalent of a successful rite to domesticate and contain the collective native soul *(dungan)* within the corpus of colonial society. As it turned out, the native soul was only partially domiciled in its colonial abode and, thus, could and did wander off.

But although many natives chose to flee, many others decided to remain within the ambit of Spanish colonialism and its orbit of power. Leaving a legacy of rural settlement patterns extant today (but not in Negros for reasons to be explained in Chapter 5), some natives moved right into the *cabecera,* the capital of the parish or town, while numerous others struck a compromise by settling in hamlets of varying sizes (the *visitas*, and the even smaller *sitios* and *rancherias*), which were widely dispersed but still "bajo de la campana" or within hearing of the church bells.

By deciding to reside within the friar-dominated settlement, *indios* placed themselves in a situation of having to appease two spheres of power, the indigenous and the Hispanic. Trapped at the center of a clash of spirits, the colonial subjects were confronted with the competing claims to loyalty and identity pressed by two similar but opposed realms, both of which were seen as authoritative and valid. Because of these historical exigencies, the natives learned to negotiate between two cultures by adhering to two religious systems, openly imitating one and concealing the other, cultivating in the process a social practice of cultural ambivalence. Fearing both power sources and seeking to appease the spirits and shamans of both domains, the natives practiced colonial Catholicism at the overt level and the indigenous religion at the covert. As a Spanish priest lamented, the *indios* managed to "embrace the matters of the faith in such a manner that they should not become separated from the ancient worship" (V. de Salazar 1906, 51). Far from being syncretistic, the religion of the colonized native epitomized what it meant to live in two colliding worlds.

The equal appeasement of two conflicting spiritual powers was not always possible, however. There were unavoidable situations when natives were confronted with the choice of which power realm to follow. Whenever the indigenous spirits became compelling, some natives were reported to have been wont to "surrender their rosaries" to the *tikbalang* or the *bibit* and, in exchange, were given indigenous magical objects "such as hairs, grass, stones, and other things, in order to obtain all their intents and free themselves from all the dangers" (Ortiz 1906, 107; de San Antonio 1906, 342). In situations when the friar's orders had to be followed to the displeasure of the indigenous spirits, the native had no recourse but to implore the latter's mercy and plead that they withhold retribution.[16] As a friar incisively observed in the early eighteenth century,

> When they are obliged to cut any tree, or not to observe the things or ceremonies which they imagine to be pleasing to the genii [*sic*] or *nonos*, they ask pardon of them, and excuse themselves to those beings by saying, among many other things, that the [friar] commanded them to do it, and that they are not willingly lacking in respect to the genii, or that they do not willingly opposed [*sic*] their will. (Ortiz 1906, 105)

Surely, in such painful betrayals of native culture and meaning, the sentiments of the political underdog began to form.

The perceived clash of powers and the *indio* response of wandering between two realms nurtured the native's gambling outlook on life. Gambling, of course, is universal and of antediluvian origin, but its local character is the product of specific contingent histories. In the Philippines, the colonial epoch gave rise to gambling as an articulation of the subjugated natives' ambivalent response that concomitantly accepted and rejected colonial realities. Given the highly spiritualized texture of Spanish colonialism and native gambling's links to the spirit-world, gambling became the idiom that expressed the *indio*'s contradictory relationship to colonial power.

An external manifestation of the subjectively felt cultural entrapment, social gambling for the *indio* was a wavering form of wagering on the odds of power. If one was caught in an inescapable situation where equal appeasement of the realms was not possible, it became a sheer case of bad luck. Otherwise, the *indios* moved back and forth between the overlapping worlds constituted by the indigenous and the colonial in a gamble that they would not be caught in either one. The simultaneous avoidance and acceptance of the clash of spirits was graphically encoded in the various forms of gambling that flourished during the epoch of Spanish colonial rule.[17]

Foremost among the games of chance was cockfighting—*bulang, sabong*, or, as the Spaniards called it, *juego de gallos*—a source of fun said to have been used to entice recalcitrant *indios* to join the colonial settlements. In cockfighting, the native could be entertained by witnessing what was essentially a cosmic battle. For the gambling contest was not confined to the participating individuals but connectively involved imagined preternatural entities who were divided by the granting of spiritual favor to the contending participants in the game. Ultimately, the shrewdness of one's gamecock or one's smart handling of a card game was reckoned as emanating from the superior otherworldly support extended to the winner. Gambling, especially in the cockfight, was a visual and thrilling display of the clash of power realms.

As a rule, only cocks of equal prowess are matched in any fight, and the opposing center bets are equalized before the fight begins. However, the assumption of parity is reserved for the liminal period, from the matching of fowls and into the fight, during which moment the idea of superiority and hierarchy is both affirmed and disbelieved, only to be confirmed anew after the fight.[18] During this liminal period when the birds are believed to have an even fighting chance, one of them is nonetheless invariably perceived as the superior, hence favorite, cock while the other is considered the underdog. Based on contemporary beliefs and practices, it is my conjecture that in the early Spanish period, regardless of the plumage and hue of the birds in the ring, the language of the ritual game simplified the cocks' colors into either red *(pula)* or white *(puti),* the first referring to the superior bird, the second to the inferior.

That the color red should connote superiority was rooted in the ancient preconquest belief in that color's potency, which signified life and courage, in contrast to white, which symbolized death and defeat. In the precolonial age, islanders who were the most valiant and had killed the most enemies in war wore, as a badge of honor, a red kerchief wrapped around the head (L. de Jesus 1904, 213; de Morga 1904, 76). Today, red continues to stand for life and strength, and the color itself is fetishistically believed to exude power that can augment one's bravery in combat.

If we take liberty of generalizing Pigafetta's observations in Palawan, it could be said that prior to the Spanish conquest certain venerated, hence spiritually linked, cocks were already made to fight for a prize: "each one puts up a certain amount on his cock, and the prize goes to him whose cock is the victor" (Pigafetta 1969, 55). It might then be speculated that, even in pre-Hispanic times, the clash of cosmic powers was already the game's message, albeit its story was that of warring, supernaturally gifted *datus.*[19] And since the *datus* valued the color red, we might say that the superior cock was even then classified as red. Corollarily, it should be noted that during the colonial epoch, the shamanic groups that resisted Spanish rule continued to use red on their persons, that color forming the basis for one of the labels by which they were known: *pulahanes,* or "the red ones."

Under Spanish colonial rule, the popularity of cockfighting (like the prevalence of the *anting-anting* amulets widely used in the pit) could be attributed to the game's subtle subversion of the dominant colonial order. The *indios* who were trapped between submission and resistance would have read into the cockfight's red–white binary codes a political significance. As red stood for indigenous prowess—as well as the shamanic resistance that posed a perennial challenge to colonial authority—it is not farfetched to assume that white was made to signify the white Iberian colonizer. As spectators vicariously involved through identification with the fowls, the *indios* could wager on either side. The equal division of the center bet between red and white reflected the social cleavage in indigenous society between resistance and accommodation to colonial rule, as well as the feelings of submission and resistance that tore apart the individual *indio.*

Moreover, the cockpit's message was contradictory. On the one hand, hierarchy and dominance were omnipresent in cockfighting, as the outcome validated the native concept of power as being the rule of the spiritually mighty; on the other, cockfighting allowed for the inversion of hierarchy in colonial society. The internal message of the cockpit was counterhegemonic. The indigenous red was not the underdog: it could be asserted and bet on as the favorite by the real underdogs outside the cockpit. Red could win, but so could white. Since the outcome was never truly predictable, the native at least had an imaginary fifty-fifty chance. And so whenever red and white clashed in the cockpit arena, the power encounter between the indigenous and the Hispanic realms was reenacted all over again—much like the perpetual reenactment of Christ's sacrifice in the Catholic mass—as though the historical outcome was totally unknown.

By the nineteenth century, with the routinization of colonial practice as well as the increased monetization of the economy, the earlier color signification appears to have become interchangeable, at least in the Tagalog cockpit. Jose Rizal (1958, 256), for instance, in his great nationalist novel, referred to the white cock as *llamado,* superior, and the red as *dejado,* the underdog. In this context, the underdog red's victory became even more emotionally charged and imbued with patriotic fervor: "A wild shouting greets the *sentencia* (the winner's proclamation), a shouting that is heard all over town, prolonged, uniform, and lasting for some time," so that everyone, including women and children, would know and share in the rejoicing that the underdog had won over the dominant power (259). The noise that burst through the rafters was noted by a Russian visitor in the 1850s, who wrote that "For a foreign spectator it was this uproar that was noteworthy" (Goncharov 1974, 203).[20] Today, despite another transformation of signs, the cheering is always louder when the underdog defeats the favorite. Although social inequalities are accepted as a facet of reality, underdog victories are seen as suggesting that "the poor farmer also has a chance" (Guggenheim 1982, 26), allowing for social catharsis at least in the fictive world of gaming. In the cockpit, history and social structure can be momentarily suspended and phenomenologically forgotten as the players—all males—indulge in an infinite series of counterfactualities that make for pure fantastic entertainment.[21]

In the 1770s, the cockpit began to fall under colonial state regulation and to be administered through licensing mechanisms and rules governing the days and times of play.[22] Despite formal supervision, the meanings generated by the *indios* in the ritual game were beyond colonial ken and control. It must be noted, however, that it was the Spanish colonial state itself that lent the conceptual framework for the cockpit's system of inversion. For although cockfighting existed prior to colonialism, it was under Spain that the game's unwritten codes were systematized, as the cockpit's argot would attest: the reading of omens *(señal)* hidden in the cock's scales on the leg that might reveal an auspicious sign of *baston* (the staff carried by the colonial elite), the *regla* or "rule" determining the trend of luck, the *logro* or odds, the *parada* or inside bet, the *casador* or betting master, the *kristo* or bookie, the *largador* or cockhandler, the *asentista* or cockpit promoter or manager, the *tasador* or matchmaker, the *pago* or payment by the cockpit management to equalize the center bets, and the *sentenciador* (referee) as though the pit were a court of law where sentence was passed (cf. Anima 1977, 1972; Lansang n.d.).

So, in the cockpit, Spanish and indigenous forms and concepts melded, allowing the experience of colonial domination to be both accepted and rejected, inverted and reinverted, objectified and internalized by the subjugated natives. Thus, within the very space seemingly under full Spanish control, the natives could enjoy a subtle subversion of the colonial order, although, paradoxically, the same event also legitimated the colonial power structure. Cockfighting was a celebration of both fact and fiction.

Males as Shamans, Imitation as Resistance

But the colonial cockpit properly belonged to only one sphere, the sphere of the church bells. Beyond the hearing of the bells, native shamans endeavored to adapt their resistance to the reconfigured social reality. One fundamental feature of this shamanic adjustment pertained to gender. With spiritual mediumship becoming a contested terrain, the perceived superiority of Friar Power resulted in the development of native male shamanism in imitation of the Spanish friarship. As males began to predominate among local shamans and as the colonizers carved out an exclusively male public sphere, the overt social role of females was eclipsed.[23] It must be noted, however, that, despite Spanish efforts to inculcate their sexual mores in the natives, local reformulations of gender remained more egalitarian than in the Mediterranean world (Blanc-Szanton 1990). Women also persisted in their role as arbiters with the spiritual, becoming devoted to Catholicism generally and, with little stigma attached to it, to friar concubinage, in the case of a few native women.

To resist more effectively, native male shamans began to alter their practices to put them on a level with the friars. A ritual "invented . . . after the Spaniards had come here" was reportedly concocted in which coconut oil and a crocodile's tooth (the crocodile having become a multivalent figure) were consecrated to local spirits, who were invoked to bestow upon the oil the power to kill (de Loarca 1903, 163). Subsequently, native shamans started furtively to steal—imitate—the magical words used by the Spanish shamans who recited Latin verses to ward off evil spirits or conquer their putative abodes. The appropriation of Latin constituted the *babaylan*'s decisive riposte to Spanish shamanism.

Prior to conquest, indigenous shamans apparently had already employed recondite words as special channels for negotiating with cosmic forces. According to early Spanish accounts, they resorted to foreign magical words that had the power to heal, what the accounts termed as "certain superstitious words" derived from "the Burneyan language which they all highly regard" (Anon. 1979, 320, 344, 349). Other accounts noted "badly-pronounced words" that were used for divination as well as the invocation of spirits (Anon. 1979, 335; L. de Jesus 1904, 204–205; de San Nicolas 1904, 137–138).

With the advent of the colonial epoch, the friars unwittingly demonstrated the heuristic value of Latin as the language of power, evinced by the frequently recited prayers such as the Pater Noster (which was mandatory for baptism) and the prayers said at exorcisms. Latin being the official language of the Roman Catholic Church, it was believed that the devil understood it and could be commanded only through that official language (Defourneaux 1970, 121). Consistent, then, with an earlier emphasis on the power of spoken cryptic words, Latin or Latin-sounding words and phrases began to compose the native shaman's formularies and incantations, which came to be known as *orasyon* or *urasyon* (from the Spanish *oración*, prayer). As unlawful knowledge, the appropriation of Latin proved to be the native's first truly subversive act. Emblematic of this subversion, still practiced by today's shamans, is the recitation of the Pater Noster in reverse order, starting with the Amen.

The *urasyon* was recited during healing rituals and, in imitation of European missals and breviaries, copied onto paper in tiny and easily concealable booklets (measuring about 1 by 1.5 inches) known in the Visayas as *libritu* (a small Spanish *libro*). Pig Latin words were also written and attached to pendants, or the booklet itself was carried on one's person to serve as an amulet or talisman, known as *anting-anting*. The booklet of Latin phrases was believed to make the native invincible and to endow him/her with the ability to negotiate or enter into alliances with spirits. Through shamanic Latin, the *engkantos* could be coaxed to effect healing, restore a wandering soul, produce a bountiful harvest, or bestow luck, as in the colonial cockpit.

Indicative of the natives' desire to augment their potency and intrude into the colonizer's power domain, the use of Latin in the form of *urasyon* and *anting-anting* became widespread among them. Along with indigenous magical objects, shamanic Latin was used to obtain luck and success, as natives negotiated between two power realms and quietly challenged, even defied, the colonial state. Reversing the magical specialization of the preconquest age, native shamans evidently shared their newfound source of power and resistance with ordinary *indios*. By the 1730s it was observed that

> It is very usual for the [natives] to carry about them various things in order that they might obtain marvellous effects: for example, written formulas, prayers, vitiated or interspersed with words arranged for their evil intent, herbs, roots, bark, hairs, skin, bones, stones, etc., so that they may not be killed, or apprehended by justice, or to obtain wealth, women, or other things. They are also very much inclined to believe in omens and in unlucky days, in regard to which they are wont to keep various books of manuscripts.... (Ortiz 1906, 109–110)

That nobody understood pig Latin did not matter, as those words were meant to address the unseen spirit-beings. What mattered was that the natives could tap into the cosmic source of colonial power. Unwittingly, the friars further confirmed the efficacy of shamanic Latin through their efforts to confiscate and destroy all the *libritu* and *anting-anting* they could lay their hands on.

As the only effective counterpoise to the friars, native shamans relied on their newfound war chest of magical power to animate the revolts and uprisings against Spanish colonial rule.[24] Armed with nothing but machetes and amulets, native male shamans and local chiefs fought to reassert their former source of meaning, power, and identity. Mustering mystical prowess from both power realms, they sought an end to the tyranny of having to contend with two opposing spirit-worlds and cultures. In seeking to reestablish the preconquest religion, they could be seen as attempting to restore the social order they once knew.

But preconquest society and culture did not remain insulated, pristine, and unaffected by the colonial epoch. On the contrary, the practices of the natives, even of those who rebelled, and those of the colonizers had become mutually determining. On the one hand, indigenous culture had become indelibly transformed by native imitation and appropriation of friar magic, particularly the shamanic gender switch; even the Hispanic spirits were also transformed into localized entities. On the other hand, the hegemonic rulers made concessions that affected their colonial practice. Gauging from its prevalence today and observations made in the nineteenth century, it would appear that the Spanish priests effectively ignored (or were incapable of monitoring) male circumcision *(turi, tuli)*, despite its initial inclusion among the so-called heathen practices targeted for suppression.[25] The colonial rulers also allowed the cockfighting so much loved by the natives to flourish. Apparently, they even indulged the native male by legislating against the entry of females to the cockpit.[26] In so doing, colonial authority empathized with the native cockfighter and invented and formalized a prohibition without precedent in the Spanish bullfight. Given the interpenetration of cultures and the interlocking alliances that emerged in colonial society, it was not surprising that organized uprisings led by the shamans were quelled by the might of Spanish firepower as well as by the intervention of *indio* soldiers.

The social order the shamans once knew was beyond restoration. But resistance persisted. Within the parish enclave, it was expressed through the covert practice of a suitably altered native religion. Outside the colonial centers, the marginalized native shamans, mostly but not exclusively male, continued to quietly draw a clientele. More importantly, the native shamans continued to enjoy a reputable status among the *indios*, still few of whom could penetrate the deepest secrets of the mystical to become masters of magical prowess. The power of the native shamans became that of the amulet-bearing fugitives and social bandits whom the natives called "good men" (*mabuting tao* or *maayo nga lalaki*, and probably *magaling na lalaki*).[27] As anticolonial fighters, fugitives and bandits became the embodiment of the idealized good (*ayo, buti*), for they possessed the enviable qualities of the strong *dungan:* a brave soul, an indomitable spirit, an invincible body, even ruthlessness toward one's enemies. The red-kerchiefed rebels engaged the colonial state in a perennial battle. Failing to exterminate them, the colonial establishment sought to establish their deviant nature by branding them as criminals and evildoers (*malhechores*). But what to the state was a lawbreaker was to the people a fascinating risk taker who transgressed colonial rules with impunity.

Albeit only from a distance, the natives admired and respected the "good men," the men who personified their longing for an unvanquished past. However, because they settled within hearing of the church bells, most natives had to put up with a life of cultural entrapment, of conflicting demands for identity that engendered a deep-seated ambivalence and inferiority. Deeply sentimental, they sang the dramas of their lives and became adept at music. But gambling was their passion. Informed by a gambling approach to life, the native elites took advantage of the circumstances enforced by colonialism to devise mechanisms for circumventing the dilemmas of culture and power, but their solutions only deepened the cultural ambivalence they had hoped to overcome.

Atrophied Charisma and the Making of Native Elites

In the early 1800s the friars were credited with having assumed "the major part in the pacification of all instances of disquietude" during the more than two centuries of Spanish rule in Filipinas.[28] By this time, as a perceptive Englishman wrote, "In the most distant provinces, with no other safeguard than the respect with which he has inspired the [natives], the *Padre* exercises the most unlimited authority, and administers the whole of the civil and ecclesiastical government, not only of a parish, but often of a whole province" (Anon. 1907, 113). Notwithstanding efforts to assert civilian supremacy, "No order from the *Alcalde* [provincial governor], or even the government is executed without [the friar's] counsel and approbation." Thus the colonial state was established not so much as a complex of institutions that formed a formal legal order but as the personal, and often arbitrary, rule of the friar. What emerged in the colony was not an explicitly political community, as politics was subsumed under religion. Spanish political hegemony was fully indebted to the regime of Friar Power.

And to the friars were subordinated the native elites who became the private landowning class of *caciques*.[29] The fundamental change came with the petrification of the ancient *datu*ship—conferred with the new title of *cabeza de barangay* or "village head"—into an inert hereditary institution. At the same time, "the village" (the barrio or today's *barangay*) was itself undergoing colonial invention as a formal political unit and as a standardized and spatially delimited social organization.[30] The recourse to heredity was a legal imposition based upon the Iberian rulers' preferred but mistaken view of the preconquest social order.

Drawing upon European legal constructs in which he was well versed, Antonio de Morga (1904, 119, 127) described the *datu*ship, in his famed *Sucesos de las Islas Filipinas,* as hereditary along the male line. However, other sources indicate that the claim to *datu*ship had to be supported by acts of valor and might. In a setting with no formal legal institutions, magical achievements were relied upon as the primary criterion to confirm, in the manner of the empiricist, the authenticity of such a claim. This latter view was not accorded official hearing, because leadership on the basis of individual prowess impressed the European as tyrannical and haphazard. The reports describing the reign of brute force can nonetheless help us understand the essence of the ancient chiefship.

Writing in the late sixteenth century, Chirino (1969, 390) observed that among the islanders "whoever was powerful enough prevailed and ruled, and not one man alone but almost anyone could come to exercise such power and authority." The Boxer Codex (Anon. 1979, 310) corroborated that view by stating that "These (chiefs) are largely brave [natives] whom they have made lords because of their deeds." A Recollect missionary in early-seventeenth-century Mindanao also noticed that "The government of those people was neither elective nor hereditary; for he who had the greatest valor or tyranny in defending himself was lord. Consequently, everything was reduced to violence, he who was most powerful dominating the others" (L. de Jesus 1904, 212).

Written in the late seventeenth century, the Jesuit Francisco Colin's *Labor Evangelica* provides what I consider the most incisive statement on the path to *datu*ship. That position was attained, he said, "through their blood; or, if not that, because of their energy and strength" in creating "some wealth," usually "by robbery and tyranny." Colin emphasized that the *datu*

> gains authority and reputation, and increases it the more he practices tyranny and violence. With these beginnings, he takes the name of *dato;* and others, whether his relatives or not, come to him, and . . . make him a leader. Thus there is no superior who gives him authority or title, beyond his own efforts and power. . . . If his children continued those tyrannies, they conserved that grandeur. If on the contrary, they were men of little ability, who allowed themselves to be subjugated, or were reduced either by misfortunes and disastrous happenings, or by sickness and losses, they lost their grandeur . . . and the fact that they had honored parents or relatives was of no avail to them. . . . In this way it has happened that the father might be a chief, and the son or brother a slave—and worse, even a slave to his own brother. (Colin 1906, 86–87)

The *datu*ship, therefore, was not governed by an absolute rule of male succession. Although a son could well prove himself a worthy successor, and might have the social advantages of becoming one, the "energy and strength" and the "grandeur" ended with the *datu*'s death. The departure of the *dungan*, unleashed from the body (*kalag*) or literally vomited or thrown out of the mouth (*kaluluwa*), terminated the force, the will, and the power of the *datu*. As is suggested in the description given by Francisco Alcina, also a Jesuit, in his *Historia de Las Islas é Indios de Bisaya*, *datus* were buried with "all the wealth that they had when they were alive" (Alcina 1960, chap. 16). The successor, therefore, had to establish his own credentials, create his own wealth, and build his own grandeur. He had to embark on his own magical journey, for the route to becoming a *datu* was contingent upon individual feats and personal exploits and upon continually increasing those feats and exploits.[31]

In the absence of ostensible procedures for selection by a higher legal authority, the petty rulerships appeared to Spanish observers (who could not comprehend the indigenous ideological context of power) as a matter of naked force, a primitive contest for individual supremacy based on violence. However, in a situation of intermittent warfare, bravery was a sine qua non of leadership. Indeed, as far as the natives were concerned, the path to *datu*ship followed its own logic, for leadership and the whole compendium of village life were intimately related to the indigenous view of the cosmos, the source of coherent meaning to the world for both chiefs and followers.[32]

As an endowment from the spirit-world, bravery was both a personal quality inside a person (the *dungan*) as well as a tangible object, a charm or talisman, which equipped that person with powers of invincibility and with abilities to perform extraordinary deeds. But the favored individual must prove his otherworldly election, a practice almost akin to "spiritual positivism." Once proven by actual deeds, the attributes of power separated the truly valorous from the rest, who were tantalized and magnetized to form around the *datu* a community of warriors and dependents. Thus, bravery, not just theoretically but as proven by unquestionable feats of valor, confirmed the *datu*'s personal worthiness and the goodwill of the spirits.

Recently, Vicente Rafael (1988, chap. 5) has cogently argued that Spanish colonialism relocated the *datu*ship into a divinely ordained system of patron–client relationships, and that the *datu*'s position finally found a stable source of authority by being linked to a centralized spiritual-cum-political realm. Earlier in this chapter, I argued for the fundamental similarity of the indigenous and Spanish worldviews in the period of contact. Testimony to the quality of this historic encounter is the structure of the spirit-world believed in by Filipinos in the late twentieth century, which, along with the localization of Hispanic spirit-beings, continues to retain its essentially decentralized character. In addition, it is my view that the preconquest *datu*ship already rested upon the highest possible source of authority meaningful to the islanders, the spirit-world. A thoroughly spiritual affair that abided by rules which, to those involved, formed a consistent and nonarbitrary order, a *datu*'s reign was intelligible to the islanders who "knew" its stability, as well as the dawning of periods of fluidity that accompanied a ruler's downfall or the death of the *datu*.

In cases where the legitimacy of an established *datu* was challenged by another claimant to cosmic prowess, which required a large following, the dispute was settled either through warfare or the formation of another relatively isolated settlement, which was possible given the expanse of land then available. The dispersion of settlements became a spatial expression of relative *dungan* strength among the native chiefs. Followers, especially warriors, shifted allegiance according to whomever they found more attractive among contending *datus*.[33] At any one time, however, especially in the Visayas, no single *datu* possessed undisputed charisma that might have justified elevating him to the position of a superordinate leader who commanded the loyalty of lesser chiefs. The seemingly inexhaustible reserve of supernatural prowess accessible to a multiplicity of spiritually endowed Big Men manifested itself in political fragmentation.

In establishing Spanish sovereignty over the islands, the colonial state transformed the preconquest elites into a fixed institution characterized by hereditary succession but bereft of their preconquest prestige and magic, and devoid of the prerogative to rule singlehandedly their individual settlements. Forced to conform to a system of political primogeniture, the families of the old chiefs held their positions in the imperial administration uninterruptedly for some two hundred and twenty years.

The transformed elites relished their honorific titles of *Don* and *Doña* and enjoyed exemption from tribute and corvee labor. They rose to social prominence as town magistrates known as *gobernadorcillos*[34] and as *fiscales* (sacristans) and cantors in the church.[35] Not until 1786, long after the Chinese mestizos had become a distinct element in colonial society, was village headship (the *cabeza de barangay*) made elective, and it retained the same perquisites as the hereditary *cabezas*. After more than two centuries, the descendants of the *datus* of old had intermarried and multiplied into a select circle of "leading families" who comprised the exclusive pool of candidates in the friar-controlled local elections.[36]

After more than two hundred years, the descendants of the *datus* had lost the ability to conjure magic but nevertheless retained their formal positions of leadership. The *datu*'s charisma was not routinized: it merely atrophied and grew stale. Through the imposition of hereditary succession, Spanish colonial rule introduced the concept of an institutional position of power that, in being separated from personal accomplishments and extraordinary feats as a sign of favor from the spirits, was thoroughly corrupted. Gone was the magical journey of achievement as the basis of exalted rank and status.

Power in colonial society thus became ascriptive and closely intertwined with the colonial construct of "the family." An imported concept that today's Filipinos denote by using the borrowed terms *pamilya* or *familia*, the family was invented, not so much as a set of identifiable relationships in a kin group, but as a conscious ideological category denoting a monogamous institution with corporate boundaries framed by parenthood rather than siblingship (as in the Tagalog *mag-anak*).[37] Marriages fell under the jurisdiction of the friar-qua-state, and the resulting union became bound with a fixed identity revolving around the colonially imposed and paternal surname—the "family name"—which became the primary criterion for the natives' social identity.[38] The former *datu*'s broad services to his kin and non-kin followers were narrowed to the family, which became the channel for the purposeful advancement of interests and the intergenerational transmission of property, power, and status—a reversal of the ancient Southeast Asian "indifference towards lineage descent" (Wolters 1982, 9). The Catholic prohibition of nuptial union between cousins[39] and the introduction of a legal inheritance system fostered the tactical use of marriage by elite families to preserve wealth "within the family." Customary law (or its elements that were not suppressed), as well as Spanish legal norms, were subordinated to the pursuit of family interests, a private sphere hardly distinguishable from the public, as the latter was itself governed by the personalistic rule of the friar.

Nonetheless, the native elite's power had to be exercised in the context of a colonial setting marked by contending social forces. Organized as political families, the native elite continually had to court the local friar to earn his favor, which they did by providing services and monetary contributions to the local church. In return, they enjoyed prominent roles in Catholic ceremonies and rituals. It also became easy for them to obtain from the priest a favorable letter of reference, required by the central government at Manila in the appointment of town magistrates. The friar became the native elite's protector against the felt abuses of civilian administrators. The markers of colonial prestige and protection, which the elite constantly had to seek and augment, seemed like signs of approval from the dominant power realm personified by the friar/*engkanto*.

At the same time, because the colonial state retained the preconquest chiefship, at least in its outward form, as a means of indirect rule, native elites were compelled to contrive a system of affirmation of their continuing legitimacy as local leaders. One mechanism was the largesse that flowed through their sponsorship of the feast of the town's patron saint, a shift in the flow of resources given that, as we shall further see in the next chapter, the *datu*'s control of the surplus had been eroded and taken over by the friar. Not predisposed to recognizing their leadership, however, were the rebel segments of indigenous society. The latter, who contested colonial authority using their otherworldly prowess, could easily terrorize the native elites who, though nominal Catholics, were awed and frightened by magic—as even the mestizo elite of Negros would be in the late nineteenth century (cf. Worcester 1898, 272–273). The elites also had to devise ways of coexisting peacefully with the unsubmissive upland settlers who had pecuniary importance for the old elite's petty commerce.[40] The descendants of the *datus,* therefore, endured further aggravation of their cultural ambivalence through the rigidified structural position they occupied, which required them to negotiate through a world dominated by the friar but also inhabited by the overtly loyal *indios* as well as the rebels, fugitives, and shamans lurking in the colonial shadows.

The old native elite's structural difficulty was heightened by the fact that the substance of the role they performed had departed from its preconquest meaning, for the *cabeza* had become a mere cog in the colonial administration. While the *datu* had rights of collection and disposition over a portion of his followers' produce, the *cabeza* was a mere tribute gatherer, a position not always enviable since, under a ruthless Spanish *encomendero* or tax farmer, the full yearly tribute was demanded on pain of torture.[41] Whereas the *datu* had full command over the labor services of his followers, the *cabeza* under the colonial system of draft labor acted essentially as a foreman for the colonial state.

Deprived of the chance of being respected and revered as an ancestor *(papu)* in the afterlife, but also excluded from the basically European Catholic sainthood, the native *cacique* saw power from a more temporal and this-worldly perspective. The native elite's solution to the dilemma of power and culture was to use its structural position as a vehicle for the opportunistic exercise of a hereditary post that, as we shall see, was an extension of the art of gambling. Probably to compensate for their social demotion, the *cabeza* became involved in illicit activities, such as pocketing the villagers' nominal wages for draft labor, with few qualms of conscience (Phelan 1959, 99–101, 115, 156–157). The wealth gained thereby was conspicuously displayed in dress and personal ornamentation, which in preconquest times would have been a sign of spiritual approval, but which under colonialism became a symbol of their insecure status and questionable role as an aristocratic class.

The native elite's vitiated view of power was passed on to later generations, no doubt abetted by the Spaniards' own exercise of power from a similar mold. The Chinese mestizos, who replaced the old native elite, were groomed in an even more ambivalent and opportunistic environment shaped by their ethnic Chinese fathers and native mothers. As merchants and artisans, a sizeable segment of the migrant Chinese male population in Filipinas went through the ritual of baptism as a shrewd legal tactic. Besides entitling them to a Spanish godparent, nominal conversion to Catholicism reduced the head tax of the Chinese, entitled them to land grants, and lifted the restrictions on residence and travel to the countryside (Wickberg 1965, 16).

The Chinese man who converted to the colonial religion married a native woman, who, as Edgar Wickberg conjectures, most likely had "some business sense herself and could help him run his business" (Wickberg

1965, 33). Many of these women probably came from the old elite, whose fortunes were dwindling and who sought to take advantage of the emerging market economy in the eighteenth century by marrying an entrepreneurial person. But for the native women it meant crossing a cultural divide, for the *indio* did not have a high regard for the ethnic Chinese who, on several occasions, were massacred or expelled from the colony by the Spanish government at Manila. On the other hand, the Chinese who saw the *indio* as an inferior creature decided on a pragmatic course of action in view of the absence in the colony of Chinese women they could marry.

By the 1740s, as Wickberg's classic study shows, the progeny of those mixed marriages, the Chinese mestizos, were numerous enough to be classified by the colonial state as a separate entity within native society. Constituting a distinct legal category, the Chinese mestizos were levied tribute higher than that of the *indio* but lower than that of the Chinese; they were required to render a fixed amount of forced labor service every year like the *indio,* an exaction to which the Chinese were not subjected; but, unlike the Chinese, the mestizos were free to change residence and participate in local government (Wickberg 1965, 63–65). Thus, the Chinese mestizos straddled a formally recognized middle position in colonial society.

Socialized into their middling status, the Chinese mestizos not surprisingly learned to be masterful opportunists who, as contemporary observers suggested, instrumentalized roles, norms, and values. They became experts at the learned imitation of religion, language (Spanish and the vernaculars), mode of dress, and other aspects they found desirable—thus fostering the modern Filipino penchant for the copying of form, thinking it equivalent to substance. Although mestizo imitation was radically different from shamanic imitation, the mestizo strategy made possible a level of adeptness at practicing both Spanish and native cultures that turned the Chinese mestizo into a skillfully versatile trader. To safeguard their interests, they had to be able to identify with accuracy individuals on whom they would place their bets. From their vantage point, the mestizos learned to stand back and become acute spectators within colonial society. The mestizo therefore became the consummate middle person as well as the gambler who adroitly profited from the manipulation of risks and intermediary functions.

Structural opportunism, however, did not resolve the socially marginal person's dilemmas of power and culture. Witness the grotesque imitation of Spanish culture and the scorn heaped upon the *indio* by the inimitable *Doctora Doña* Victorina de los Reyes *de* de Espadaña in Rizal's *Noli Me Tangere.* Fostered by the Chinese father's feelings of superiority over the *indio,* on the one hand, and *indio* society's lack of regard for the Chinese, on the other, the mestizos became overzealous in their mimicry of the Spaniard, who generally held both *indio* and mestizo in contempt. In this manner, the mestizos, whose *dungan* were in virtual limbo, added another layer of ambivalence to the contradictions absorbed from received *indio* culture. As shown in the next chapter, notwithstanding the affluence they were beginning to reap from trade and agriculture, the Chinese mestizos did not enjoy any residual prestige from the ancient *datu*ship, nor did they sustain whatever prestige was left of the old native elite. The Chinese mestizos, who later comprised the ruling classes of the Philippines, were certainly not a traditional aristocracy. But, as the high point of the colonial invention of the family, the mestizo elites began a tradition of conjuring up genealogies of illustrious forebears with many a connection to an imaginary royalty.

Although intermittently bothered by "evildoers," the friars by the early nineteenth century retained undisputed power in colonial society—until the merchant capitalists began to pose a serious challenge to the monopoly of Friar Power. The cholera epidemic that struck in October 1820 was an unprecedented disturbance of the health of the capital and suburbs, which, just a year earlier, had been described by an American sailor as "proverbial" (White 1962, 104). More significantly, in the midst of the epidemic, holy water and the friars' magical ensemble failed to neutralize the poison believed to have been sown by foreigners. It seemed as though only through massacre could Friar Power avenge itself against the enemy.

The subsequent growth in the presence of foreign merchant capitalists must have impressed the natives, who could well have arrived at the conclusion that, in the white man's spiritual realm where the Hispanic spirits also had their enemies, those in support of the Protestants/Masons were acquiring a strength sufficient, in a new clash of spirits, to allow them to challenge Friar Power. The wedding of George Sturgis at about the middle of the nineteenth century proved to be the crucial test. The specific attraction of the wedding episode to the natives was that it bore the mark of a spectator event on which those anxiously gathered at the Manila harbor could bet as to its eventual outcome.

Highlighting the dialectical inseparability of economics, politics, and culture, the pivotal conjuncture represented by Sturgis' wedding led to the accelerated incorporation of local agriculture into the circuits of global capitalism. In the wake of a triumphant Masonic Capitalism, the presence of Catholic Spain's enemies in the colony had to be tolerated, with the native reading of the situation fully in accord with the realities of Spain's twilight empire. Applauded by cheering natives, the successful Protestant rite added impetus to native participation in the export economy. By protecting Sturgis' Catholic wife from the clergy's forebodings of infernal damnation, the foreigners demonstrated the sort of mystical strength that could serve as a counterpoint to Spanish shamanism. Condemned as evil but able to ward off friar opposition and Catholic reprisal, the Protestant/Masonic capitalists began to signify an alternative storehouse of power that could be tapped for luck as the natives negotiated their way through a changing world.

For the enterprising Chinese mestizos, the wedding's cosmic significance might well have meant that colonial categories could be transgressed and stakes safely wagered with the formerly denigrated "merchants of evil," from whose hands money—as in a gambling den—flowed to the winners. To the mestizos, the foreign merchants represented a clear opportunity for economic advancement as well as a cultural pole of identification to deal with the dilemmas of cultural ambivalence. There would prove to be no dissociation of the Chinese mestizo from the contradictions of received *indio* culture, but class and cultural differentiation was accentuated later in the nineteenth century by the liberal education wealthier mestizo children acquired in Europe, where the more activist became Masons. Education and travel were afforded by gains from export agriculture, particularly sugar.

By the time of its full incorporation into global capitalism in the early nineteenth century, Philippine agriculture had already undergone a series of changes that became integral to the founding of export-crop production in various parts of the colony, including Negros. The changes in the economic structure of colonial society must be understood as inseparable from the same historical dynamic that produced the social and cultural transformations discussed in this chapter. The historical sociology of colonial agriculture is presented in the next chapter through a reconstruction of the relations of production prior to conquest and during the colonial era.

3

Elusive Peasant, Weak State:
Sharecropping and the Changing Meaning of Debt

Cosmology and Preconquest Production Relations

As we saw in the previous chapter, the islanders of the preconquest world had configured a hierarchized social order according to the distribution of charisma and prowess and the economy of prestige as ordained by the spirits. An islander who could not claim otherworldly prowess to be reckoned as *datu* entered the penumbra of one whose claim to individual supremacy was validated empirically by deeds of valor. Either as warriors or dependents, the *datus'* followers were grouped in settlements, conventionally known as *barangay* in Tagalog (*hapon* or *haop* in Visayan), which were of highly variable character and size, some encompassing thirty, others about a hundred households. In a world splintered into dispersed bands and settlements locked in intermittent conflict, group membership was a requisite for individual survival. In that context, the *datu* managed the *barangay* as his protectorate. If the *datu* was not waging war on other settlements, he was entering or cementing alliances with other leaders, often by giving daughters in marriage. Such alliances were useful in confronting a common enemy or undertaking a common war effort, and in establishing a conducive environment for trade.[1] While at one level ensuring the safety of followers and dependents, the management of external relations at another level was crucial to a *datu*'s personal competition with other men of renown and the maintenance of power within his own settlement.

Within the micropolity he ruled, the *datu*'s spiritually sanctioned authority covered such aspects of life as we would today segregate into the legal, the military, the political, and the economic. Standing at the vortex of human society, the *datu* was the broker of goodwill from the spirit-world to his followers and dependents. As the repository of special qualities, skills, and knowledge, the *datu*, with the priestly consort of the *babaylan*, rendered invaluable services to the settlement, ranging from ensuring its defense and the individual members' protection to settling disputes and adjudicating difficult criminal cases, hosting ceremonial rites, and being responsible for the overall welfare of kinsmen and dependents (Chirino 1969, 384, 430; de Loarca 1903, 141, 147–149, 151; de San Antonio 1906, 356–358). The legalist Antonio de Morga (1904, 119) described *datus* as having the "duty . . . to rule and govern their subjects and followers, and to assist them in their interests and necessities." Hardly ever completely recompensable, the *datu*'s services were like gifts to his followers, which amounted to the socially recognized debt *(utang)* the latter owed their headman.

As conduit, hence receiver, of favor from the spirit-world, the *datu* was in turn the giver to his community, his ostensible services finding embodiment in tangible gifts. Even in the vestigial *datu*ship of Bukidnon in the late 1960s, "informants appeared to regard the principal sign of *datu*ship as the ability to give things away freely" (Biernatzki 1973, 43). In fact, the Bukidnon *datus* were "called *tatay,* 'father,' by their willing followers" (40)—not dissimilar from observations proffered by early Spanish chroniclers that old chiefs were "esteemed as a father," and that if followers "wanted any small trifle, they begged the head chief of their *barangay* for it, and he gave it to them" (Chirino 1969, 430; Colin 1906, 96). Another European observer described *datu*–follower relations as one of "friendship" (de San Antonio 1906, 348). Certainly, the anthropological ideal type of Big People is one of generous gift giving to juniors and subordinates, for which the former were accorded great respect (Collier 1988, 92–105, 114–119; Sahlins 1963, 289–293).

In return, the *datu*'s leadership was expected to be reciprocated by followers and dependents with loyalty and gratitude expressed in the form of obeisance, deference, labor services, and crop sharing. As Colin (1906, 96) noted, the *datu* "summoned" his followers, who were "obliged to go to him to work in his fields or to row in his boats." Whenever a feast was to be held, members of the settlement "all came together," bringing with them "a jar of wine, so much rice, and to assist in such feasts" (Anon. 1979, 353; Colin 1906, 96). Followers of *datus* "served in their wars and voyages, and in their tilling, sowing, fishing, and the building of their houses. To these duties the natives attended very promptly, whenever summoned by their chief" (de Morga 1904, 119). The islanders were wont to "give their chief a little of everything," and in turn, were held by the *datu* "in great veneration and respect" (Anon. 1979, 353; de Morga 1904, 119).

Because the whole compendium of *barangay* existence was seamlessly intertwined with the indigenous worldview, *datu* and followers could be seen as linked in a social nexus consisting of norms prescribed by the indigenous understanding of the cosmos. These norms entailed economic as well as political relations, and in a real way affected the flow of material resources. In return for the *datu*'s nonquantifiable and overfulfilled giving, his followers engaged in infinite expressions of gratitude because of their necessarily underfulfilled reciprocal obligation. The preconquest settlement, however, should not be romanticized as an idyllic gift economy. The *datu*'s concern for his followers' general welfare was profoundly indistinguishable from the self-centered calculation of his own personal power and social standing. Moreover, the acts of reciprocity on the part of followers were inseparable from a tacit fear that the magical *datu* could, if he chose to, exact retribution for noncompliance with norms.

Fear, however, was muted by a more explicit set of social expectations deemed legitimate by everyone in the settlement. The soothing of this fear, an indirect acknowledgment of its existence, was initiated by the *datu* himself through the courtesy of not treating his followers with contempt, but rather relating to them as a "father" or "friend." Neither was the *datu* dismissive of the things given him by his dependents; rather, he treated gift offerings "with veneration and respect." Through this exchange of deference, which seemed to erase hierarchy but in fact was its very affirmation, the *datu* evinced wisdom from the spirits and became even more respected and "esteemed as a father."

Let me point out that the apparent paternalism and salience of kinship ties among members of the band or settlement do not justify the description of the *barangay* as a "family-based community" often given in Philippine historiography.[2] It has been asserted, for example, that "The social order was an extension of the family with chiefs embodying the higher unity of the community" (Constantino 1975, 38). Such statements assume that "the family," employed in its modern ideological sense, was already in existence. On the contrary, the path to *datu*ship discussed in the previous chapter and, as shown below, the practice of debt peonage evinced an underlying disregard for what today would be revered as "family ties." Moreover, warriors could transfer loyalty and move from one *barangay* to another without encumbrance from the kinship system, a practice that would be considered incongruent with contemporary notions of family.[3]

In a similar vein, to describe the preconquest settlement as governed by "communalism" is problematic (cf. Constantino 1975, 38–39). Although the institution of private property was absent, personal possession even in the non-Islamized areas was socially recognized. Both chiefs and ordinary people were entitled to rights of possession. Especially in the matter of dwelling units and personal effects, individual possession was respected to the extent that "no one was able to usurp" something claimed as belonging to another (de San Antonio 1906, 348). Owing to the inseparable unity of possession and personal honor, personal ownership of things was valued, and stealing within the settlement was considered a most serious offense (de Loarca 1903, 145, 147; de Morga 1904, 128–129). In addition, everyone had a piece of land to till, which, in the case of a swidden patch, an individual selected following a ritual that served to divine the fertility of the soil (Alcina 1960, chap. 5). After having planted the field with coconut, banana, fruit trees, upland dry rice, and other crops, one was considered "master" of the land one had cultivated. Within the settlement, labor exchanges, which were crucial to speed up work on sowing to take advantage of the seasonal rains, were rewarded with "the invitation to eat on that day, the owner of the field bearing the expense."[4] The "master" of a piece of tilled land was socially recognized.

In the interest of safeguarding his monopolistic dominance over gift giving and the concentration on his person of respect, the *datu* controlled comparatively larger and more strategic holdings that were worked for him by dependents. To mark status boundaries, ordinary members of the settlement observed spatial boundaries: they were not allowed to enter or walk "in the water in the field belonging to the chief" (Anon. 1979, 330).[5] Furthermore, because of his superior magical status, the *datu* could engage in "removing and giving lands" (338). Evidently, the final decision on the use of land within the *datu*'s protectorate was vested in the *datu*, in keeping with the deep knowledge he had acquired from spirit guides. Seven days prior to harvest, the *datu* led the observance of the rites for a bountiful harvest (de Loarca 1903, 165). From their own fields, the islanders gave a portion of the produce "in varying quantities" to the *datu*, what de Morga and the writer of the Boxer Codex conceived as "tribute," but which the natives called *buwis* (Anon. 1979, 353; de Morga 1904, 119).

In the complex world of exchanges, material and magical, the settlement's surplus product was appropriated primarily by the *datu*. Because of his control over the surplus, which was small but sizeable enough to make trade with peripatetic merchants from distant places possible, the *datu* served as the main conduit through which trade goods flowed in and out of the settlement. Thus magnified was the chief as channel of blessing and goods. From the perspective of indigenous cosmology, the mighty chief deserved the act of appreciation expressed through the *buwis*. Given this cultural system, it was to the *datu*'s advantage that more people be dazzled by his courage and magic, as there would then be more givers to the chief who, consequently, was the wealthier. On the whole, preconquest production of material goods, whether in the field or sea or generated through warfare, was under the overall aegis and sponsorship of the *datu*, for which he was entitled to a share of both crop and booty. Where greater spiritual assistance was crucial, as in plundering expeditions, the *datu* claimed a larger proportion of the objects gathered (de Loarca 1903, 151).

Thus, not only were preconquest relations of production inseparable from, if not totally subsumed by, the islanders' worldview, but, following Maurice Godelier (1978a, 1978b), preconquest cosmological principles could be said to have constituted the preconquest relations of production. Cosmology determined social control over resources such as land and labor and the disposition of the surplus. Wrapped in fear and reverence, preconquest crop production and distribution did not constitute a form of sharecropping, as some have referred to it (Cushner 1971, 72; McLennan 1969, 655; Phelan 1959, 20), as though the produce given to the *datu* "in varying quantities" had been stipulated by a land rental contract.

Moreover, contrary to what the Spaniards perceived as the *datus*' high-handedness and tyranny from which Spanish colonialism would rescue the natives, the people in each settlement, all of whom shared the same belief system, generally did not feel oppressed. They willingly submitted to the *datu* and cheerfully consented to the socially accepted exactions of labor and produce. Indeed, so enjoyable and intoxicating were the feasts and celebrations they shared that, in the end, entertainment was inconceivable apart from the *datu*, and the islanders must have felt that, in serving him, they ultimately were serving themselves. Individual survival and day-to-day existence were inconceivable apart from the *barangay*.

Certainly, there were culturally defined limits to the *datu*'s prerogatives that, when exceeded, were felt as oppressive.[6] A *datu* who transgressed those limits became the object of "insulting words," considered a grave personal affront, hence a serious crime (de Morga 1904, 128–129). Under normal circumstances, however, there was no reason for a *datu* to impose his will upon the weak. If that happened, followers could strike against an unreasonable head, their last recourse being the offering of their allegiance to another Big Man or to one with such a potential. But although the weak could play off the powerful against one another, they could also be refused the support desired from another Big Man, which would place the disloyal one under the betrayed *datu*'s mercy. In the end, loyalty was prized and the *datu* sought to minimize internal disputes or resolved them immediately. With chiefly domination consented to by members of the settlement and continuously affirmed by the spirit-world, physical coercion was needless.

Among the nonchiefly islanders, it was the practice to lend to one another as part of the social norm of belonging to the same settlement, what Chirino characterized as the "spirit of good neighborliness" (Chirino 1969, 310). By virtue of the social relations in the settlement that placed the *datu* at its center (with an underlying, but euphemized, fear), everyone in the *barangay* was expected "to aid one another mutually" (de San Antonio 1906, 348). Alcina (1960, chap. 20), while decrying the "avarice" of the Visayans, also offered the concomitant observation that "among themselves" they were "very liberal and helpful of one another." In the period following the rice harvest, the pot was always full of the prestige food ready to serve any visitor, a "prodigality" that fast depleted the rice supply, "because on these occasions they eat without moderation" (Alcina, chap. 20). If rice was needed for ritual purposes but the supply had been exhausted, one had to borrow it, which was returned during the next year's harvest doubled in quantity to express gratitude and ward off anger and retaliation, a custom Spanish chroniclers interpreted as avarice and usury (Anon. 1979, 354; de Loarca 1903, 151, 153; de Morga 1904, 124, 127–128).

In the precolonial epoch, the early Spanish accounts suggested a level of abundance and a relative ease in the matter of acquiring sustenance. If rice or millet had run out before harvesttime, there were tubers and fruits to subsist on (Scott 1992b). In this context, to view debt in the pre-Hispanic settlement as arising from the present-day sense of a subsistence level of living—that is, poverty—would be inappropriate, just as using the same meaning of "debt" in the context of a hunting and gathering society, what Marshall Sahlins (1972) calls the "original affluent society," would be untenable. But in a famous essay, William Henry Scott (1982) viewed precolonial debt using the poverty lens and even conjured reified patron–client relationships as though the preconquest settlement had been a rigidly structured feudal society,[7] a point clarified by Rafael's arresting critique (1988, chap. 5). Rafael (chap. 4), nonetheless, has similarly juxtaposed twentieth-century relationships of debt and reciprocity with those obtaining during the period of conquest. To use Sahlins' (1972, 3) eclectic phrase, both Scott and Rafael have therefore fallen prey to "bourgeois ethnocentrism." But, as E. P. Thompson

(1977, 256–258) reminds us, there is no constant "act of indebtedness" that can be isolated from particular social and historical contexts.

What debt there was did not emanate from economic deprivation, as in a class-structured society. Rather, debt could have been induced by the lack of ritually prescribed goods, particularly marriage gifts, prompting some bachelors to obtain those goods from seniors with larger families and more family labor and, hence, more marriage-validating resources.[8] Big People, especially those from smaller settlements, could also have incurred debts in the process of hosting alliance-building ceremonies or compensating military allies from other settlements.[9] Indebtedness was also frequently triggered by legal infractions or acts violative of personal honor and prestige, for which reason the payment of gold, the highest symbol of prestige, was the usual penalty (de Loarca 1903, 143, 181).[10] If a violator did not have the required quantity of gold, it had to be borrowed from someone—a parent, sibling, or another person in the settlement, the *datu* especially—who was then repaid by the offender in terms of obligatory labor. Occurring irrespective of kinship relations, debt peonage existed as an indirect act of appeasement, via the lender, to soothe the wounded honor of a third party. Judging from the early Spanish accounts and the known *dungan* rivalries among lesser mortals, offenses against personal honor must have been rather rampant, which would have accounted for the sizeable number of debt peons at the time of the Spanish conquest in the late sixteenth century.[11] Debt was the equivalent of *dungan* defeat.

Loosening Ties: Land, Gambling, and Individual Autonomy

A hundred or so years after the establishment of colonial rule and the atrophy of the preconquest *datu*'s charisma in the face of Friar Power, a fundamental change had occurred in native society: the decline of the ancient debt peonage. Although at the outset of the colonial epoch the *datus* and their immediate descendants had innumerable peons working for them, by the mid-seventeenth century the number in servitude to a native elite family had been reduced to only about two or three (Cushner 1976, 18).[12] The ties that bound had been loosened.

In marked contrast to observations made during the period of conquest that labor was at the *datu*'s beck and call (Anon. 1979, 323, 330, 342; Colin 1906, 96; de Morga 1904, 119, 155), debt peons thence, as Scott (1982, 107) noted from an early dictionary, had to be begged to render "help with something" and only "from time to time," not "too often." Alcina's (1960, chap. 5) observation in the 1660s that the native elite celebrated the end of farm work in a special way by reveling with "a kind of vanity and ostentation" in being able to serve "food and drink in great abundance" probably pointed to a new mechanism for mobilizing labor. No longer were *indios* under compulsion to bring contributions of food and drink to feasts and celebrations; somehow the tables had been turned. The *"conquista espiritual"* had broken the unitary canopy that, in the precolonial world, subsumed economic relations of production under cosmological tenets.

The social expectations of the preconquest *barangay* had lost their force, and it became possible to see the *datu*'s demands as excessive or even oppressive.[13] With the waning of the *datu*'s traditional magic and prestige, and the concomitant imposition of a new social and legal system under the friar's personalistic rule, the *indio*'s release from debt bondage revitalized the *dungan* and freed surplus labor from the elite's control. But, in simultaneously being subordinated to an external overlord, the *indio* became a peasant. The evolution of a new societal arrangement found the *indio*-as-peasant in a new contested terrain. In colonial society, the old norms safeguarding personal honor were replaced by expectations organized around conformity to the friars' religion and personal dictates and to fulfillment of the colonial state's demands for head tax, corvee labor, and a quantity of produce known as the *vandala*. The new rules of the social game affected the totality of interpersonal relations, and one colonial edict was particularly pertinent to the issue of debt.

In the 1640s, Spain passed a law prohibiting loans in excess of five Spanish pesos, not even "under pretext" that the money was advance pay for rice or some other product: "If more is given, it shall be lost, and the [native] receiving it cannot be made to pay it" (Corcuera and Cruzat 1907, 199). In the 1780s, an edict reiterated that imprisonment or coercive payment of personal debts in excess of five pesos was illegal (Basco y Vargas 1907, 292). In conjunction with the sometimes forcible emancipation of peons bound by debt obligations, the law on personal debts must have in time gained wide recognition across the colonized territories (cf. Cushner 1976, 18; Phelan 1959, 129–131). By the mid-seventeenth century, *indios* who had inherited the preconquest peonage status could have ignored the calls of the by then hereditary village chiefs to abide by customary norms. With the loss of the *datu*'s magical mystique and with the law on debts, the old native elite lost legitimate grounds for the imposition of debt bondage. For the first time, the *indio*, though subservient to Friar Power, experienced a liberation of sorts from tradition.

Colonial society and its relative peace made room for the real possibility that the native could stake a claim on a parcel of land to become an independent cultivator. Although most of those who formalized land claims were from the native elite, the introduction of legal titles to land and the emergence of a land market as early as the 1590s (Cushner 1976, 18, 85), or some thirty years after conquest, built upon the indigenous concept of personal possession and transformed it into a more solid concept of individual property. The idea of land as exchangeable for cash, hence an alienable private commodity, was early on fostered, following colonial state directives, by the use of the pulpit to announce all land sales.[14] Despite the persistent confusion over land titles during the whole Spanish era, the natives did imbibe the concept of private proprietorship and of land as inheritable property. Consequently, even before the commercial revolution of the late eighteenth century, the smallholdings of *indios* were demarcated from lands owned by the native elite. Native peasants also engaged in land disputes.[15] By the early 1800s, Tomas de Comyn reported that natives "generally possess a small strip of land situated round their dwellings, or at the extremities of the various towns and settlements," landholdings he differentiated from those owned by the elite mestizo class (de Comyn 1969, 21).

The idea of peasant autonomy had been born. Though many times overbearing, the exactions of the colonial state came at predictable intervals, and no special relationship need have been cultivated with the village *cabeza* who, while exercising some power in his own right, had become a mere tribute gatherer and foreman of the state. Often paid in kind, the tribute could be avoided by natives whose geographic movements were beyond colonial state control (Cushner 1971, 104–108).[16] Moreover, the natives learned that, as long as they complied with the routine performances of attending mass and the compulsory rituals at life's passages, Friar Power could somehow be held at bay.

But what was the native to do to ensure that the land under his possession yielded a good harvest? With the marginalization of the *babaylan*, the friar's disapproval of indigenous practices, and the disappearance of the magical *datu* from the scene to lead the customary rites of agriculture, but with the continued salience of the spirit-world, the individuation of the *indio* was forged. Peasants were left to their own devices to negotiate with spirits, a process deemed imperative to enhance good fortune in agricultural production. Deploying their ingenuity to domesticate an alien source of power, peasants individually began to use Palm Sunday leaves, pieces scraped off saintly statues, and other Catholic artifacts furtively obtained during a procession or some other ritual as magical objects on the farm and at sea. The popular access to the colonizer's supernatural power realm intersected with what in Chapter 2 was portrayed as the quiet spread of new practices fostered by the shamanic subversion of Friar Power and the appropriation of the Catholic magical ensemble. But, though magic was possible, the direct intervention of a priestly caste in a collective agricultural ritual was not. Therefore, magic became highly individualized, frequently a personal secret, in agriculture and in other areas of life.

As the epitome of the atomized native's strategy of negotiating with the spirit-world, the *anting-anting* was marshaled by colonial subjects, as the previous chapter has shown, in their individual and largely uncoordinated attempts to defy the Spanish colonial state. The *anting-anting* was also relied upon in the cockpit, the arena that expressed the *indio*'s contradictory relationship to the colonial state. Because of the *indio*'s cultivated penchant for cockfighting, the game attracted natives to the colonial centers that housed the gaming structures, thus facilitating the state's use of the game to effect native incorporation into the expanding money economy, particularly from the eighteenth century onward.

The florescence of gambling during the Spanish colonial era can be deduced from the fact that the contemporary Tagalog word for gambling is *sugal,* from the Spanish *jugar* (to play or gamble), while the Spanish *tahur,* meaning gambling as well as cheating, is the Ilonggo noun for gambler (as well as a "deep" Tagalog word for gambling). As appropriated words, they suggest radical changes in the *indio*'s views about destiny, fortune, and success in a cosmologically altered world.[17]

In the precolonial Visayas, sacrifices were offered to the *diwata* in supplication of good fortune for an undertaking such as rice planting or a sea expedition (Alcina 1960, chap. 15). However, one's overall destiny was already foreordained and etched in the *palad,* the palm of one's hand, which showed whether one was to become a *datu* or an ordinary person. One could attempt to reverse a specific misfortune through *sibit,* by piercing the lines that told of the foreboding and drawing blood from it, but the palm remained the supreme text that disclosed one's destiny *(kapalaran),* the verdict of the spirit-world.

During the colonial epoch, the reconfiguration of indigenous society and the *indio*'s own gambling response gave currency to concepts of *suwerte* (or *swerti*) for good luck, from the Spanish *suerte* (fortune or luck), and *malas,* or bad luck, from the Spanish *de malas* (out of luck). Ironically, under colonialism, the palm lines were liberated from signifying immutable events bound to happen. Instead, *suwerte,* in being closely associated with gambling, became individually negotiable with the spirit-world and with the changing structural features of colonial society. As in the imagery of the modern Tagalog phrase *gulong ng palad,* the palm itself could rotate *(gulong),* literally like a wheel *(gulóng).* Somehow, despite the dictates of fate, the individual bore some responsibility for the

Sharecropping and Debt

turning of the wheel; one was answerable for the *suwerte* or *malas* encountered in life.

Moreover, the appropriation of the Spanish *tahur* was suggestive of the varied routes to success made possible by colonial society, where corruption was endemic. Many beliefs and practices flourished as individual devices for negotiating with the supernatural, for manipulating the spirits' volition, a very early expression of which was analyzed in Chapter 2: the romantic liaisons by native women with the *engkanto*/friar. In due course, gambling itself became the avenue for taking risks with the colonial state. In 1701, the Spanish *Alcaldes Mayores* (provincial governors), who were charged with penalizing gamblers, were denounced by some friars as "the very ones who secretly give full license and permission for gambling games, in consideration of the money [read: *suerte*] which they receive every month for the said license" (Vila et al. 1906, 135). Corruption on the part of Spanish officials continued until the end of Spanish rule. In the 1890s, Amadeo Valdes, governor of Negros Occidental, masterfully amassed what then was a staggering 70,000 pesos during one year in office (Simonena 1974, 36). This phenomenon in the colonial state, of flouting the law with impunity, became a principal strategy of the native planter class that emerged in Negros.

In addition to outright cheating, *suwerte* was believed to emanate from a variety of sources, such as the disabilities of children. Because congenital handicaps were nonexistent among the preconquest islanders or, as Alcina (1960, chap. 23) hypothesized, were banished through infanticide,[18] the Iberian colonizers introduced the value of physical features culturally defined as deformities in relation to the attainment of *suwerte*. To discourage infanticide, considered morally reprehensible in Christendom, physical oddities would appear to have been redefined by the friars, especially as the association of such phenomena with luck was well established in the Peninsula.[19] Various forms of mental and physical handicaps came to be believed by local parents to be magnets that attracted *suwerte* to their families. What had earlier been seen as misfortune became convertible to good luck, in the same way that subjugation could be put to advantage depending on how well one played the social game.

As we shall see in Chapter 6, the negotiation with the spirit-world was later to be seen as also attainable by "selling one's soul to the devil," which people deduced from wealth accumulated from export agriculture. But transacting with spirits could also be conducted through rather innocuous practices that flourished in time, many still prevalent today, such as the manner of sweeping dirt out of one's house, which can bring either *suwerte* or *malas*. A home-lot owner who desists from disturbing a mound of earth and its spirit-dwellers *(dwende)* will be lucky some fine day. The number of steps in a staircase are counted in sets of threes, *Oro, Plata, Mata,* signifying Gold, Silver and Death, to ensure that the last step does not end in *Mata,* the harbinger of misfortune.

The central part played by coins in gambling—for access to the cockpit and for making bets, a practice firmly in place by the mid-eighteenth century[20]—has lent coinage the quality of mediumship for negotiating with the spirit-world. Catholic icons, the Santo Niño especially, are garbed in finery and "given money" to bring *suwerte* to the idol's owner. During house construction, coins (sometimes foreign currencies) are buried at corner posts and beneath the main entrance to appease unseen spirits who, because of the offering, are expected to bestow luck to the homeowner—the money being expected to literally "enter" and "step into" the house, as well as serve as its foundation. In such a fetishistic context, money begets money. The principle finds its generalized expression in the *balato,* the distribution of small amounts (to passersby as well as to the victor's kin and associates) from one's winnings in gambling, an integral aspect of the ritual of gaming founded on the belief that to share one's luck augments future chances of victory.

Thus, systematic generosity in allowing money to leave will encourage it to return in bigger quantities. The willingness to bet repeatedly and part with money (and other valuables convertible to money) can be a sign of trust in one's friendly relations with certain preternatural beings who are expected to reciprocate in the form of financial windfalls. The bet is an investment in both the unseen and the social world, the act of lending, a bet and an investment. In turn, as part of the complex of wagers and *suwerte,* the act of being in debt can entail betting one's farmland, which was resorted to by substantial numbers as early as the 1760s (Raon 1907, 241). Regardless of the risks involved, including betting one's very station in life, every potential avenue to *suwerte* is explored.

Within the gambling milieu, transactions with the spirit-world are a form of propitiation but also of contractual exchange, a seeking of favor but also a quid pro quo, the offering somehow obligating the spirits to respond favorably. Fatalism, with its unchangeable givens, nonetheless makes room for human cunning and manipulation in order to avoid the snares of fate. With gambling constituting the organizing principle of society as well as its joy, social existence and personal life chances have come to be seen as the outcomes of how shrewdly individuals pursue their *suwerte,* conceived as somehow preordained yet amenable to purposeful circumvention.

The early emergence of a land market, particularly in and around the colonial capital, provided a means for the pursuit of *suwerte.* Relatively easy access to money could be had by selling off small land parcels. Following colonial regulations, the native was required to state the reason for parting with the land—the owner being too old, the land being too hilly, the seller having other plots to till—to initiate the procedural conversion of property into cash. The reasons were often fabricated to consummate the deal. Through this process, many small parcels of land for grazing or already devoted to rice farming were alienated by and from the natives in the Tagalog area. These patches of land were initially acquired by Spanish civilians, then subsequently resold to the religious estates (Cushner 1976, 21, 26–27; Roth 1982, 134). Speculation in land had very early beginnings.

The Transition to Sharecropping

Perhaps because they no longer controlled the surplus and their resources were increasingly strained by societal demands, the native elites proved most ingenious in their use of the new land market. They sold off lands they had usurped from within the protectorates of the deposed *datus* as well as uncultivated land parcels belonging to the public domain, which they had claimed as their personal property (Cushner 1976, 18–20; Larkin 1972, 53; Phelan 1959, 117). Even where sale of land was not intended, the native elite used the new legal system opportunistically to extend claims over land parcels possessed and cultivated by other natives. In the latter case, land ownership became a means to secure control over native labor "freed" from preconquest norms.

By the end of the seventeenth century, the old protectorate concept had been utilized to claim control over *barangay* lands, and the *buwis* were reduced into a contractual form of land rent.[21] The legacy of such native elite machinations is the equation of the word *buwis* with farm-lease payment and, in present-day usage, with taxes paid to the state. The collection of land rent was justified in terms of the elite's responsibility for making "legal safeguards" to protect what was deemed to be the local community's corporate property. By recourse to legal formalism, the reemergence of the preconquest protectorate concept in a new guise allowed the native elite to gain control over a portion of the surplus product generated by the native peasantry.

In this context, leaseholding and farm tenancy, as well as economic debt owing to failure to pay rent, had their formal genesis. As functionaries of the colonial state, the elite also found a pretext, illegal of course, for seizure of private property in case of noncompliance with regulations on tribute. Although at the level of appearances they seem akin to custom, social relations were gradually but profoundly altered between the native elite-turned-landowner and the disenfranchised *indio* cultivator-turned-sharecropper.

A probable model for the leaseholding system devised by the native elite was the monastic estate in the Tagalog region. To raise revenue for the religious coffers, the estates began to lease land in the 1590s to sharecroppers (called *inquilinos*), initially ethnic Chinese who, later in the seventeenth century, were replaced by *indios* as the Chinese moved on to trade. Rice was the principal crop, but sugarcane and tobacco were also raised. For the right to cultivate small plots of land, the *inquilinos* paid the ecclesiastical landowners a fixed fee, with a four-year grace period should the land need clearing of forest growth (Cushner 1976, 42–45, 49–50). The monastic orders made sure they did not suffer from fluctuations in farm yield by stipulating the rent in fixed monetary terms but paid for in kind, according to the price of rice prevailing during harvesttime (Roth 1982, 137).

The terms of the sharecropping contract, on the face of it, ran counter to what James Scott (1976) has referred to as "the moral economy of the peasant," as the cultivator's right to subsistence was not assured but the cleric's right to ground rent was. Why, then, did some natives choose to work for the monastic estates when they could have farmed on their own in other areas? It would seem that, for such *indios*, to be in the employ of Friar Power was to bask in the magic and protection of the Hispanic shamans, as though the monastic estates were a reincarnation of the *barangay* under the leadership of men of prowess. Historical sources suggest, for instance, that native *inquilinos* felt arrogant toward and superior to other natives who worked casually as day laborers or who tilled their own lands bordering the friar estates (Cushner 1976, 49). To be within the penumbra of Spanish magical men also meant added protection from colonial state exactions, because the religious shielded their wards from tribute collection and had them exempted from the harsh corvee labor of timber cutting, log hauling, and shipbuilding in the face of the Dutch threat during the first half of the seventeenth century (50–54). Attachment to the monastic estates was also a source of pride and privilege, as the clerics provided loans and cash advances and introduced innovations in farm technology. Their seemingly esoteric knowledge led to the spread of irrigated wet-rice cultivation, the use of the plow (consigning to oblivion the digging stick), and the harnessing of the indigenous water buffalo, or carabao, to pull the plow (11). (Not coincidentally, the Philippine word for plow, *arado*, is the same as the Spanish.) A sign of monastic prowess, the rise in land productivity could thus have made the *inquilino* arrangement even more attractive.

There were natives, however, who could not endure conditions in the friar estates, particularly the landless, who were supposed to have been hired and paid a nominal wage but were not. They rebelled by fleeing to the hinterland beyond the reach of the colonial state, there to pursue their own distinctive anticolonial style of life. Because of such outright desertion by countless natives, a new social category had to be introduced into the tribute lists of the late seventeenth century: the *vagamundos*, who were "without fixed residence" (Cushner 1976, 68). These so-called vagabonds formed the core of bandits that proliferated during the eighteenth and nineteenth centuries, who "occasionally attacked" the Tagalog towns and estates of Cavite, Tondo, and Laguna (Roth 1982, 140).

Outside the Tagalog region and its ecclesiastical estates, friar dominance and command over local economies were felt in a different manner. In their respective parishes, priests extracted alms and cheap produce; kept part of the tribute as well as a hundred bushels of rice, along with fish, as the parishioners' annual support for each of them; mobilized domestic servants, rowers, and porters at will; called on native labor to construct and repair ecclesiastical buildings; and demanded excessive sacramental fees (Phelan 1959, 102–103). Through these everyday occurrences, the friars evinced they had taken over the functions of both *datu* and *babaylan*. With the mystical aura they were perceived to possess, the Spanish clergy dominated the social relations in colonial society and became the most visible appropriator of the peasantry's surplus product.

Later generations of natives who decided not to flee to the interior but live within hearing of the church bells found themselves in an increasingly contradictory relationship with the friars. As the colonial economy became more monetized and as shamanic resistance persisted amid the *indios*' own ambivalence toward colonialism, natives began to doubt the legitimacy of clerical surplus extraction—especially as there was hardly any tangible gift giving on the part of the friar. As a later elucidation of folklore in Negros would show, the Catholic Church was depicted as singularly exploitative, in that it "ate money." The friars' proverbial gluttony and greed overshadowed the transformation of production relations between the *datus*' descendants and their sharecroppers. Yet such relations were also important and had undergone fundamental changes such that, by the eve of the commercial revolution, a complexly differentiated native society had taken shape. There had emerged a distinct native landowning class, a group of individuated tenants and lessees paying ground rent to monastic and private landowners, a group of independent cultivators, and a floating stratum of natives. But the contradictions implied in the relations internal to indigenous society were largely eclipsed by the primary axis of contradiction and exploitation revolving around relations with the religious colonials.

Chinese Mestizos and their *Kasama* and *Agsa* Tenants

The old native elite eventually lost whatever lands they had usurped because of their inability to take advantage of the commercial revolution that commenced in the late eighteenth century. By the 1740s the Chinese mestizos, those masterful opportunists discussed in the previous chapter, were numerous enough to form a distinct category within indigenous society and were on their way to supplanting the increasingly inactive, impoverished, and uncharismatic old elite. With the expulsion of the ethnic Chinese who as a group had dominated domestic trade, conditions during the second half of the eighteenth till the middle of the next century were propitious for the economic ascendancy of the mestizos.

Engaging aggressively in commerce, the mestizos supplied Manila with foodstuffs and exportable products obtained from the provinces. In the 1820s, despite institutional difficulties in equilibrating supply and demand in the Manila market, "the peripatetic capital of the mestizo, who buys only in the years when he calculates that he must . . . make a profit" nonetheless provided "the irregular and transient stimulus" to colonial agriculture (Bernaldez 1907, 245). In the Visayas, particularly on Panay Island, the districts of Molo and Jaro in Iloilo Province became the site of a vibrant native textile industry that thrived under mestizo traders (McCoy 1982a, 301–302).

The protracted ban on the Chinese also made room for mestizos to pursue their *suwerte* by taking advantage of the onset of commercialized agriculture, impelled by the embryonic urbanization of the domestic economy. Land acquisition became an attractive form of investment and gamble. In the second half of the eighteenth century, mestizos began to lease farmlots from the monastic estates that far exceeded the average of about 1.4 hectares rented by *indios* during the previous century (Cushner 1976, 50; Roth 1982, 142). In the nineteenth century, the mestizos sustained their leaseholding activity despite the cumbersome three-year contracts and the rising ground rent, which doubled from the 1840s to the 1890s (Roth 1982, 147–148). In turn, the mestizos subdivided the leased property and entered into share tenancy *(kasamahan)* contracts with *indio* peasants. The mestizos also took over the estate's role of providing cash advances to tenants. Rice and sugarcane were planted on many such *inquilino* lands.

With the commercialization of agriculture, the mestizo leaseholders raised, on the aggregate, a substantial amount of products to be able to pay the fixed annual rent to the religious landowners, provide the tenant *(kasama)* the contractual half-share of the crop net of the rent, and in the end still realize a profit. A number of mestizo lessees, however, could not recover from their mounting debts and so lost the gamble (Roth 1982, 145). To forestall failure, the mestizos sought to extract full advantage from the subletting arrangement by requiring the tenant, treated as production partner, to provide seedlings, the carabao, and working implements. If he had no work animals or tools, the tenant was charged high rental fees for their use (137–138).[22]

As they accumulated capital, the mestizos acquired more land through the moneylending device known as *sanglang-bili* in Tagalog, or *pacto de retroventa* in Spanish, which resulted in a pattern of landownership characterized by scattered smallholdings (McLennan 1969; Wickberg 1964, 1965). In the contract of retrocession, the *indio* cultivator who pawned his property for a determinate time period—often to raise money for cockfighting and gambling (Jagor 1907, 305)—was converted temporarily into a sharecropper. With the latter's inability to redeem the land, the moneylender became full owner of the property for a third or half its market value while the peasant became a full-fledged contractual sharecropper.

A process of perpetual indebtedness commenced with the tenant's request that the mestizo landowner or lessee furnish an advance on production, the money being earmarked for farm expenses (especially when switching from rice to sugarcane [Larkin 1972, 51]), entertainment, and food to tide over the household until the next harvest season. Lent at usurious rates of interest, loans made on the proceeds of the next harvest (called *alili* in the Visayas) were difficult to repay. Riding on the cycle of debts, land mortgages, and cash advances, share tenancy spread to more areas in the settled portions of the archipelago. Wanting to enjoy the benefits of the unfolding commercial era, the sharecroppers for their part generally remained on land owned by mestizos, honoring their debts, cultivating the land—and acquiring more debts.

Sharecropping, known in Ilonggo-speaking areas as the *agsa* system, probably emerged under different circumstances, or at least the entry of the word *agsa* into Ilonggo occurred in a somewhat different context. The original moneylenders would appear to have been the ethnic Chinese who, because of nominal conversion to Catholicism, could reside in the weaving districts of Molo and Jaro and profit from the lucrative textile trade. Such beginnings may be inferred from the Hokkien word *acsa*, which was used interchangeably with *agsa* during the Spanish period. In Hokkien, *acsa* is an old curse word, the equivalent of the American term "shit," with all its connotations.[23] The use of *acsa* would therefore seem to be suggestive of Chinese contempt for the *indio*, even as it spoke of the disdain of the merchant class for the peasant masses in the

Chinese mainland.[24] Not unaware of its derogatoriness, the Chinese mestizo descendants who observably formalized the sharecropping contract continued to use the term *acsa* to refer to their indebted tenantry. If the speculative sketch presented here captures some aspects of a complex process the history of which still needs to be written, the emergence of sharecropping in Ilonggo-speaking areas would, on the face of it, have been more contentious than in Luzon. At any event, what seems rather clear is that the term *agsa* or *acsa* is inseparable from the historically conflictive relations between *indios* and mestizos.

Indeed, from the *indio* vantage point, the culturally confused and imitative mestizos were wanting in prestige and status; they had none of the aura of the pre-Hispanic *datu*ship (glimmers of which were seen among native shamans) nor that of the friar. In some places, as in Cebu in the 1810s, the friars influenced the *indio*'s negative perception of the mestizo (Cullinane 1982). The wealthier mestizos, on the other hand, felt superior to the *indio*. Many towns with both *indio* and mestizo guilds *(gremios)* were rent by innumerable disputes along ethnic lines (Wickberg 1964, 1965). Interethnic antagonism was felt by the late eighteenth century when old Tagalog elites, reacting against mestizo ascendancy, protested that mestizos had "no right" to acquire land (Roth 1982, 145). In the early 1820s, an Englishman observed that the *indio* "is pinched or cheated by the Mestizos, a forestalling, avaricious, and tyrannical race," and in return the *indio* "repays them with a keen contempt, not unmixed with hatred" (Anon. 1907, 94, 105). In the 1840s, the French traveler Jean Mallat (1983, 515) reiterated the observation that the *indio* was "an enemy of the mestizo and vice versa." A peasant religious movement with a large following in the southern Tagalog area in the 1830s, the Cofradia de San Jose had no restriction on membership but for one: mestizos could not be admitted, as the association saw itself as exclusively for "the poor" (Ikehata 1990, 127–131; cf. Ileto 1979, 32, 42, 55). These comments and observations suggest a social tension that rendered either wholesale *indio* admiration for or subservience to the mestizo of this period implausible.

The contested social status of the mestizos, their paucity of what Pierre Bourdieu (1977) has termed "symbolic capital," was evident in the nomenclature applied to the sharecropping contract in Tagalog areas. By calling the *indio* sharecropper a *kasama,* literally a companion (which in Spanish documents was translated as *socio,* or partner), the mestizo lessee attempted to project an acceptable partnership based on equality where in reality there was none. Living in comparatively better houses in the towns, the mestizos avoided farm work by using capital amassed from trading activities for investment in commercial rice production. On the other hand, the *kasama* tenant lived in a friable hut on the farm with the tenant household's labor power as principal contribution to the production process. Notwithstanding the ostensible difference in class positions—aeons before the Green Revolution and the reliance it spawned on capital-intensive inputs—labor and land would have appeared at par, which justified the fifty-fifty division of the harvest between the so-called partners.

There was nonetheless an aspect of parity in the *kasamahan*, in that, as far as the *indio* was concerned, there was no compulsion to work as a tenant. In this differentiated but negligibly commodified society, the mestizo had no power, either economic or extra-economic, to compel the *indio* to labor. The immense frontier and the lack of any insurmountable barriers to mobility made subsistence achievable in a variety of ways. However, the native who chose to live within hearing of the church bells and who desired ready access to the cash economy and the world of gambling could be enjoined to enter what would have been seen as a mutually beneficial relationship. (Incidentally, the "pooling of resources" in the tenancy contract resembled in form the trading arrangements of precolonial times.)[25] When originally used in the eighteenth century, *kasamahan* would thus appear to have been deployed as a linguistic evocation of equity to make the historic introduction of a formal sharecropping contract palatable to the *indio* whose labor power had to be wooed to serve the distrusted mestizo.

Moreover, the mestizo could not and did not dictate the rudiments of farm work and the labor process, which belonged to the *indio*'s autonomous space. At the core of that space with which the mestizo could not interfere, nor knew how to, were the individualized rites of negotiation with the spirit-world. As long as they tilled the land in the manner they preferred and exchanged labor with other cultivators at the time of their own choosing, the *indios,* whether called *kasama* or *agsa,* felt they retained autonomy in a partnership where they also acted as personal negotiators with the spiritual realm. Where there was no economic equality, there was nevertheless parity of status, a relative matching of strength of soul stuff.

The prevalence of the *kasamahan* contract prompted Captain-General Jose Basco y Vargas (1907, 294) to outlaw it in 1784, "on account of the burden which ensues from it to the poor, and . . . the many usurious acts which are committed therein." But by granting the elderly and the infirm the right to contract a *kasama,* evasion of Basco's prohibitory edict was easy. The colonial state also did not have the means to enforce the law.

In any case, where Basco's decree was known, the sharecropping contract was entered into and maintained by mestizo and *indio* partners knowingly in contravention of a colonial state ruling.

Economic contracts represented by the *kasamahan* and the *agsa* system emerged and flourished according to the rules of interpersonal agreements, and were hence anarchic in not being dependent upon legal enforceability and the state's authority. Amid this anarchy and a feeble state, peasant autonomy was confirmed by the *indio* who, despite insufficient resources, could exercise a real decision in bargaining through the sharecropping agreement.[26] It was the *indio* who, individually, had to decide whether to work as a tenant or not, how to work the land, how to propitiate the spirits, whether to honor debts or not, and in general how to relate to the mestizos. On the other hand, the mestizos rose in economic prominence, but they occupied a structural position marked by incomplete hegemony, the transcendence of which would later become an object of pursuit by the sugar-planter class of Negros and other areas.

Meanwhile, the religious remained the biggest landowners. By the 1850s, apart from the regionally restricted production of tobacco, which was under colonial state monopoly, the friar lands were the only locus of large-scale agricultural production in the Spanish Philippines. The emergence of a monetized economy and incipient commercialized agriculture might have given the impression of a colony primed for economic change. Contrary to the dream of some Spanish officials, however, a wage-labor plantation regime was nowhere installed in the colony. There was an intransigence to sharecropping by smallholders that the colonial state found difficult to overturn. Even in the monastic estates, which long predated capitalist penetration, lands devoted to rice and sugarcane had to be subdivided in a chain of subcontracting arrangements based upon sharecropping.

Problematizing Sugar Production before 1860

The entrenchment of small-scale agriculture dependent upon *indio* labor and mestizo petty capital was deemed problematic by a number of Spanish civilian authorities. Developments in international trade, centered on the debate in Britain concerning tariff preference for sugar made by nonslave labor (which led to the subsequent establishment of a British consular office in Manila in 1844), made it plain that the increased cultivation and manufacture of sugar in Filipinas would be extremely beneficial for Madrid's royal exchequer. But there were major obstacles to the expansion of sugar production. For one, the colonial administrators found it worrisome that Spaniards displayed hardly any interest in agriculture—a situation that, it must be stressed, arose from the corrupt use of official positions and from decades of commercial restrictions surrounding the galleon trade. If wealth could be amassed without productive labor, what incentive was there to engage in agriculture? In addition, the few who did farm had little to contribute, for, as Tomas de Comyn reported in 1810, "the Spanish proprietors, whose number possibly does not exceed a dozen persons, and even they labor under such disadvantages . . . [are not] in a situation to give to agriculture the variety and extent desired, or to attain any progress in a pursuit which in other colonies rapidly leads to riches" (de Comyn 1969, 21). From de Comyn onward, it became a popular recommendation to encourage more Spaniards to migrate to Filipinas to develop large plantations devoted to export crops. These estates were to be provided, following Spanish American lessons, with batches of *indio* labor who would work there for a certain time, "provided only they are taken care of during their journeys, maintained, and the price of their daily labor, as fixed by the civil authorities, regularly paid to them" (26). In 1842 Sinibaldo de Mas was still grappling with the problem of how to motivate Spaniards to devote themselves to colonial agriculture (de Mas 1963, 131–132).

An appropriate model of sugar production presented itself to the Spanish ambassador at The Hague, who was captivated by the case of Java. The forced cultivation scheme decreed in 1830 had created a sugar industry that was "grand in scope, starkly simple in design, and ruthless in intent" (Elson 1984, 32), but it prospered and rescued Holland from its financial troubles. The diplomat was prompted to address Madrid in a memoir dated 16 August 1842, suggesting that Manila imitate Batavia, which he claimed was a particularly apt model, as it conformed to the pattern of smallholdings prevalent in Filipinas.[27] He failed to note, however, that in Dutch Java the native elites had been fully incorporated into the colonial state apparatus and had authority to mobilize village labor, whereas in the Spanish Philippines the colonial power would not countenance a "forced" system, and neither was the local elite positioned to command labor at will on behalf of the colonial state.

With magisterial haughtiness, Madrid's Royal Palace dismissed the suggestion and gloried in the fine sugar from Cuba and Puerto Rico, asserting that production from its West Indies colonies was more than adequate to supply the Peninsula, and a sizeable portion of the European market besides. Internally, the palace reproachingly observed that the Spanish Philippines produced inferior-quality sugar that was difficult to sell in Europe, as freight costs made the product even more uncompetitive.[28] But the ambassador's letter served as occasion for Madrid to question Manila on the condition of sugar production in the colony and the measures being taken to promote its growth.[29] Though perfunctory, the injunction for a report was an appreciable improvement over Madrid's disregard of a captain-general's report written from Manila about a decade earlier concerning the backward condition of sugar production in the colony.[30]

Madrid's query generated responses from three entities: the *Ayuntamiento de Manila* (1843), the *Real Tribunal de Comercio de Manila* (1843), and the *Real Sociedad de Amigos de Filipinas* (1844).[31] In a collective defense of their actions, the memoranda from these three bodies argued that it was impossible to isolate the state of sugar production from the overall backwardness of colonial agriculture. But in their attempt to explain the deep reluctance of Spaniards to invest in agricultural estates despite likely profits, the responses from Manila unwittingly brought into sharp relief the weaknesses of the colonial state and the growing internal decay of the Spanish empire.

There was no doubt, according to them, that sugar production was profitable. Even in Tondo Province where land leases were costlier, the cultivation of sugarcane and other cash crops was guaranteed to yield substantial returns.[32] However, the extant system of sugar production did not fulfill its economic potential. The three *haciendas* run by "whites" *(blancos)* in Jalajala, Calauan, and Calatagan devoted portions of these private estates to sugar production but did not prove to be financial success stories, for reasons to be discussed shortly.[33] The other sugar producers operated on such a tiny scale that each harvested from 5 to at most 30 *pilones* of sugar for the whole year, a *pilon* weighing about 110 pounds. The colony had no more than thirty planters who each produced an annual sugar harvest of at least a thousand *pilones*. The small-scale sugar producers were dispersed in nine provinces in Luzon and the Visayas, with Pampanga as the lead producer, followed by Cebu and Bulacan.[34]

Extreme fragmentation hindered the growth of the sugar industry. Not only was it difficult to foresee aggregate inventory levels, but the small gains from limited production constrained capital accumulation and the acquisition of more efficient and modern means of sugar manufacture. The market had been able to improvise in sourcing produce from innumerable small growers by relying upon middlemen: Chinese until the 1760s and, after that until the 1850s, mestizos. Middlemen networks helped raise the colony's annual sugar exports, which averaged roughly 300,000 piculs, or 18,750 tons, by the early 1840s.[35] The shortage of capital, however, was a more recalcitrant obstacle. Cash-strapped Spain was in no position to capitalize the Philippine sugar industry. Quite the contrary, the colony's surplus was used to alleviate the hardships of the Peninsula,[36] leaving the colonial state with no funds to lay down the basic infrastructure that might have more effectively linked small-scale sugar producers to the market.

The colonial state also had no funds to loan out to the industry, which had to continue its reliance upon expensive noninstitutional credit sources. Apart from the cash advance provided by merchant houses in Manila, sugarcane growers in the countryside raised funds through moneylenders, principally until the 1840s the Spanish provincial governors. Although the loan agreement gave the creditor a claim over the produce, credit was obtainable only at interest rates pegged at very high levels to offset the grave risks faced by moneylenders. These risks ranged from losses arising from natural calamities to the possibility that the borrower might flee without a trace to another town or province. The inability of the colonial state, including its village functionaries, to control the geographical movement of *indios* was an important factor in the cycle of backwardness. Moreover, because of the lack of legal safeguards, the producer might decide to sell the sugar to another trader before the creditor could collect the produce. Given these risk factors, interest rates hovered from 25 percent to as much as 100 percent over the sugar price prevailing at the time of harvest, financial impositions that reduced the small producers' net gain.[37]

The meager capital investment and the comparatively lower sugar prices in the colony[38] resulted in a modicum of returns, which prevented the formation of a respectable asset base. The small-scale producers had no surplus (and the information) to seek a more advanced technology to make higher-grade sugar. Local manufacturing methods remained antiquated. Borrowed from ancient Chinese techniques of claying molasses, the sugar mill was ordinarily nothing more than a carabao-driven wooden cane crusher plus a cauldron or two for cooking the sugarcane juice. To produce muscovado, the boiled juice was poured into large earthen pots called *pilones*. As the molasses drained through a hole at the bottom of the pot, the mixture hardened. The manufacturing process was under the individual cane grower's supervision, although bigger producers hired a Chinese expert (called *maestro*) to oversee the process. Use of these rudimentary methods was estimated to cut productivity to an estimated one-twelfth that of Cuba.

The solution to the ills of the Philippine sugar industry, the three bodies argued, lay in developing large plantations equipped with modern mills. Economies of scale should lead to better, cheaper, and more competitive sugar. Going beyond a narrow Hispanic protectionism, especially because few, if any, Spaniards possessed the requisite capital, the memoranda suggested that "whites," in general, be encouraged to invest in colonial plantations. Momentarily absent from their discourse was any derogatory reference to "foreigners." However, large capitalist *blancos* did not develop agricultural interests in the Spanish Philippines, where colonial policies toward foreign capital remained ambivalent and unpredictable, if not restrictive, and where Spaniards themselves did not feel safe.

Flight, Banditry, and the Cultural Solvent

Identified by the three corporate entities as the principal stumbling blocks to agriculture were *(a)* the insecurity of life and property *(falta de seguridad personal y de la propiedad)*, and *(b)* the scarcity of field hands *(falta de brazos)*—problems perennially cited because the colonial state aimed but failed to turn the colony into a moneymaking venture. In 1843, the anxieties over personal safety were heightened by the murder of the son of a Spanish farmland owner within the *hacienda*'s premises not far from Manila. The *Tribunal de Comercio* claimed that the slaying was the latest in a lengthening list of crimes against estate owners victimized by natives. The situation was probably overdramatized, but the massacre of October 1820 had remained vivid in the memory of Manila's expatriates,[39] whose fears were not about to be dispelled by periodic robberies and murders of foreigners.[40]

Moreover, Spaniards felt that plantations had no protection: "either from negligence or malice," cattle and draft animals frequently escaped from their corrals and destroyed crops. Moreover, native bandits ruled by what the *Tribunal de Comercio* called "a miserable passion against the landowners," robbed and pillaged huts belonging to the estates, and "not a few times" killed cattle and left their carcasses to rot in the fields with no other apparent motive than to irk the estate owners.[41] Banditry, which drew on the magical power of the quietly subversive *anting-anting*, had raged for some time, but its problematization in the nineteenth century was emblematic of the colonial state's awareness of its inability to construct a profitable enterprise in the blossoming age of world capitalism.

Early in the 1820s, an English observer noted that natives wanted "to exonerate themselves from paying tributes and taxes, in return for which no protection is granted. In many provinces . . . whole districts are rendered impassable by the robbers, who even lay villages under contribution!" (Anon. 1907, 93). In the coastal towns, the danger of Moro attacks was also omnipresent. Aggravating the situation was the fact that "The very soldiers and sailors sent for their protection plunder them. An [*indio*] in whose neighborhood troops are posted, or who sees the gunboats approaching, can no longer consider his property safe; and in the very vicinity of Manila, soldiers ramble about with their loaded muskets, and pilfer all they lay hands upon at midday!" (93–94). Brigandage was everywhere in colonial society: the perpetrators were external to the colonial state as well as right within the apparatus of power.

The problem concerning the security of life and property seemed insoluble, but the authorities felt that the difficulty of mobilizing *indio* labor for regimented work in plantations—equally symptomatic of colonial state weakness as was the existence of banditry—was even more intransigent. The "slothfulness" of the *indios,* as another official termed it, had so vexed de Comyn that he recommended in 1810 the proscription of "idleness as a crime" and the mandating of "labor as a duty" (Bernaldez 1907, 249; de Comyn 1969, 24). He also argued that Basco's program of using indirect stimulants to foster agricultural development was unsuccessful: the subsistence-oriented natives were not moved by economic incentives. The handful of private-estate owners, de Comyn observed, could mobilize labor only by acquiescing to *indio* preferences, which "compelled [the owners] to divide their lands into rice [farms], in consequence of this being the species of culture to which the [*indios*] are most inclined" (21). The *Tribunal de Comercio* lamented the impossibility of disciplining the *indio* into wage work: "should you pay them on a daily basis, they work the least they could, and it is impossible to exercise the necessary vigilance to remedy this abuse."[42]

Native resistance to regimented work was highly effective. The inability of the colonial state to buttress the position of estate owners vis-à-vis labor left them with no option but to resort to sharecropping. But the sharecropping arrangement preferred by the peasantry was distasteful to Spaniards, as it required them continually to make provision for the tenant household's subsistence on account of the half-share. This aspect of the social game, however, was masterfully played by the Chinese mestizos, who after all were part of indigenous society, to cultivate social ties to their advantage. But as non-natives Spaniards were more fearful—a tacit admission of their ultimately *extrangero* position. They felt sharecropping was too risky, as native demands might be limitless and they could end up with huge losses. They were afraid of being "abused," strangely projecting onto the subjugated *indio* the intentions of the colonial power.

But the Spaniards' fears were also born of the awareness of the colonial state's effeteness: in case of disagreements and notwithstanding unpaid debts, there was no stopping a native from leaving the farm and moving to another location, undetected by the colonial state. The unrestrained movement of natives, particularly in the monopoly-tobacco-growing Province of Ilocos Sur, triggered the Manila government to issue a decree in 1843 making transfer of residence without prior notification illegal.[43] But in general the state could not deter the flight of its native subjects. In this regard, "avoidance migration" (Adas 1980, 537) in the face of an unbearable situation was a strategy shared by *indios* with other Southeast Asians whose movements across vast tracts of unoccupied land were beyond the power holder's control.[44]

In the unlikely event that the *indio* was caught, a protracted process of procedural entanglements was required before the native could be made to pay any debts owed, as it was possible to take refuge in the old law that limited recoverable debts to five pesos. Moreover, the native could seek cover in the prohibition against contracts authorizing monetary advances on production, another old ruling that outlawed, in effect, the very cornerstone of the export trade. The legal system evidently did not encumber the mestizos, but neither did it help them. Local cultural norms were more effective guarantors of the safety and flexibility of the mestizos' "peripatetic capital." Spaniards however, with their meager capital and uncertainty in dealing with the native, found the colonial legal framework an impediment to their involvement in agricultural production.

Feeling powerless to put the native under full control, the *Tribunal de Comercio* bewailed the "most vicious order of things": antiquated laws meant to protect the natives had the unintended effect of being cunningly manipulated by colonial subjects who had learned to become "fraudsters" and "swindlers."[45] The situation was made more deplorable by the colonial bureaucracy the Spaniards themselves had created, in which a court case could not be won without routine bribes. In the perceptive comment of an Englishman in the 1820s:

> The imperfect mode of trial, both in civil and criminal cases (by *written* declarations and the decisions of judges alone), lays them open to a thousand frauds; for if the magistrate be supposed incorruptible, his notaries or writers *(escribanos* and *escribientes)* are not so; and from their knavery, declarations are often falsified, or one paper is exchanged for another whilst in the act of or before signing them. (Anon. 1907, 94; italics in original)

The intramural games spawned by the convoluted legal and judicial system, like insult added to the injury of an impoverished and feeble colonial state, were a barrier to Spaniards and other *blancos* who might have intended to develop large plantations in the colony.

The sharecropping system thus became the entrenched mode of mobilizing a stable work force. The state's lame military arm could not be helped by local elite functionaries, who were also unable to control the movement of people across the archipelago and its vast hinterland. Apart from Friar Power, itself increasingly detached from the *indio* peasant by the wedge of mestizo intermediaries, the Spanish colonial state had no power to coerce or cajole a wage-labor agricultural regime to come into being.

The difficulties elaborated on by Spanish civilian authorities mirrored the tenuous character of the social relations involved in the sharecropping contract, which could easily be abrogated by the peasant through flight. That the *kasamahan* and *agsa* schemes entered into by *indios* and mestizos worked with acceptable predictability suggested that both parties shaped it to be mutually beneficial from their respective points of view. To stabilize the overtly instrumental partnership with the tenant, the mestizo deliberately offered and provided subsistence and monetary advances to the *indio*, a behavior facilitated by the fetishistic belief that to let money depart in the form of a bet would allow it to return in even bigger quantities. To the mestizo, the system of advances was a gamble, but it was worth it despite a few possible losses because the debt became a tool to control the elusive peasant and thus ensure even bigger winnings in the long run. However, there was nothing in debt itself and its social context that made it inviolable.

But a social construction of meaning around debt was possible. A systematic investment, the mestizo's loan was made to represent a willed trust in the farm partner, almost an appeasement that was also a quid pro quo: a gambler's gift. Like a mollified spirit, the *indio* saw in the flow of resources from the mestizo an expression of sociability and of "fair treatment" in a colonial setting where fairness seemed wanting and the surplus was largely usurped by the friar. The return flow of resources to the mestizo in terms of crop share and high rental dues was thus deemed legitimate, and so was the honoring of debt.

The idea of mutual understanding expressed by the Tagalog phrase *nagkahulihan ng loob* suggests that the partners in a stable sharecropping agreement could be said to have "caught" *(huli)* each others' literal "inside" *(loob)* or inner being. In a power encounter involving the soul stuff *(dungan)* of two parties, stability in interpersonal relationships entailed, as it were, the trapping of each other's soul, a necessary stage that made an agreement in a common undertaking possible. The Ilonggo counterpart, *nagkahulugan sang boot,* suggests a mutual "pushing" of wills *(boot)* until both "fall" *(hulug)* comfortably into place. The instability and fluidity of debt relations would imply a continuous process of pushing and catching and trapping until the intermeshing of wills was reached. The culmination of the testing period would be the traversal of the referent of debt from the money-object to the person, so that the lender caught "the inside" of a debtor, who thereby had an *utang na loob* or *utang nga boot,* a debt of the inside, a debt of the will. Flight, which would have made debts and cash advances irrecoverable, was obviated, as the parties had already "caught" each other's will before the debtor could abscond. Spatial movement was discounted, as the partners' wills had "fallen into place," much like a *dungan* that had been fully domesticated into one's body.

In the sharecropping agreement, the parties of rather equal soul stuff sought to stabilize the persistent "catching" and "pushing" and disguise the materiality of the social game by entering into other forms of social bonding, such as godparenthood (e.g., Larkin 1972, 84), that ritually transformed the social relationship into a permanent fictive one. That the mestizo was prepared to become godfather (*compadre*) or godmother (*comadre*) to the *indio* sharecropper's offspring, which required periodic gifts to the godchild, meant that ethnic differences could be transcended for instrumental purposes. But instrumentality was camouflaged, because godparenthood meant the elevation of a relationship to a firmer ground of "friendship." The sharecropping relationship was thus wrapped in several layers of meaning generated in the course of a social game the mestizos could not avoid, given their incomplete dominance over the elusive *indio*.

Despite the prevailing climate of ethnic differences, the compounding of debt at the individual level gradually incorporated the *indios* into the mestizos' gamble at the societal level. The *indio* tenants were "won over" by the mestizos and gradually tied to what became an economically unequal but workable tenancy contract. The cycle of indebtedness deepened, and was passed from one generation to the next. As generalized commodity production spread, as population density increased, and as the broader social and political circumstances in which sharecropping was embedded changed, the nature and quality of the share-tenancy system were transformed across different parts of the country.[46]

Fantasies of Policy and Colonial Agriculture

By the time Madrid had examined the three memoranda from Manila in mid-1846, the metropole's perception of the colony's sugar industry had changed to one more appreciative of its potential. This view resulted from an appreciation of the colony's proximity to China, whose capacious market had been fully opened to free trade by virtue of the treaty system instituted at the end of the Opium War of 1842. Nevertheless, Madrid's *Consejo Real* failed to comprehend the import of the three memoranda. The Royal Council reacted in disbelief to the suggestion that the colonial state could not protect estate owners or mobilize native labor. Madrid maintained its illusion that the *indio* was "too timid" to jeopardize agriculture. However, it did concede that, if it would satisfy estate owners, the colonial state might consider fielding "small detachments of troops to reconnoiter the nearby areas."[47]

Choosing to ignore the voices from Manila, the council advanced its own three recommendations, namely: Chinese and Spanish immigration to augment labor supply, the distribution of Crown lands, and agricultural education. Only the last of these measures was appraised by the *Ministerio de Ultramar* as deserving immediate implementation, as it would provide technical skills to "obviate the litigations and frauds" committed in connection with land measurements.[48] But not until 1887 did Manila's *Escuela de Agricultura* (School of Agriculture) receive formal approval.[49] In the same year, plans for a model farm *(la Granja-modelo)* in La Carlota town on Negros Island were finally approved by Madrid following a similar scheme in Pampanga; preparations for the collection of agricultural statistics in Negros were also made.[50] Apparently, some concerted policies to boost colonial agriculture were finally coming off the drawing board very late in the nineteenth century.

The plan to distribute uncultivated portions of the public domain materialized starting in 1875.[51] However, the regulations for public land sales belatedly decreed in 1883 prohibited foreign corporations and associations from owning plantations, thereby contradicting the concession of 1870 that allowed foreigners the right to acquire landed property.[52] The imperial resistance to foreign capital in the Spanish Philippines stood in marked contrast to the open-door policy that drew corporate capital to the east coast of Sumatra (Stoler 1985, chap. 2). As Spanish antagonism toward foreigners was once more in resurgence, natives became the preferred policy target of the 1884 law on *colonias agricolas*, which offered tax exemptions to encourage natives to form plantations ten kilometers or farther away from major population centers. However, mestizo and European *hacienda* owners, especially in Negros, vigorously protested that the law was iniquitous to existing farms.[53] The colonial state was far from adopting a coherent set of policies to attain the goal of establishing large plantations.

On the issue of labor shortage, one of the solutions contemplated was the purchase of children from China, a plan that included cost-benefit calculations for each child.[54] But the large-scale importation of Chinese coolie laborers was not pursued due to fear of the political instability that might arise from greater numbers of Chinese and fear of displacement among local traders who dreaded Chinese competition.[55] The other stream of migration, that of Spaniards, was made doubly difficult because Madrid (prior to the 1869 opening of the Suez Canal) had given the navigationally distant colony a reputation as a place of banishment for political offenders, not only from Spain, but even from Cuba.[56] Racist pride also obstructed the flow of migrants. Madrid would not countenance poor Spaniards working as farmhands alongside the "racially inferior" *indios,* as an implicit equality among the laboring classes would damage what was assumed to be the native's veneration for the colonial *hombres superiores* (superior men).[57]

Effectively ignored by the Royal Council were the suggestions concerning the judicial and legal system. To the very end of the Spanish regime, for example, the regulation that contracts in excess of five pesos were null and void was never rescinded, an issue the first British vice-consul at Iloilo denounced as a hindrance to agriculture.[58] Also disregarded by Madrid was the proposal that each town form a corps of "vigilantes" who would be paid adequate salaries and equipped with arms and horses to apprehend missing workers. Also, for a long time, Madrid shied away from using an institutional network of native spies that colonial powers elsewhere had effectively used. Although committed in principle to the development of agriculture, in particular the sugar industry, Madrid, unlike The Hague, was hesitant to take decisive steps to extend unequivocal support to agriculturists, as by doing so it would obliterate its complacent image of itself as a benevolent imperialist in full command of a submissive and inferior race.

Only after Manila had filed repeated complaints about the uncontrollable incidence of thievery did Madrid, after meticulous calculation of expenses, agree to create the *Guardia Civil* (Civil Guards) for Tondo Province in 1855; this elite military corps was extended to other Luzon provinces in 1867, but a similar one for the Visayas was not organized until as late as 1882.[59] The *Guardia Civil* failed, on the whole, to curb what colonial officials saw as the rising tide of brigandage and criminality. In Negros the authorities thought the island's administration could be improved through the creation of two provinces demarcated by the mountain range along the island's middle spine, a suggestion originally proposed in the 1860s but approved by Madrid only in 1889.[60]

In many respects, Filipinas reflected the turmoil and volatility of politics that had overtaken the Peninsula for most of the nineteenth century and beyond. Following the political convulsions that resulted in the loss of the Spanish-American empire by the 1820s, Spain experienced from 1833 to 1875 the age of the *pronunciamiento,* literally the proclamation of a new government, which came in quick succession as political fortunes seesawed in the fight for state power between the Carlist absolutists and the fractious "liberal" camp. The political discord in the metropole was echoed in the colony through a succession of officials whose terms of office were truncated, predictably, by events in Spain. During the last quarter of the nineteenth century, the colony reflected even more the demoralization and confusion in Madrid's colonial administration.[61]

But it was precisely in the context of a debilitated, destitute, and drifting empire and a colonial state with an inept military, a ravenous judicial and administrative system, and a corpus of antiquated laws to which Spain stubbornly clung that export agriculture emerged after mid-century in Negros. The immediate trigger was the opening of provincial ports to direct foreign trade in 1855. The declaration of Iloilo, Zamboanga, and Sual (in Pangasinan Province) as ports directly accessible to international commerce occurred, as I have shown elsewhere (Aguilar 1994c), in the crevices of colonial administrative inefficiency and without prior sanction from Madrid. The opening of provincial ports was staunchly opposed by Spaniards with interests in the coasting trade, but fissures in the colonial state apparatus allowed one interest group to triumph in having the outports opened. The internal decay of the Spanish empire resulted in a policy coup by a short-lived governorship in Manila: a fundamental change toward a liberal economic direction resting upon an unstable political structure. The direct admission of foreign capital to the Visayas was an exemplar of ad hoc policy-making in a crumbling empire.

A year after the pivotal opening of Iloilo to world trade, a British vice-consulate was established in Iloilo with Nicholas Loney at the helm. After an initial period of hesitation, dispelled in part by Loney's aggressive strategy, foreign capital began gradually to trek to adjacent Negros Island.[62] From about the same time, rapid increases in per capita consumption of sugar in advanced capitalist countries as sugar became a necessity (Mintz 1985, 197) saw overseas markets buying increasingly large quantities of sugar directly from the Visayas. Through Loney's brokerage, modern machinery acquired by easy credit from foreign merchant houses began to be used in the manufacture of sugar. Lured by the speculative profits obtainable from the sale of sugar to the export market, mestizos in Iloilo and a number of foreigners moved to Negros to engage in agricultural production. *Indio* labor, elusive but amenable to enticement, began to move as well. Built upon sharecropping and debt relations, the sugar *haciendas* of Negros began to take shape. After 1860 sugar production steadily accelerated, and Negros sugar exported via the Iloilo Port would finally surpass Pampanga sugar from the mid- and late 1880s. Overcoming old while creating new problems of social control in the face of an ambivalent, almost irrelevant, colonial state, migrant mestizos, *indios,* friars, Chinese, colonial officials, and foreign merchant capitalists gambled their way through the terra incognita that was Negros, in time crafting a peculiar world of Masonic Capitalism on this central Philippine island. To that story we now turn.

Part II

The World of
Negros Sugar
after 1855

The Formation of a Landed *Hacendero* Class in Negros

The Race of Races and the World Market's Allure

At the time of its opening to world trade in 1855, the Iloilo port area was the center of a thriving piece-goods trade controlled by the Chinese mestizos of Molo and Jaro. The native textile industry, which produced finely crafted *sinamay* and *piña* fabrics woven from cotton, silk, and pineapple and abaca fibers, had exported its products to other parts of the Spanish Philippines and to overseas markets since the mid-eighteenth century (McCoy 1982a, 301–302). In the 1850s local textiles worth an estimated $400,000 (Mexican) were annually exported by mestizo traders, who embarked "in numbers from 6 to 10, 15 and sometimes twenty in the coasting vessels leaving for the capital."[1] On their return, European, chiefly British, goods amounting to about $30,000 to $40,000 (Mexican) per month were brought to Iloilo and sold in the larger markets of Jaro, Molo, and other towns whence some goods gradually found their way to the interior.

But the mestizos soon found themselves facing stiff competition from the ethnic Chinese, generally males, most of whom had migrated from the Hokkien language area (Doeppers 1986; Wickberg 1965, 48–61). The readmission of Chinese sojourners and their participation in economic activities outside Manila were made possible by the colonial state's liberalization of immigration, residential, and travel restrictions commencing in the 1830s. By the 1850s, there were about thirty Chinese in Molo and about two or three in Jaro who had established permanent shops as agents or partners of Manila Chinese.[2] In 1861, Nicholas Loney reported that mestizo traders were being squeezed by the Chinese who procured manufactured goods "from first hands in Manila."[3] The Iloilo Chinese were positively advantaged by their participation in the so-called *cabecilla* system, an ethnically based trading network that appeared in the 1840s and lasted until the 1880s. This network enabled the Chinese to present a united front in bargaining with the European importers whenever a cargo of textiles arrived in Manila, and it assured all Chinese wholesalers, including smaller Manila dealers and those based in provincial capitals like Iloilo and Cebu, a supply of textiles for sale (Wickberg 1965, 74). On their part, foreign commercial houses found the Chinese network advantageous, as the imported goods obtained the broadest possible distribution. The Chinese traders also doubled as a network of purchasing agents that provided the Manila-based foreign merchants a steady supply of exportable produce. Faced with such formidable competition, the Iloilo mestizos could not but lose out to the Chinese.

It must be noted that in spite of the opening of Iloilo to world trade the direct flow of imported goods to the region did not commence immediately, in part because of the unsettled market conditions and the "cotton famine" triggered by the American Civil War in the first half of the 1860s.[4] Moreover, British shipmasters continued to complain about the unsatisfactory and dangerous state of the Iloilo port's entrance.[5] In the early 1870s, the direct import trade between Iloilo and Britain was considered "very limited" and did not exceed £20,000.[6] The direct importation of cotton goods and textiles from Manchester and Glasgow into Iloilo did not reach significant levels until the 1880s; and only in 1888 did the direct trade in textiles peak, its value exceeding 3.3 million pesos (cf. McCoy 1977, 62–63). The sluggish growth of direct trade notwithstanding, the flow of imported goods via Manila through the Chinese trade network continued, this indirect trade (rather than the opening of the Iloilo port per se) dealing the fatal blow to the trading activities of the Iloilo mestizos. Spaniards with interests in the mestizo-led coasting trade were also badly hit.

The financial decline of the mestizos resulted in racial tension, manifested in "a more or less subdued feeling of hostility towards [the Chinese] and a tendency both among mestizos and Spanish to regard them as interlopers."[7] Viewed as naturally vicious, Chinese shopkeepers, as their numbers rose, began to push the mestizos out of retail trade, which the latter increasingly found "much more difficult, precarious and unremunerative."[8] Ejected from "a branch of commerce where they had become superfluous,"[9] the mestizos began to shift their attention to agriculture. The displacement of mestizos in Iloilo and in other parts of the colony bred a general resentment against the Chinese, which prompted the mestizos to oppose vehemently the idea of importing Chinese coolie laborers to solve the perceived problem of labor shortage (a position they staunchly clung to even during the American colonial period). However, as far as Loney was concerned, that the mestizos diverted themselves to sugar production suited his schema that an appropriate racial category develop the region's agricultural potential, race and class being conjoined in his frame of mind (cf. De Rooy 1990). Although the Chinese occupied a somewhat higher position in Loney's racial hierarchy, he reported that "their numbers in the interior are too few to enable them to cultivate the ground on a large scale, and in small isolated bodies they would not have sufficient security against the ill will of the natives."[10] Loney's desire that "more capital and intelligence" be applied to agriculture saw partial fulfillment when the mestizos turned to sugar production, especially on adjacent Negros Island. In setting their sights upon sugar, the mestizos deliberately abandoned the native textile industry to rapid destruction.

By 1861, Loney reported that, along with a few Europeans, a "great number of Yloilo mestizos" who "hitherto [had] been engaged in the piece goods (wholesale and retail) trade in this province" had already "invested in the large tracts of fertile and well-situated land on the coast of Negros."[11] In contrast to the *indios,* whom Loney depicted insultingly as "a set of deceitful, lazy vagabonds," the mestizos he regarded as "a remarkable commercial, industrial and speculative race" who were "increasing yearly in social and political importance, and though not so fully possessed as the Chinese of the persevering and commercial qualities necessary for continued success under the pressure of great competition, are not without pre-vision, energy and enterprise."[12] Sharing the social Darwinism sweeping Europe at that time, Loney applauded the formation of a sugar-planter class extracted from mestizo stock.

Clearly, the advance of capitalist free trade in the Iloilo-Negros area was accompanied by social reorganization and economic sectoral transfers caused by the wedge between mestizo and Chinese. Given impetus by the dynamics of the local cultural milieu, the resulting changes were more attuned, as far as the British vice-consul was concerned, to the requirements of the world market. Capitalist penetration converged structurally with the ethnic tensions peculiar to the plural society of the Spanish Philippines. In other words, despite his racism, the institutional linkages to the world market he sought to build, and the initial capital he lent to interested parties, Loney did not engineer the mestizo–Chinese competition that facilitated the emergence of a sugar-planter class in

Negros. Local actors as human agents in dialectical interaction with changing structures adopted strategies that resulted in the willful neglect of the mestizos' economic interests in Iloilo in favor of sugar production in Negros.

As early as 1857, the speculative mestizos, according to Loney, were favorably responding to the "very high prices which have been ruling in Europe for sugar for some time past" due to the Crimean War, the prices serving as "a great inducement" for natives to cultivate "any amount of sugar" (Loney 1964, 71). Moreover, the "high prices" of the initially trifling amount of sugar exported to Australia from the Iloilo port in 1859, compared with what planters would obtain from the Manila market, "led planters and others to see how much they benefit by the direct trade."[13] Along with remunerative prices, the establishment of commercial houses in Iloilo that handled Negros sugar afforded the advantages of a secure market as well as "prompt cash transactions in place of the delay and risk of forwarding consignments to Manila."[14]

The allure of windfalls, indeed of *suwerte,* from export agriculture attracted the nascent capitalist mestizos to gamble on cane growing and raw-sugar production. The Iloilo price per picul of sugar was definitely on the rise: from a range of 1.25 to 1.50 pesos in 1854 to 3.25 pesos in 1866, and still further to from 4.75 to 5.00 pesos in 1868 (Jagor 1965, 220). The initial direct shipment in 1859 raised Loney's expectation that the following year's crop would be *"ten* times larger" than that of 1856 (Loney 1964, 97; italics in original). Rising world market prices induced greater sugar production. The Iloilo port saw its direct sugar exports rise sharply from 584 tons in 1859, to 1,555 tons in 1860, to 3,722 tons during the first half of 1861; these figures represented 11 percent, 22 percent, and 95 percent, respectively, of all sugar leaving Iloilo.[15] As direct sugar exports rose, sugar shipments to Manila declined. Notwithstanding erratic prices from 1880 to 1889, the Iloilo port exported sugar at an annual average of 78,345 tons, registering a peak of 165,407 tons in 1892 (Foreman 1899, 295–296).

In the early 1850s there were in Negros only four notable *hacienda* owners: the creole Spaniards Agustin Montilla in the *sitio* of Pulupandan in the town of Bago[16] and Eusebio Ruiz de Luzuriaga in Bacolod,[17] the Frenchman Ives Leopold de Germain Gaston in Silay, and someone called Tia Sipa in Minuluan (present-day Talisay). Their combined annual sugar production reportedly totaled 3,000 piculs (Echauz 1894, 22–23). With the opening of the Iloilo port to world trade in 1855, the ranks of Negros *hacienda* owners—or *hacenderos* in local parlance—soon expanded to include the mestizos who decided to leave their Iloilo-based trading activities to relocate to Negros. Also moving to Negros were a

Formation of a *Hacendero* Class

number of *blancos*, with few exceptions peninsular and creole Spaniards, who, as later discussions in this chapter will show, had virtually no capital for investment in sugar production but who nonetheless dreamed of building a fortune by cashing in on the island's economic boom.

The admixture of different racial and ethnic groups was potentially problematic. Europeans of all shades did not look kindly upon the native, and even among Spaniards the peninsular and creole categories formed a major divide. In the 1860s, Jagor made the trenchant observation that, in Manila:

> Life in the city proper cannot be very pleasant; pride, envy, place-hunting, and caste hatred are the order of the day; the [peninsulars] consider themselves superior to the creoles, who, in their turn, reproach the former with the taunt that they have only come to the colony to save themselves from starvation. A similar hatred and envy exists between the whites and the *mestizos*. (Jagor 1965, 16)

The creoles were known to have disdained agriculture and commerce, so they relied primarily on their posts as colonial state employees. Within the bureaucracy, however, they could not aspire to any position of significance because the best sinecures were assigned to Spaniards born in the Peninsula. Despite cultural and physical similarities, the accident of birth outside the metropole and the bureaucratic structure of empire created deep feelings of separateness; indeed, the creoles of Latin America became the pioneers of modern nationalism (Anderson 1991, chap. 4). National differences were articulated at the same time in economic terms, so that cultural and nationalist consciousness blended with class awareness and economic conflict, in the same way that the racial and caste categories of the Inquisition discussed in Chapter 1 had given way to cultural differentiation along class lines within the emerging national societies.

But the animosity between creole and peninsular did not hinder members of both groups from trying their luck in Negros. In 1861 there were a total of thirteen *blancos* in Negros in various occupations, a number that rose to twenty in 1866; by 1880 the count was placed at sixty.[18] In 1888 as many as 254 *blancos*, all but five Spanish, reportedly resided on the island, along with 631 Chinese, most of whom were devoted to trade.[19] The magnetic pull of Negros contrasted sharply with Pampanga where, also in 1888, resident Spaniards numbered fifty-eight only, of whom only a small number (twenty-seven) were classified as agriculturists (Larkin 1972, 77).

The 1888 data for Negros indicated that the majority of *blancos* were clustered in eight towns, with the largest concentrations found on the eastern side of the island.[20] The Recollect historian, Fr. Angel Martinez Cuesta, has observed a spatial distribution among the three ethnic categories of *hacenderos* (Cuesta 1980, 376). For instance, the peninsulars congregated in the town of La Carlota, the creoles in Jimamaylan. Mestizos, on the other hand, predominated in Minuluan, Silay, and Saravia (present-day E. B. Magalona). Apparently, the location of *haciendas* followed a pattern of clustering and segregation along the cultural and somatic lines that stratified colonial society.

Nevertheless, social cleavages no longer posed insurmountable barriers. In Bais, creoles and peninsulars were mixed together, while mestizo-dominated Minuluan had twenty *blancos*. The Negros export economy helped blur, to a degree, the social divisions that previously were all-important in colonial society. Whether mestizo, creole, or peninsular, everyone had embarked upon the same journey with a common object. On the island of Negros they faced the same vicissitudes and hopes. However, once they had settled and carved their own niches and *hacienda* enclaves, the journey turned into a social game that simultaneously emphasized as well as deemphasized racial and status differences.

As migrants starting from about the same footing, as no one possessed spectacular capital such as to outshine the rest, the *hacenderos* were comparatively even, rather equally matched—an essential rule for combatants in the gaming arena of the cockpit. As in any contest, the sugar game produced winners and losers, the differences becoming more apparent after a series of matches. In the race of races, the mestizos were in the lead of sugar production. Excluding Minuluan, seven towns with a sizeable number of *blancos* in 1888 had an estimated annual sugar production of 187,500 piculs. In contrast, the two overwhelmingly mestizo towns of Saravia and Silay produced 245,000 piculs, which, when combined with the annual output of Minuluan, totaled 405,000 piculs, or 42 percent of Negros's aggregate sugar production (Table 1).

Earlier in 1881, the local elite *(principalia)* composed of officeholders in the different Negros townships assembled to list the residents in their respective towns from whom at least 200 pesos could be collected in property and business taxes.[21] This exercise yielded 1,009 names, a monied group that no longer coincided with the extant political elite of the island. The Negros export economy had produced a new economic elite, one composed primarily of migrant mestizos. The towns with a concentration of at least forty such individuals had no sizeable *blanco* population, except for Bais and to some extent Minuluan.[22] Numerically, the mestizos were the most visible beneficiaries of export agriculture.

Table 1.
Principal Agricultural Production per Year, by Town, Negros Island, 1880s

Town	Sugar (Piculs)	Rice (Cavans)	Corn (Cavans)	Tobacco (Fardos)
Negros Occidental				
Argüelles	10,000	—	—	—
Bacolod	19,500	—	—	—
Bago	40,000	18,000	—	—
Binalbagan	10,000	3,000	—	—
Cabancalan	41,000	15,000	—	—
Cadiz	35,000	—	—	—
Calatrava	—	—	4,000	500
Cauayan	—	5,000	—	—
Dancalan	—	4,000	—	—
Escalante	—	—	1,500	2,900
Ginigaran	12,000	10,000	—	—
Granada	14,000	500	—	—
Guiljuñgan	—	3,000	—	—
Ilog	50,000	25,000	—	—
Isabela	12,000	120,000	—	—
Isiu	—	3,000	—	—
Jimamaylan	2,000	10,000	—	—
La Carlota	48,000	4,000	—	—
Manapla	40,000	—	—	—
Minuluan	160,000	—	—	—
Murcia	3,000	2,000	—	—
N.S. de las Victorias	40,000	—	—	—
Pontevedra	40,000	35,000	—	—
San Enrique	30,000	20,000	—	—
Saravia	125,000	—	—	—
Silay	120,000	—	—	—
Suay	3,000	9,000	—	—
Sumag	4,000	3,000	—	—
Valladolid	15,000	14,000	—	—
	873,500	303,500	5,500	3,400

Table 1. (continued)

Town	Sugar (Piculs)	Rice (Cavans)	Corn (Cavans)	Tobacco (Fardos)
Negros Oriental				
Amblan	1,600	—	1,500	—
Ayuñgon	—	—	4,000	200
Ayuquitan	—	—	800	—
Bacon[a]	—	—	12,000	—
Bais	50,000	—	—	—
Dauin	—	1,000	6,000	—
Dumaguete	22,000	4,000	—	—
Guihulñgan	—	—	14,000	600
Jimalalud	—	—	10,450	500
Manjuyod	4,000	—	1,000	—
Nueva Valencia[a]	—	—	—	—
Siaton	—	600	1,500	—
Sibulan	5,000	—	2,500	—
Tanjay	6,000	2,000	2,000	—
Tayasan	—	—	6,000	200
Tolong	3,000	4,000	—	—
Zamboanguita	—	800	1,000	—
	91,600	12,400	62,750	1,500
Totals	965,100	315,900	68,250	4,900

Source: Datos estadisticos conocidos, Memoria de la Ysla de Negros, 31 May 1888, PNA *Memoria de Negros, Oriental y Occidental.*
[a]Bacon and Nueva Valencia also produced 1,000 and 4,000 piculs of abaca, respectively.

Further data disaggregation showed that 187 individuals, or 18.5 percent of the 1,009 listed names, were deemed liable to pay more than 1,000 pesos in taxes.[23] Although not an exclusively mestizo group, the wealthiest *hacenderos* were found in sizeable numbers in predominantly mestizo towns, particularly Ginigaran and Saravia.[24] In Bacolod, the mestizo Teodoro Benedicto was the only one who made it to the highest tax category, while the creole Jose Ruiz de Luzuriaga was classified along with six mestizos/natives in the over-600-peso category.[25] Evidently, mestizos had surpassed the *blancos* in the contest of wealth, a fact made even more salient in elite consciousness by the very exercise of ranking one's peers for tax purposes.

With profits from sugar making, the mestizos who had mimicked the Spaniards to deal with the cultural ambivalence of their middling position found grounds to feel on a par with, if not superior to, the *blancos*. The impressive economic showing by individual mestizos, which allowed them to purchase the trappings of a European lifestyle, must have taught them that they could truly aspire to the status of the colonial *hombres superiores*. The imperial fear that the colonial subjects would imagine themselves equal to the power holders had been realized, but not in the leveling field where *indios* would toil alongside poor Spanish peasants, as Madrid had earlier imagined. On the contrary, the perception of equality came with the attainment of economic success. Money ignited the mestizo nationalist imagination: henceforth, money and nation would dance around each other. The mestizos began to see themselves as akin to the "Filipinos," the creole label designating Spaniards born in Filipinas. This same label, which mestizos would later appropriate to designate a newfound sense of national identity, would be used to emblematize the commonality of birthplace and, in due course, to homogenize the mestizo and the *indio*.

The imaginings of equality notwithstanding, the mestizos' cultural ambivalence persisted. Unlike their role as liminars in the cockpit, in the broader arena of life the mestizos were marginals who had no assurance of a final resolution to their deep feelings of cultural ambiguity (cf. Turner 1974). The status anxiety was provoked, I believe, by the fact that the mestizos who aspired to "Filipino" status were not unaware of the foreigner's taunt of the Spanish. In 1850s Iloilo, Loney himself had described Spaniards in the colony as, "from a social point of view, of a third or fourth rate sort, many of them ex-mates of vessels and that sort of thing who have left the sea and taken to small trading operations" (Loney 1964, 60). Such class contempt could only be thinly disguised by the upwardly mobile British, although Loney seemed particularly skillful at dissimulation.[26] Because they also tried to imitate the foreigners, the mestizos immersed themselves in a convoluted mix of ranking systems.

But although interaction with foreign merchant capitalists accented feelings of cultural ambiguity and problems of identity, the mestizos found it necessary to ally with them in order to gain access to the capital that lubricated the sugar economy. Moreover, it was precisely in the reliance upon foreign capital that mestizos were on a level with the similarly capital-starved Spaniards. Regardless of race or ethnicity, *hacenderos* were ostensibly even in their equal dependence upon the foreign commercial houses at Iloilo for the funds that defrayed *hacienda* production. The social distinctions that energized the sugar contest were simultaneously overshadowed by a sense of uniformity that *hacenderos* were all players in a game which, despite British and American dominance, could reward astute players with money that somehow knew no race, even as money held different meanings for the various players. A document about the state of Negros Island in the decade ending 1887 provides graphic evidence that sugar capitalism could transcend racial barriers: the signatures of peninsulars like Miguel Perez and Domingo Tejido appeared side by side with those of mestizos like M. Locsin and J[uan] A. Araneta.[27]

Negros had become like a giant colonial cockpit where social categories were intermingled. As Jean Mallat incisively observed elsewhere in the 1830s, "On holidays, the curious people flock from all sides to these [cock]fights: not only Indios are to be found there but also mestizos, Chinese and even Spaniards; all ranks and status are mingled and confounded; the priests sit beside the military" (Mallat 1983, 301). Like the imaginary egalitarian world of the cockpit, Negros had become an arena for the phenomenological forgetting as well as affirmation of the social classification scheme of colonial society.[28] Moreover, like the colonial cockpit where law enforcers mixed with law offenders, the distinction between the legal sphere and the illegal was progressively blurred. Thus pursuing their *suwerte*—a concept shared by mestizos and Spaniards alike—the *hacenderos* were all gamblers in the sugar economy where they had been lured by the prices and profits afforded by linkage to the world market.

Hacendero Gambling in the Negros Sugar Cockpit

Occupying the time and saturating the thoughts of *hacenderos*, gambling bloomed into a worldview that served as the operative framework in the farms of Negros. In a way, *hacienda* production was pitted in a contest with the elements, the unpredictable rains and the periodic infestations of locusts. The risks were heightened by the uncertainty of whether a sufficient number of workers could be found in time to harvest the cane (Echauz 1894, 75–85). The unforseen and uncontrollable factors surrounding sugar production legitimated a gambling principle encapsulated in the *hacenderos*' "faith in an unwritten law of crops, which they often quote, i.e., that the good and bad crop years alternate" (Hord 1910, 14), like the flipping of a coin.

In a deeper way, the *hacenderos* deployed the gambling mentality to make systematically calibrated moves that contravened colonial state regulations concerning labor, as the next chapter will show, and concerning land, as this chapter's discussion suggests. Gambling confounded the appearance of dualism that seemingly separated the loyal colonial towns from the unsubjugated hinterlands, the religious subjects from the heathen rebels, the orderly and law-abiding insiders from the troublesome outlaws. The putative alliance between sugar *hacenderos* and the colonial state was less straightforward than had previously been portrayed in Philippine historiography (Ileto 1988; McCoy 1982b, 173; McCoy 1992). On the contrary, in pursuit of their sugar interests, *hacenderos* gambled with the state for the chance that they would neither be caught nor prosecuted by a weak colonial state in the process furthering its enfeeblement.

Moreover, as a speculative gamble, the sugar industry required enormous funds to prepare the land; purchase cane points, work animals, and machinery; construct lodgings and warehouses; and build access points. Although on their own they did not have sufficient resources, the *hacenderos* could utilize somebody else's money—the foreign merchants'—as stakes in the game of sugar. The chain of borrowings initially pivoted around Loney who, using funds supplied by Russel and Sturgis, built a warehouse on the Iloilo waterfront and began the practice of lending capital to *hacenderos* at 8 percent annual interest; at the same time, Loney extended credit payable in four to twelve months for the express purpose of acquiring modern machinery.[29] Paying their loans in the form of raw sugar, the *hacenderos* still garnered high net returns because of the rising prices and the technical changes that augmented sugar yields.

Dependence on credit became deeply ingrained in the minds of the sugar players. The system of cash advances practiced by foreign commercial houses as a mechanism to secure access to exportable produce was a business strategy that became the *hacenderos'* addiction. Instead of channeling their surplus into savings and investments for the next crop year, the *hacenderos* were wont to gamble away a huge chunk of their profits and to spend the rest in conspicuous consumption. As John Foreman observed in the 1890s,

> although [the native planter] may owe thousands of dollars, he will spend money in feasts, and undertake fresh obligations of a most worthless nature. He will buy on credit, to be paid for after the next crop, an amount of paltry jewellery from the first hawker who passes his way, or let the cash slip out of his hands at the cock-pit or the gambling table. (Foreman 1899, 307)

Habits forged by earlier generations of natives in the wake of colonial subjugation, particularly gambling and the ostentatious display of prestige goods as unstable status markers, forestalled capital accumulation. Confused over the position they occupied in colonial society, the mestizo planters were further influenced by the reputed flamboyance of the merchant capitalist lifestyle exemplified by Russel and Sturgis. Amid the ambience of easy money and accessible *suwerte* came the constant need for capital, thus perpetuating the yearly cycle of cash advances and loans from the commercial houses—a practice that did not differ much from the indebtedness of *indio* sharecroppers. Eventually, as we shall see in Chapter 7, the annual running-into-debt habit would be institutionalized by the American colonial state. Meanwhile, given the outstanding loans they had incurred, Foreman projected that, if forced to liquidate within twelve months, "certainly 50% of [all the Philippine planters] would have been insolvent" (ibid.).

But the risk of insolvency and failure was just that: a risk among many risks. Cane cultivation and sugar making, after all, constituted a foreign-funded game in its entirety, the production phases being a staggered and protracted liminal period when all forces, from weather to locusts, from spirits to labor, could affect the outcome in unexpected ways. In every agricultural cycle, the mystery deepened after the weeding period, when the growing canes were purposely left unattended, when the fields were said to be "closed" *(cerrado)*, at least to human intervention. Like the last phase of a struggle, the harvesting came, followed immediately by the caking of sugar. Then, as in cockfighting, the *sentencia* began: the counting of the number of *pilones* made, the grading of sugars, and, above all, the revelation of prices and the rough estimation of gains. Then Negros was abuzz, as it was about the same time each year, preoccupied with ranking winners and losers according to their greater or lesser *suwerte*.

Marketing soon became a strategic game in itself, especially after sugar prices became highly erratic in the mid-1880s. One was never sure what the gains would be from one year to the next. But unless the planter was completely bound to a credit source, as the industry's periodical noted in the 1920s of an old *hacendero* practice, "it was his habit to gamble with the market or to await an auspicious moment before selling his sugar," such as the meaningfully impregnated time of Holy Week (*Sugar News* 1924b). As reported at the turn of the century, the foreign merchant houses at Iloilo indeed saw sugar planters as "gambling in their product," for instance, by delivering their sugar stocks to the traders without actually selling until a given day when, following a lucky date or a prophetic dream, the price of sugar was expected to rise (Stewart 1908, 200). Thus, the *hacenderos* dabbled in the market; and not surprisingly, as prices constituted the aspect of the industry best known to most *hacenderos* who, as a rule, were hardly conversant with the technical or financial intricacies of sugar production. As former textile traders, playing with prices was what they knew.

In the 1850s, Loney observed that "a superior class of sugar" could

Formation of a *Hacendero* Class 109

not be produced in Negros because of the "very defective nature of the process employed by the native and 'mestizo' planters."[30] In the late 1880s, Spanish officials in Negros observed that the *hacenderos* were "absolutely devoid of agricultural instruction" and that 95 percent of them were extraneous to the sugar industry.[31] The lack of agricultural expertise was similarly observed in Negros as well as in Pampanga during the early American period.[32] In financial management the *hacenderos* also had negligible technical knowhow. In the 1920s, it was reported that planters did not keep financial books and had no precise financial picture of the *hacienda* (*Sugar News* 1920b). As Carlos Locsin observed of his fellow mestizos, who by then had developed the "Negrense" identity, it was "exceptional to find *hacenderos* who keep cost account records or even a segregation of accounts in order to determine the actual field expenses as distinguished from other running expenses connected with the management of a sugar cane plantation" (Locsin 1923, 73). In cognizance of this problem, the Philippine Sugar Association (PSA) began to make a concerted effort to disseminate a system of accounting for sugar plantations.[33]

Still, the American-dominated PSA, notwithstanding the by then illegal status of gambling in the United States,[34] relied upon the gambling mentality to prod the sugar planters in the American Philippines. In the early 1920s, at a time of dropping sugar prices, an editorial of the *Sugar News* (1922i) reaffirmed that "The gambling instinct is strong in most of us. We like to take a chance. Surely the Filipino *hacendero*, at the moment, has an excellent opportunity to gamble on the future." For the possibility that prices might suddenly rise, the editorial asked, "Isn't it worth the gamble?" Official policy pronouncements to the contrary, elite gambling was exalted.

In the present century, as in the past, regardless of the earnings for the year the *hacenderos* have been consistent in the avidity they have shown toward the cockpit, card games like *monte* and *panguingui*, mahjong, and other forms of gambling—a behavior analogous to that of the slaveholders of 1680s Virginia who sought to underscore their freshly acquired gentility (Breen 1977). By the late nineteenth century, high-stakes gambling had become a desirable outlet for the display of *hacendero* wealth. For wagers they reportedly risked titles to *haciendas* and sometimes sugar by the shipload once the wads of cash they had brought along to the gambling dens had been exhausted. A law passed in 1861 limited the amount that could be staked by any person on a cockfight to a maximum of fifty (Mexican) dollars, but the *hacenderos* could not be bothered by such petty legalism (Artigas 1894, 152; Foreman 1899, 406; Worcester 1898, 286). After all, as Francisco Varona wrote in the 1930s, the speculative spirit was "the first manifestation of opulence and good taste" for the region's "legendary magnates," such as Isidro de la Rama, who philosophically considered the "calculation" involved in gambling as a suitable "sign of intelligence and of mastery over self."[35] In 1939, a Bacolod newspaper editorial admitted that the gambler "is as respectable as a saint or a reformer" (*The Commoner* 1939b). As a system to stratify the ranks of sugar planters, gambling gave the ancient Visayan notion of interpersonal *dungan* rivalries a modern expression as *hacenderos* engaged in their contests of acumen and grit, skill and otherworldly favor, strength and superiority. Gambling was the foundation of the sugar-export economy and society.

Betting on Land, Gambling on Legalism

In the pursuit of *suwerte*, control of cultivable land, like casino chips, was integral to the broader game of sugar. To acquire more land for sugarcane cultivation *hacenderos* adroitly manipulated the legal system to their advantage.[36] In particular, because the Laws of the Indies recognized the right of usufruct, natives could reclaim land by judicial recourse to the court injunction known as the *interdicto de despojo*. Combined with other means, this legal provision was utilized by Teodoro Benedicto, the mestizo cause célèbre from Jaro, to acquire vast tracts of land in the central interior of west Negros.

Benedicto's alleged machinations were roundly protested by eleven Spaniards from the town of La Carlota in a petition to the governor-general dated 12 August 1876.[37] In clamoring that Benedicto's *interdictos* be nullified, the petitioners argued that these types of legal document were easily falsifiable: *(a)* the written testimony of a complainant pretending to have once been the possessor of a piece of land now occupied by certain John Does *(Fulano y Mengano)*; *(b)* the complicity of three witnesses; and *(c)* the posting of a bond, all of which were sufficient grounds for a judge to issue an injunction. In turn, with the assistance of local officials, particularly the town magistrate *(gobernadorcillo)*, the document could be used to summarily order legitimate occupants to vacate the land.[38]

Benedicto's detractors vouched that he had seized "perhaps no less than" the equivalent of seven thousand hectares in Negros.[39] "With no law other than his own caprice," Benedicto, according to the Spaniards, had also diverted the river waters of Najalin, in the process depriving *haciendas* in La Carlota of irrigation. Alleging that Benedicto was not "without imitators," the peninsulars identified the creole Bonifacio Montilla of Bago as having similarly usurped the two barrios of Santa Teresa and Zaragoza and "extended his property toward the slopes of Kanlaon."[40]

Using colonial discourse for their own ends, the Spaniards lamented the injustices inflicted upon displaced *indio* settlers. Although the complaint about misuse of legal procedures seemed valid, the La Carlota peninsulars' act of publicly denouncing Benedicto was more suggestive of the ethnic conflict that fractionalized Negros' emergent *hacendero* class than of genuine concern for native welfare.

A few days before the peninsulars wrote their statement, two of the petitioners, Lucas Rubin and Manuel Pacheco of La Carlota, and a third person, Alejandra Bagallon of the town of San Enrique, were implicated on 1 August 1876 by fifty-five residents from the barrio of Borja in Pontevedra. In this earlier instance the Spaniards were accused of appropriating "immense" parcels of land in *sitio* Panlopgasan, which were converted into exclusive pasture lands where *indios* were barred from entering even if merely to gather forest products for subsistence.[41] A fortnight after the La Carlota petition, another testimony dated 27 August 1876 was executed by ten native *principales* from Pontevedra with the stamp *(Visto Bueno)* of the parish priest. In it Lucas Rubin was charged with constructing a canal that "diverted the waters of the river Bujangin from its natural course, depriving everyone of water in the barrio especially in years of scarce rainfall."[42]

Obliquely responding to the peninsulars' allegations, these last two cited documents also reported that many in Borja had been ejected from the 300 hectares which Crisostomo Ramos and "his associate Teodoro Benedicto" purchased legally for 200 pesos on 3 June 1876 from a previous owner, Juan Espinosa, who in turn had acquired the property for 81 pesos in 1866.[43] Espinosa was blamed for the disorganized land situation as not more than 5 or 6 hectares of his property were tilled, the rest being abandoned for occupancy by migrant settlers who formed the nucleus of barrio Borja. A similar pattern emerged in the case of another piece of property purchased by Benedicto. The vendor, Micaela Herrera, had earlier requested a survey of the land "conceded by the *principalia* of the town" because, "with the passage of time the old boundaries had been confused, although the land appeared to have measured 300 *cavans.*"[44] Arguing that after two years one's claim to ownership could be forfeited through noncultivation, which implied "abandonment" (cf. Corcuera and Cruzat 1907, 198), the Pontevedra elite expressed their own view of the matter by calling for "equitable distribution" by restoring the contested property to "the legitimate owners," the settlers who had occupied the land before Espinosa and Herrera entered into transactions to sell those properties.[45]

Meanwhile, in October 1876 the plot thickened when the parish priest sought the Negros governor's assistance on behalf of the residents of La

Carlota's barrio San Miguel, citing that they had been twice expelled from the land after Benedicto had illicitly extended the boundaries of his Hacienda San Bernardino.[46] Claiming to have bought the farm from the Custodio family in 1871, Benedicto later claimed that the legal papers of that purchase had been destroyed in a fire that gutted his Iloilo residence. For his part, Mauro Custodio, *gobernadorcillo* of Pontevedra, was identified in a complaint filed in 1872 as having "usurped" thirteen hectares in Binigsian, La Carlota, accomplished in all likelihood with the connivance of Joaquin Custodio, the local *juez de sementeras* in charge of adjudicating agricultural conflicts.[47]

In evaluating the case against Benedicto, Negros governor Ramon Pastor found that the mestizo had been the recipient of two *interdictos*, one for 30 hectares in La Carlota and another involving 44 hectares in Isabela.[48] But Benedicto had also used other tactics. Testimonies of thirteen Spaniards and twenty-one displaced migrant *indios* in the investigation conducted by the *Inspección General de Montes* in March 1877 revealed that Benedicto had resorted to intimidation. Threatened by armed men, occupants of lots adjoining Hacienda San Bernardino were made to sign documents that relinquished a total of about 272 hectares in exchange for compensation amounting to 1,029 pesos, purportedly for clearing the land of forest growth.[49] By coercing the occupants to sell, Benedicto enlarged his *hacienda* from about 800 hectares to over 1,100 hectares and could even produce documents supporting his contention of a legal purchase, a tool that assured him of judicial victory.[50]

While providing a glimpse of Benedicto's nefarious activities, the investigation was also suggestive of a serious conflict between Benedicto and the peninsulars, for the latter evidently had sought to acquire some of the contested land parcels. One of the eleven petitioners, Teodoro Gurrea, for instance, allegedly offered 200 pesos for the 18 hectares tilled by the *indio* Vicente Agatis, but the latter refused, only to be pressured by Benedicto into accepting 100 pesos for the land.[51] (Incidentally, just a year before the peninsulars filed their protest, Gurrea had borrowed 4,500 pesos from Benedicto, with Gurrea's 119-hectare *hacienda* in La Carlota as collateral.)[52] The investigation concluded that Benedicto "and his accomplices" were criminally liable for the methods they had employed, but the legality of the land acquisition itself could not be ascertained. In a decision penned in June 1883, the *Real Audiencia de Manila* exonerated the local officials involved and endorsed the case back to the Court of First Instance in Negros, deemed by the *Audiencia* as the only competent body to resolve the issue.[53] The collusion of local functionaries with land grabbers went unpunished.

To the envy of his peninsular opponents and the misery of migrant

Formation of a *Hacendero* Class 113

indio settlers, Benedicto thrived on winnable court suits brought about by the ambiguities of colonial land laws, the anarchic land market in which transactions could be consummated *sin formalidades* (without formal procedures), and the absence of an orderly system of property titling and registration. Benedicto, however, was not the only one who exploited the malleable legal system. As some Spanish officials were prepared to admit, there were "countless" cases resulting "continually" in dispossession because of the misuse of the *interdicto de despojo*.[54] In response to complaints about the *interdicto*, officials in Manila reacted by defending the judicial instrument as well "within the existing legal framework."[55] By allowing formal legalism to triumph over sound governance, the colonial bureaucracy perpetuated the twilight zone of the law. All that the governor-general did was issue a banal instruction to parish priests and local officials to inform the people and "make them understand" that legal remedies existed in case they were wrongly subjected to an *interdicto de despojo*.[56] In 1880 the *Real Sociedad Economica de Amigos del Pais* appealed to Madrid to legislate reforms on the *interdicto*, but the request was met with the almost predictable fate of imperial inaction and apathy.[57]

In the meantime, countless peasants were dispossessed even as the *interdicto* figured in the incessant squabbles among *hacenderos*. For instance, it was central in a case involving Antonio Buenafe, leaseholder since 1879 of Hacienda Dauimon in Victorias. Some 50 hectares of the leased property were encroached upon by Manuel Suarez in 1887 and restituted to Buenafe in the same year; but two years later 13 hectares were again usurped by Manuel's son, Felix Suarez.[58] *Interdictos* and counter-claims were also employed in a litigation that lasted a decade involving 25 hectares in Tangcong, Saravia, which Julian Giguiento claimed he had legally purchased in 1873 but which, in 1881, Pedro Montino claimed as his inherited property.[59] Paradoxically, despite elite gambles against the colonial state, the *hacenderos* nonetheless continued to appear before the colonial judiciary to sort out their differences, suggesting differential use of the state apparatus. The settlement of intra-*hacendero* disputes over land seemed to be one of those instances when the parties concerned regarded state intervention as useful. After all, in the cockpit as well as in a court of law, the *sentencia* was passed by an individual appointed by the colonial state (cf. Artigas 1894, 152–153).

In addition to sensational cases of land grabbing *(usurpacion)*, the *hacenderos* resorted to various other means of acquiring land. Particularly in the more settled zone along Negros' western coast (Figure 2), appreciable changes of property ownership entailed the purchase of farm lots already cleared of forest growth—belying the image of planters as pioneers romantically painted by the hagiographer of Negros *hacenderos*,

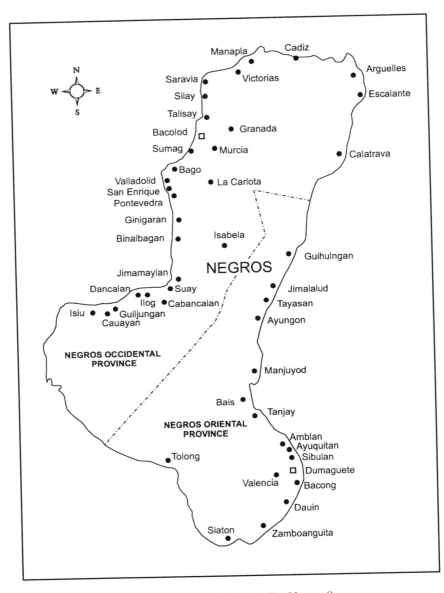

FIGURE 2. Map of Negros Island with its *Pueblos* c.1890

Francisco Varona (1965). Because small *indio* landholders were willing to receive cash for land long made productive, probably thinking there were no serious obstacles to opening up new land for cultivation, they sold off their fields, usually without the legal formalities. This pattern of land acquisition accounted for the large concentration of sugar production by the 1880s in the established coastal towns along the northwestern portion of the island (cf. Table 1).

The farmland transactions notarized in 1861 (the first year for which a fairly complete record is available) revealed that thirty-one of the thirty-six legible cases involved the sale of land that ranged from small plots of about 2 hectares to at most 60 hectares.[60] Nearly three-fifths of the plots sold were located in Bacolod, and slightly over a fifth in Pontevedra. The average land sale involved approximately 16 hectares valued at 246 pesos, or an average of roughly 15.5 pesos per hectare. Close to three-fifths of the land sales involved a Spanish buyer.

By 1875 some 800 hectares changed recorded ownership, the largest parcel measuring 200 hectares. This time the farmlands were acquired mostly by mestizo buyers (Spanish buyers formed only one-fifth) and were somewhat more evenly distributed from Cadiz in the north to Ilog in the southwest.[61] Reflecting the profitability of sugar, the average land sale involved about 35 hectares worth 1,039 pesos, each hectare costing an estimated 30 pesos, or about double the 1861 figure. Of the other forty-five notarized farmland transactions in 1875, most were either requests for the "legitimate title" of land "long held without interruption" or the mortgaging of property.[62] In absolute terms, a greater area was involved in requests for titles, as the average case measured nearly 120 hectares, which the predominantly mestizo petitioners invariably claimed as the aggrupation of small plots bought over the years from several previous owners.

The land market continued to thrive, with an unprecedented 3,117 hectares being sold in thirty-one documented transactions in 1890.[63] Four transactions involved *haciendas* at least 300 hectares in size, with another five involving 100 hectares or more. In general, the 1890 public notary records reflected an enlargement in the average farm size sold (about 100.5 hectares) and mortgaged (123 hectares), whereas in 1875 the corresponding figures were only 35 hectares and 45.5 hectares, respectively. Three and a half decades after the region's direct incorporation to the world market, the pattern of land distribution had become more skewed: as *haciendas* expanded in size, more and more land areas were falling under the control of fewer individuals. The increasing concentration of land was suggestive of a trend toward the centralization of wealth in the hands of large (and luckier?) *hacenderos*.

Despite the sugar crisis that began in 1884, the average sale price per hectare in 1890 was computed at 28.3 pesos, only slightly lower than 1875's 30 pesos, while the average value per hectare of mortgaged property amounted to 25.5 pesos, nearly the same as 1875's 26.5 pesos. But the fact that entire *haciendas* were being sold or leased meant that those more averse to taking risks were seeking to temper the sugar gamble. However, the crisis did not drastically destabilize the sugar industry, and the properties for lease or sale were not without takers, who believed that market woes were amenable to "calculation." Certainly, land speculation was rife when uncultivated portions of the public domain were auctioned off in 1890. In seven such auctions, about 874 hectares were sold by the colonial state, averaging 270 hectares at a giveaway price of just over 2 pesos per hectare.[64]

A land registration scheme introduced in 1889 provided yet another facile means to claim and control land legitimately. Data from 1894 to 1896 indicated a total of 538 land parcels registered for the first time in Negros.[65] Land parcels not exceeding 20 hectares comprised over half of the properties registered; in fact, plots 10 hectares in area or even less comprised 37 percent, suggesting that smallholding peasants also took advantage of the chance to formalize their claim to land. But the most cunning players were the bigger *hacenderos*. During this three-year period, they registered 128 land parcels ranging from 21 to 50 hectares; they also registered another 51 parcels that varied from 50 to 100 hectares, and as many as 76 parcels that exceeded 100 hectares.

The resulting pattern of landholding was characterized by a few relatively large *haciendas* and a preponderance of small and medium-sized farms.[66] Through various means, legal and extralegal, an inequitable land distribution pattern became entrenched in Negros. Nevertheless, even the island's largest nineteenth-century *haciendas* were comparably small by the standards of Latin America's *latifundios,* which ran into several thousands of hectares.[67] Probably in recognition that they fell markedly short of the scale and grandeur of Latin America's landowners, the local sugar planters desisted from calling themselves *hacendados*, the Spanish word for landowner. Instead, they settled for the humbler term that had long been used in the Spanish Philippines: *hacendero*. In reports and memoranda about Negros, Spanish officials, especially when addressing Madrid, normally explained *hacendero* in an aside as the local custom: *"como se titulase en la localidad."*

The *hacendero* tag, however, was strangely inappropriate. As a noun, it originally referred to a daily wage worker in the mines of Almaden. In contrast, the Negros *hacenderos* saw themselves as belonging, economically and culturally, to a class category far superior to that of a wage worker. Indeed, they fancied themselves as constituting an aristocratic class, a new oligarchy pretending to be old, with self-consciously crafted images of gentility supported by a life of leisure and gambling. Their style of life resonated oddly with the adjectival meaning of *hacendero,* which referred to someone thrifty and industrious, for thrift was furthest from the language of the sugar *hacenderos,* who approached the growing of cane and the marketing of clayed sugars as one big gamble.

Thus, *hacendero* was an oddly fashionable term, but every landowner relished being known by such a label, no matter the size of his or her farm. To belong to the elastic construct of the *hacendero* class permitted the use of the coveted titles of *Don* and *Doña,* now accessible even to those who did not come from the old native elite. The owner of 5, 50, or 500 hectares could claim the old honorifics in the new capitalist age of the late nineteenth century. Undeniably, differential adeptness at the sugar game resulted in an evident system of social stratification. Even so, it was reassuring that all were intermingled in the same colonial gambling den where to be *hacendero* meant to express unbridled individualism as each player asserted the strength of one's *dungan* in pursuit of *suwerte.*

The Colonial State and Foreign Capital in Negros

The fact that the gambling den that was Negros was lubricated by foreign capital did not please the colonial state, despite the high revenues generated by the Iloilo port where, as in 1891 and 1892, customs collections surpassed 1.3 million pesos.[68] Spanish functionaries complained that the merchant houses were the real beneficiaries of foreign trade and not the *hacenderos,* much less the colonial state. Structurally, however, sugar planters had little option but to rely upon the merchant houses.

In the matter of technology, for instance, Loney's tied loans for the purchase of British equipment to upgrade local sugar manufacturing became indispensable, inasmuch as the colonial state was "opposed to the practice of foreign importers of machinery at Manila [selling] on other terms than cash."[69] How, then, were the *hacenderos* to improve backward techniques of sugar production as they had no wherewithal to acquire machinery? As it happened, technical change began to dawn upon the region only after foreign merchants and local planters decided to bypass the colonial state's obstructionism. By 1861 Iloilo Province had three animal-drawn iron mills *(molinos de sangre)* and one steam-powered mill; even more noticeable was the change in Negros, where, in 1857, there had been only one iron mill but where, four years later, thirteen were in operation.[70] By the late 1880s, there were 500 iron cattle mills in Negros, supplemented by 200 steam engines and 30 water-powered mills.[71]

Less than a decade after the Iloilo port's opening, the island of Negros was already being referred to as *el emporio de la riqueza* (the emporium of wealth) of the Visayas. Nevertheless, the dramatic boom of Negros did not coincide with the dreams of patriotic Spaniards who, thinking themselves cheated, began to reactivate the exclusionary term *extrangero*. Among those who emphatically denounced the foreign merchants was Remigio Molto, head of the Provincial Government of the Visayas based in Cebu.[72] In the mid-1860s Molto wrote in a memorandum that the Spaniards who had ventured into Negros were poor retirees who had served either in the military or the civilian bureaucracy and whose meager salaries did not allow them to accumulate the needed capital for sugar production.[73] This "misfortune," he said, was exploited by foreign merchant capitalists, who were charging exorbitant rates of interest, forcibly making Spaniards buy machinery and other items, and requiring the Spanish debtors, whose loans were tied to the standing crops, to surrender their produce at prices lower than prevailing market prices.[74] The result, according to Molto, was prosperity for the predatory foreign commercial houses but misery for the Spaniards who—like the sugarcane—were "smothered," unable to extricate themselves from the foreigners' iron hands, which literally "squeezed" them of cane juice.

Spaniards in Negros were said to be desperately selling their properties, and the specter was raised that eventual foreign ownership of *haciendas* would make a mockery of Spanish suzerainty: "and Filipinas would become a foreigners' province with Spanish authorities" as mere figureheads.[75] Molto's fears, however, were baseless, as foreigners did not dominate landownership in Negros. Specifically mentioned among the Spaniards who were supposed to have sold out were *"los señores Coscolluela, Lusuriaga y Macunana."* Only the last name, however, does not figure among the large sugar planters in late-twentieth-century Negros, and its disappearance may have been due to reasons other than those raised by Molto. Moreover, though foreigners did acquire properties in Negros, these were eventually bought by mestizos.[76]

At any event, Molto suggested several measures that would turn the tide in favor of Spaniards, such as exempting their work force from corvee labor and deploying corvee workers to assist Spanish planters in dire need of farmhands for harvesting and other tasks.[77] He also suggested, as many others had done before and have done since, that an agricultural bank be created to lend capital at low interest rates so that Spaniards could cut their dependency ties *(dependencia)* with the foreign merchant houses. But Molto's suggestions were either politically ill-advised or financially infeasible. The "dependency" connections were not severed, even as interest rates in general soared due to the unabated demand for capital. In the notary records, interest rates jumped from

9 percent in 1861 to from 12 to as much as 36 percent in 1875, declining somewhat to from 6 to 20 percent in 1890.[78] The wealthier sugar planters also began to loan funds to smaller planters at even higher interest rates than those charged by the commercial houses. Despite ostensible problems of capital, labor, and transportation, the Negros *hacenderos* pushed ahead with their production, and sugar exports continued to rise, even as Spanish officials saw fit to continue to denounce the foreign merchants.

Reporting on conditions in the mid-1880s, Antonio Tovar, the governor of Negros, censured the commercial houses at Iloilo for being *"casas acaparadoras"* (monopoly houses) that purchased Negros sugar on grossly unfair terms by virtue of the crop loans.[79] Except for Ynchausti y Compañia, a Spanish commercial house based in Manila that loaned to a number of sugar planters in Negros, the sugar export trade in Iloilo in 1886 was dominated by four British commercial houses (Smith-Bell & Co., Ker & Co, W. F. Stevenson & Co., and Macleod & Co.), one American (Peele-Hubbell & Co.), and one Swiss (F. Luchsinger) (McCoy 1977, 85, 87). Tovar accused these Iloilo exporters of colluding on the prices paid for Negros sugar. Whether or not he was aware of the broader crisis in the global sugar market at this time did not seem to matter, for he called the prices "arbitrary" and pinned the blame on the foreign merchants' whims. Censuring the latter as *"fingidos protectores"* (deceitful protectors), the governor listed the merchants' malpractices as including: *(a)* the absence of a "legal classification" system to determine the quality and value of sugar; *(b)* tampering with weighing scales; *(c)* the practice of *tara,* by which the buyer made the sugar producer pay for packing the sugar; and *(d)* manipulations like "humidifying," which lowered sugar volume, and changing the sacks of the sugar temporarily stored in Iloilo warehouses even when not needed. Even the means of transporting sugar from Negros to Iloilo were also apparently controlled by the merchant houses (Stewart 1908, 180). The disastrous outcome, Tovar averred, was that "bit by bit the monopoly houses at Iloilo appropriate the wealth . . . leaving behind in such a fertile island nothing but the frustrated dreams of the planters."[80]

The sugar planters were probably cognizant of the delicate balance in their relationship with the foreign traders, which hovered between exploitation and mutual advantage. But the *hacendero* point of view was certainly different from that espoused by Spanish officials. The planters instead faulted the colonial state for exacerbating their problems. Not only were they deprived of technical and credit assistance and a supportive administrative system, but they were made to bear an additional burden arising from a new tax on urban property and industrial-commercial income, without the "odious" *diezmos prediales* (tithes on real estate) being abolished.[81] This imposition prompted sugar planters to devise schemes to reduce or avoid paying the new tax, a strategy at which Philippine elites have been most adept. Although some were caught, they were eventually absolved of any offense,[82] suggesting that taking gambles with the state does pay off. To the planters taxation was a mere irritant and did not constitute an insurmountable problem. What they found most offensive were the contradictory policies at the level of empire, which prompted the *hacenderos* to enter into a formal alliance with the foreign merchants despite harangues by such Spanish officials as Molto and Tovar.

The crisis of the 1880s starkly demonstrated the chaos and apathy that beset Madrid and their deleterious consequences for the Spanish Philippines. The bounty system, which subsidized beet-sugar producers in Germany and France, occasioned a precipitous decline in world sugar prices, with the year 1884 particularly critical, as it saw total Philippine sugar exports plummet by over 42 percent from that of the previous year. The decline in sugar exports, the scarcity of rice that required huge importations from French Saigon, and the withdrawal of about 4 million pesos from the colony to support Madrid left the colonial treasury with a loss of about 17 million pesos in 1884.[83] Amid this crisis, Madrid made policy decisions detrimental to Philippine sugar. First, on 25 July 1884, Madrid reduced by 5 percent the export duties on Cuban products. Then, a little over two months later, on 5 October, the sugar exports of the Spanish Antilles, Cuba, and Puerto Rico were allowed duty-free admission to the Peninsula.[84]

In objecting to such moves, the colony's most important commercial interests, Spanish and foreign alike, lamented that their needs were heard "so late and sometimes inexactly at the bosom of the Mother Country."[85] They beseeched the *Ministerio de Ultramar*, in December of that crisis-ridden year, to place Philippine sugar on an equal footing with that of Cuba and Puerto Rico to encourage fair competition. In July 1885 Madrid acceded to only one item of their petition, that of allowing Philippine sugar duty-free entry to the, in any case, almost negligible market in the Peninsula.[86] This concession paled in comparison to Spain's other market interventions. Notwithstanding formal representations by Philippine sugar interests in Madrid and the support extended to the December 1884 petition by the highest Spanish officials in Manila,[87] Spain went ahead to conclude negotiations on a treaty that granted Cuban and Puerto Rican sugar duty-free entry to the vast and lucrative United States market. In January 1885 the "businessmen and agriculturists" of Negros and Iloilo reacted in disbelief. They told the *Ultramar* that they preferred to view the "omission" of Filipinas from the coverage of that treaty as an "involuntary lapse of memory," which ought to be rectified to forestall the "complete ruination" of sugar production, the principal pillar of the colonial economy.[88]

The United States had become the major market for the sugar produced by the Spanish Philippines beginning around 1869, and its importance became pronounced in the 1880s (see Table 2). The United States absorbed over half of Philippine sugar exports in 1882 and more than two-thirds of that in 1883. Thus the Negros and Iloilo sugar interests reasoned that the U.S. market was indispensable to the colony's sugar industry, in much the same way that Cuba depended upon the U.S. market for some 80 percent of its sugar exports. Exclusion of the Spanish Philippines from the treaty was catastrophic, especially so for Iloilo's sugar exports: of the 1,518,341 piculs exported in 1883 (which comprised 45 percent of the aggregate volume sold by the colony), as much as 80 percent went to the United States.[89] Undoubtedly, the duty-free admission of Cuban sugar gravely disadvantaged local sugar.

Table 2.
Destination of Philippine Sugar Exports from 1850 to 1919 (Total Exports in Tons of 2,000 Pounds, Muscovado)

Calendar Year	% to U.S.A.	% to China-Japan	% to U.K.[a]	% to Europe/Continent	% to Other Countries[b]	Total Exports
1850–54	27.6	—		42.9	29.5	35,881
1855–59	20.0	—		48.3	31.7	8,809
1860	31.5	—		48.8	19.7	61,457
1861	14.2	—		50.6	35.2	58,122
1862	13.2	3.6	46.6	1.4	35.2	89,003
1863	10.9	31.6	35.8	1.1	20.6	82,682
1864	24.1	6.8	65.6	0.7	2.8	70,311
1865	22.6	18.0	36.8	1.7	20.9	60,821
1866	24.8	13.7	53.6	1.2	6.6	60,445
1867	17.4	18.8	49.1	2.8	11.8	71,163
1868	19.4	2.4	69.1	0.9	8.2	81,659
1869	42.2	0.1	46.6	0.6	10.5	75,867
1870	29.6	6.5	51.8	2.9	9.1	86,213
1871	46.1	0.04	39.7	4.1	10.0	96,412

Table 2. (continued)

Calendar Year	% to U.S.A.	% to China-Japan	% to U.K.[a]	% to Europe/Continent	% to Other Countries[b]	Total Exports
1872	33.7	0.5	55.2	2.9	7.6	105,298
1873	40.1	0.2	39.5	5.0	15.3	98,478
1874	49.3	0.5	40.1	2.4	7.6	114,487
1875	42.4	0.1	50.0	1.4	6.1	139,099
1876	60.2	0.02	37.8	1.2	0.7	143,773
1877	52.5	0.1	45.3	2.1	—	134,935
1878	54.2	2.2	39.4	2.6	1.4	130,100
1879	41.6	5.5	51.3	1.6	—	148,595
1880	57.0	1.6	38.6	2.5	0.3	199,238
1881	40.4	2.1	51.5	4.5	1.5	233,044
1882	51.6	1.4	43.8	2.1	1.0	166,440
1883	69.5	0.1	26.2	4.3	—	237,255
1884	72.9	5.3	15.2	6.6	—	135,500
1885	71.8	10.6	15.6	2.0	—	234,560
1886	72.6	10.2	14.2	2.9	—	204,792
1887	68.9	11.8	16.7	2.5	0.05	197,474
1888	58.5	14.3	19.0	2.5	5.7	204,208
1889	59.4	6.5	25.1	2.3	6.7	241,324
1890	27.5	24.4	27.1	2.7	18.2	162,616
1891	36.1	11.2	29.3	1.4	22.0	183,488
1892	24.9	30.8	29.9	1.2	13.2	272,493
1893	23.3	29.8	37.7	1.6	7.5	288,456
1894	18.7	37.1	31.5	2.1	10.5	214,199
1895	17.6	32.4	39.9	1.7	8.3	254,547
1896	34.1	38.5	17.1	1.5	8.8	254,433
1897	7.5	63.8	23.3	0.9	4.5	222,767
1898	15.5	58.5	25.9	0.1	—	199,318
1899	25.5	27.8	14.4	—	32.3	94,608
1900	3.3	75.6	13.6	—	7.4	71,860
1901	9.2	90.8	—	—	—	62,691
1902	5.2	71.7	6.4	—	16.6	108,682
1903	34.4	57.2	—	—	8.4	94,035

Table 2. (continued)

Calendar Year	% to U.S.A.	% to China-Japan	% to U.K.[a]	% to Europe/Continent	% to Other Countries[b]	Total Exports
1904	29.7	70.2	—	—	—	95,959
1905	40.2	59.4	0.5	—	—	19,598
1906	9.2	90.8	—	—	—	42,697
1907	8.6	82.0	9.4	—	0.03	141,003
1908	32.3	59.8	7.7	—	0.2	159,541
1909	41.0	58.9	—	—	0.02	142,559
1910	82.9	17.1	—	—	0.01	133,899
1911	89.8	7.9	2.3	—	0.00	230,429
1912	67.9	30.2	1.8	—	0.00	217,237
1913	19.5	80.5	—	—	0.01	173,430
1914	71.7	26.0	2.3	—	0.01	260,691
1915	39.2	51.1	9.6	—	0.01	232,599
1916	39.1	37.4	19.9	1.0	2.7	372,015
1917	30.3	67.6	1.9	0.00	0.1	226,973
1918	38.8	60.9	—	0.2	0.1	301,212[c]
1919	23.6	72.5	3.6	0.2	0.1	149,979[d]

Source: Palmer 1920, I-8, U-6.
[a]Exports to the United Kingdom and other parts of Europe are lumped together for the years 1850–1859.
[b]Other countries include Australia, British India, and Canada.
[c]Includes 70,567 tons of centrifugal sugar; prior to January 1918 all sugar is stated in terms of muscovado.
[d]Includes 32,914 tons of centrifugal sugar and 30 tons of refined.

In response to the petition of Philippine sugar interests, Madrid reduced sugar export duties by 20 percent, a concession belatedly granted in 1887.[90] Also in September 1887, about a year after Spain and the United States entered into a general agreement on the equalization of tonnage dues and imposts, an amendment was forged to include the Philippines in the modus vivendi that originally covered Cuba and Puerto Rico.[91] In principle, such policy changes might have served to make sugar produced in Filipinas more competitive in the world market. By the 1890s, however, Western consumer tastes had shifted markedly to refined white sugar. Because of its inferior quality, Philippine raw muscovado sugar found its market mainly in the low-priced China market, a situation that prevailed until the early 1900s (Table 2). Meanwhile, Spain and the United States continued to sign commercial agreements, such as one issued in early 1895 pertaining to the application of the lowest tariff rates possible for the products and manufactures of Cuba and Puerto Rico, with Filipinas again remaining excluded.[92]

The disjunctures within the Spanish imperial domain crystallized to local sugar interests the importance of their dependence upon the "deceitful protectors" at Iloilo. Local parties could not have marketed Negros sugar on their own internationally because of the region's relative isolation from other parts of the world. Prospective local traders did not have the connections the foreign merchants at Iloilo had established with markets worldwide. Global marketing was made doubly difficult by the absence of a direct channel of communication with the centers of sugar trade. Although by 1880 Luzon was already linked internationally by means of telegraphic connections via China,[93] the Visayas had no rapid means of communication even with Manila. Businessmen perceptively argued that telegraph lines to the Visayas would not only improve commerce but would also "establish and strengthen the criterion of unity."[94] Madrid's inaction on repeated requests to extend the telegraphic network again compelled the *hacenderos* of Panay and Negros to band together with the merchants of Iloilo in raising another clamor to the *Ultramar* in December 1892.[95] This last request was granted, but not until 29 November 1897 did the southward telegraphic network finally become operational.[96] By the time Madrid acted, the telegraph lines merely served to facilitate the relaying of messages as the revolution against Spain struck the islands.

With little support from the colonial power, the sugar economy of the Negros-Iloilo region proceeded under its own momentum. The colonial state's vision of a lucrative agriculture was finally realized, but in a way that made the state virtually irrelevant as the *hacenderos* took matters into their own hands, contravening the strictures of the colonial state with impunity. Whether on questions of land or, as discussed in the next chapter, on questions of labor, the Manila government chose to remain unmindful of the exigencies of the emerging regional economy. Ultimately, the colonial state was more concerned with upholding the frequently obsolete legal principles of Spanish imperial administration. Plagued by political and financial disarray, Madrid for its part was suffering from indifference that bred chaotic policies, while rhetorically upholding its beloved self-image as a paternalistic imperial ruler. As in an earlier era, to vituperate the foreign merchant houses as villains was all that patriotic Spanish officials in Filipinas could do. The foreign merchants, as might be expected, were attentive to the promotion of their own economic interests. In the process, the funds and market outlets they provided were seized upon by the *hacenderos* to indulge in sugar gambles in a game that intermingled different social categories and a contest where the state was an extraneous, though occasionally useful, umpire. But the mestizos were not the only gamblers; as we shall see in the next chapter, the *indios* were as well.

"Capitalists Begging for Laborers":
Hacienda Relations in Spanish Colonial Negros

Labor Recruitment and the *Indio* Advantage

Given the prevailing circumstances on Negros Island, the fledgling sugar planters resorted to a two-pronged approach to the acquisition of farm labor: they employed in the *haciendas* a permanent resident work force, mostly of migrant background, and a temporary stream of migratory laborers hired during peak periods. Labor of the permanent kind was provided by *agsa* or *acsa* sharecroppers (known in their Spanish variants as *agsadores* or, more generally, *aparceros*) who were contracted on a share-tenancy basis similar to the *kasamahan* arrangement in the Tagalog area. Labor of the temporary kind, needed especially during the cane harvesting and crude-sugar claying seasons, was provided by day laborers *(jornaleros)* imported from neighboring islands.

From the second half of the nineteenth until the early twentieth century, share tenancy was widespread in the *haciendas* of Negros. In the 1860s, Loney observed that "on all plantations of any extension there are besides day laborers a number of families or tenants, called *acsas*."[1] In the 1880s Antonio Tovar, the island's military governor, reported that a great number *(muchos)* of property owners relied upon sharecroppers *(aparceros)* to cultivate the land.[2] Writing in the 1890s, Echauz referred to sharecropping as "lo mas corriente," the most common or well-known system of working the land (Echauz 1894, 43).

In some farms, the management of plantation work was not directly handled by the *hacendero* but delegated to a salaried staff composed of overseers *(encargados)* and foremen *(cabos)* who supervised the work process of tenants and day laborers. This arrangement was known as the administration *(administracion)* system (Echauz 1894, 43). However, consistent with the perennial problems of colonial agriculture discussed in Chapter 3, a strictly wage-labor regime did not emerge in Negros. As in other parts of the country, the preponderance of sharecropping suggested that the balance of social forces in Negros made the control of labor that knew how to insist on its autonomy highly problematic. In Negros the situation was aggravated by the island's scant population, the core of which was composed of aborigines (hence the Spanish name, Isle of Blacks). The indigenes would later be displaced and dramatically outnumbered by migrants. In the early 1850s, Negros had a population of around 100,000, which rose in the late 1850s to an estimated 134,000 (Table 3). Despite this increase, Negros' population size paled in comparison to Iloilo Province's population count in 1859 of 390,833 (McCoy 1977, 28), nearly three times that of the entire island of Negros at the inception of the latter's sugar industry. Negros' deficit in labor supply was met by importing workers from surrounding islands, an unwieldy process that highlighted the incomplete class dominance of the *hacenderos,* who could not beckon labor at will.

In 1861, Loney reported that the Iloilo mestizos who became sugar planters in Negros "each [took] with them several families from Molo, Jaro, Miagao and other *pueblos* of [Iloilo] province to settle on their estates and work on the usual system of proportionate share of profits," that

Table 3.
Estimated Total Population of Negros Island, 1850–1903
(Selected Years)

Year	Population	Year	Population
1850	101,260	1882	292,688
1854	123,420	1885	401,423
1858	134,079	1888	387,640
1862	180,867	1891	394,355
1865	198,264	1894	440,944
1876	255,717	1897	461,972
1879	272,767	1903	463,743

Sources: Cuesta 1980, 257; United States 1905, 123, 126–127.

is, to work as sharecroppers.³ In 1867, only six years later, it appeared that the "taking of families" was no longer feasible, moving Loney to delve at length into the issue of labor in his consular report. The sparse and widely scattered population of Negros made it imperative, he said, for the *haciendas* "to draw nearly all their supplies of labor from the more populous province of Iloilo," with the *hacenderos* themselves "personally [going] to the villages" to engage workers "for as long a period as they can."⁴

But the legal recruitment of labor followed a complicated procedure, with the employer first having to pay the tribute money the prospective worker owed the colonial state. The employer then had to make a monetary advance to coax the *indio* to accept *hacienda* work on adjacent Negros Island. And the problems did not end there. The internal passport system in place since the 1840s required the native to obtain permission from the village head, the town magistrate, the provincial chief, and the local parish priest. The friar was "very often quite averse to the laborers leaving his curacy, as they often do not return, and thus diminish the importance of the parish and his own revenue. This increases the difficulty of getting his signature and that of the local native authorities."⁵ Should the requisite permission be obtained, the internal passport system allowed a sojourn of at most three months. State strictures on the geographic movement of colonial subjects evidently encumbered the recruitment of plantation workers.

Nonetheless, the legal restrictions were not insuperable obstacles to interisland mobility, particularly within a small contiguous area.⁶ Natives who wanted to pursue their *suwerte* by cashing in on the new opportunities in Negros decided to migrate even without proper documentation, and hence were known as *indocumentados*. Perceiving that the colonial state did not have the ability to enforce its regulations effectively, the native, as Loney reported, "very often unites with a number of companions, hires a canoe, and sails across to Negros, disembarking on an unfrequented part of the coast and, with the connivance of the planter who engages them, enters with his companions on the estate without any passport at all."⁷ One of these conniving planters was Loney himself, who hired illegal migrants on his *hacienda* in Minuluan.⁸ The illicit recruitment of labor gave currency to the term *sacada* (which today refers to transient migratory workers), suggesting that the *hacienda* employers did "draw" and "bring out" *(sacar)* illegal laborers from neighboring islands. Thus, to obtain the requisite work force the sugar planters systematically broke colonial state rules and thereby knowingly straddled the sphere of legitimate action as well as the dark underside of colonial society.

For their part, many *indio* migrants crossed the Guimaras Strait to Negros in flight from the oppressive colonial order, a process in which the planters were unavoidably implicated. But having risked their lives, many did not wish to be cooped up in the *haciendas*. While most *indios* did assume their new jobs, many used the illegal recruitment process to their advantage, and to the great annoyance of sugar planters. Echauz lent his dramatic flair to this subject:

> Twenty or thirty individuals, with their bundles of clothing, present themselves at the mill. They speak to the *cabos* expressing their willingness to work because they know the *hacienda* owner is kind and treats people well. After a lengthy and windy talk, they request seven to ten pesos per person to pay the *cabeza* for the *cedulas* [official identity papers] and the fifteen days of corvee work, and to pay for the needs of their families. The *hacienda* owner sees the heavens open and, deluding himself with countless illusions with the arrival of additional hands, surrenders the money. But the surprise is reserved for the following day when he finds himself without both money and labor because all the workers who presented themselves the previous day had fled during the night. (Echauz 1894, 84)

Resisting the impositions of the colonial state as well as subsumption to capital in the *hacienda* labor process, many migrants deliberately absconded to begin their new life as independent settlers in the unexplored portions of the island.

The undocumented persons at the time of Loney's 1867 report were estimated to number around 9,000, and their ranks swelled further as illegal migration proceeded unabated. Numbered among the generally peasant migrants were some prospective *hacenderos*, who subsequently established farms without ever appearing on the tribute list.[9] By the 1890s Echauz estimated that there were as many as 35,500 *indocumentados* who, together with about 57,600 *infieles* (infidels), accounted for some 29 percent of the island's total population (Echauz 1894, 119–136). A sizeable proportion of the native populace therefore was beyond the reach of the colonial state and the effective grip of capital.

Indicative of fissures in the Spanish administrative apparatus, local officials in Negros, who probably had personal stakes in the island's sugar industry, tolerated the streams of illegal migration. As Loney reported in 1867: "Constant reclamations are being sent by the Governor of Iloilo to the Governor of Negros requesting that the men of certain villages should be returned, but these meet with little real attention, as the authorities at Negros do not look with aversion at the augmentation of their population, tribute-money, and labor-power."[10] When queried by the central government in 1883 about the illegal transprovincial population movements, Negros governor Ramon Estevanez explained away the phenomenon by evoking what he termed "the nomadic instincts of the native,"[11] thus conjuring an abstract cultural trait as an alibi for political economic realities.

In the context of a weak and fragmenting colonial state, it can be said that the *indios* themselves posed the more severe hindrance to labor recruitment when compared to the easily flouted internal passport system. By the late 1860s, they appear to have been acutely aware that the new sugar industry on Negros Island was desperately in need of workers and that they could use their decided advantage to promote their own interests as they deemed fit. Whatever binding ties that had been established in the past through sharecropping and debt bonding elsewhere in the colony were shattered in Negros. Like spirits requiring to be propitiated before acquiescing to a healing, the *indios* began to impose conditions before agreeing to work. After all, if mestizos could assert their *dungan* in daring to move to Negros and to engage their fellow planters in the game of sugar, so could the *indios*.

The stirrings of an inchoate proletarian consciousness signaled the need for attractive cash advances in order to entice labor to a new terrain. From all indications, the emerging sugar-planter class was bereft of seigniorial rights that might have equipped them with the symbolic capital necessary to mobilize and command workers at will. In addition to the colonial state's inability to control labor, the planters' faulty hegemony was heightened by the fact that they sought to employ labor where no free labor market, in its fully commodified sense, existed.[12] To lure peasant labor to cooperate with the *hacienda* ventures on Negros Island, the sugar planters were compelled to improve upon the social game that had been played since the rise of commercial agriculture in the late eighteenth century: the giving and taking of cash advances. *Hacenderos* of various racial backgrounds had to play that game with workers deftly, in the same manner that they had to gamble skillfully with the cash advances provided by the foreign merchant capitalists at Iloilo.

Cash Advance: Debt as Appeasement of Labor's Spirit

By the late 1890s the population of Negros was about four and a half times that of 1850 (see Table 3). Despite the quadrupling of the potential labor force, John Foreman (1899, 475) observed that sugar planters often had to offer *antecipos* or cash advances to workers in order "to secure their services." Sounding perplexed, he added, "It is rather novel to see capitalists begging for laborers; sometimes there is quite a competition amongst the planters to secure hands." Also in the 1890s Dean Worcester (1898, 260) reported that "cane was suffering in the fields for want of men to cut it," explaining that "laborers cannot be had at all, unless they are imported, and in any event it is usually necessary to make them considerable advances on salary before they will do anything." A similar comment was expressed by Forbes-Lindsay (1906, 473), who noted that in Negros imported workers "demand considerable advances before they will enter upon their work."

The granting of cash advances became not only customary but mandatory in the employment of labor, hired increasingly with the aid of agents. Initially, the cash advance was 3 to 4 pesos per worker but, as more land was brought under cane cultivation and as competition among employers intensified, the advance rose to about 10 or 15 pesos per worker in the late nineteenth century (Stower and Urquijo 1924), exceeding the maximum allowable *indio* debt of five pesos. The practice of offering a cash advance was decried by Negros Governor Tovar in the 1880s as "perverting the worker" and as a "form of piracy which could not have existed if the *hacendero* had not developed the worker's greed."[13] Carried over to the first decade of the American period, the system of advances was branded as "pernicious" and sugar planters were blamed as "alone responsible" for "[o]ne of the greatest drawbacks to obtaining labor" (Leonard 1910, 20). But the *hacenderos* had few options available to them at that time.

Tolerance for labor's "piracy" reflected the sugar planters' imperfect dominance as a social class. However, they tried to disguise this weakness to themselves and to outsiders by explaining the system of advances as a concession to the *indio*'s so-called penchant for debt. John White, an American military officer stationed in Negros, reminisced that the *hacendero* "would tell you that unless the peasants were in his debt they would not work" (White 1928, 117). By their actuation, however, the *indios* evinced that they saw the cash advance from a different light—as a prerogative, a guarantee of reward, almost an appeasement of the spirit, to be enjoyed beforehand in exchange for a form of drudgery that was largely unnecessary, as the *indio* was under no absolute economic or other compulsion to sell labor power. Thus, if no cash advance was offered, no worker could be hired. And if an *indio* had agreed to work but found *hacienda* conditions disagreeable, he would simply flee and leave the debt unpaid, as countless *indios* had done before in other parts of the country. In a few exceptional cases workers were made to swear on oath not to abscond, but what apparatus of power could guarantee full compliance with a piece of paper or a verbal promise?[14]

There is little to suggest that the ties between planter and worker were strong in the late nineteenth century. On the contrary, such ties could be expected to be demonstrably tenuous in a general atmosphere of lawlessness engendered by the unrestrained illicit peopling of Negros. The problem of social control may be inferred from available judicial statistics. In 1864, long-settled and populous Iloilo reported 154 adjudicated criminal cases whereas Negros reported only 22.[15] By 1869 Negros registered 226 criminal cases, a tenfold increase in five years (Table 4). During the 1870s the number of criminal cases tapered off to an annual average of 138. During the restive 1880 to 1886 period, the annual average jumped to 249, an 80 percent increase from that of the previous decade; the Negros figure in the 1880s was also 28 percent higher than Iloilo's annual average of 194 cases. Criminal cases in Negros reached a record high of 368 in 1882, while the corresponding figure for Iloilo was only 273 cases registered in 1878.

In particular, the growth of *hacienda* society in Negros fueled a dramatic increase in "crimes against property" (Table 4), which may be considered a rough indicator of *hacendero* class weakness. While the annual average number of cases of "crimes against persons" (mainly homicide) increased only moderately from 45 in the 1870s to 59 in the 1880s (a 31 percent increase), the number of "crimes against property" soared from 48 in the 1870s to as many as 100 in the 1880s (a 108 percent increase). In contrast, the trend of crimes against property in Iloilo was reversed: the annual average declined from 105 cases in the 1870s to 87 in the 1880s. In Negros, crimes against property, a marked plurality of which were robbery and theft cases, peaked to 170 adjudicated cases in 1882 (Table 5).

Arson was almost negligible in Negros in the 1870s, with only 16 reported cases for the decade. However, the seven-year period from 1880 to 1886 saw the total number of arson cases jump to 78, 31 or two-fifths of which occurred in 1885 when, coincident with the global sugar crisis, which was aggravated in Filipinas by erratic Spanish imperial policies, many cane fields were torched.[16] The reported rise in attacks against property in Negros stood in marked contrast to the relative stability in the aggregate number of crimes against property adjudicated for the colony (which averaged 2,087 cases in the 1870s and 1,982 cases in the 1880s). In Negros the offender was generally a native *(indio)* male between eighteen and sixty years old, of peasant *(labrador)* or day-laborer *(jornalero)* background.[17] As may be glimpsed from the judicial statistics, the newly formed *haciendas* rapidly became a locus of social conflict. In that context, it was difficult for planters to use heavy-handed techniques of labor control without expecting some form of retaliation.

Thus confronted with labor recruitment problems and inadequate social control mechanisms, the *hacenderos* had virtually no option but to

Table 4.

Criminal Cases Adjudicated in Negros by Type of Crime and Total Criminal Cases in Filipinas, 1864 and 1869 to 1886

Year	Against Persons	Against Property	Public Order[a]	Flight	By Public Officials[b]	Others[c]	Total[d] Negros	Filipinas
1864	8	6	2	1	0	5	22	n.d.
1869	95	75	40	—	5	11	226	4,641
1870	52	40	25	—	6	7	130	4,114
1871	70	37	29	—	10	4	150	4,034
1872	60	35	13	3	8	4	123	4,199
1873	64	29	18	30	7	17	165	4,665
1874	33	68	18	8	15	6	148	4,272
1875	30	61	9	4	11	4	119	4,314
1876	25	38	14	1	13	6	97	5,082
1877	26	42	8	10	10	9	105	4,518
1878	39	44	21	15	9	7	135	5,125
1879	53	86	31	9	19	9	207	5,260
1880	76	81	54	20	18	17	266	4,260
1881	89	130	36	16	16	13	300	4,616
1882	72	170	56	22	39	9	368	4,413
1883	23	70	39	11	15	12	170	4,376
1884	76	111	57	35	12	19	310	5,412
1885	57	103	37	21	13	14	245	5,170
1886	19	37	6	6	3	14	85	3,781

Sources: Estadistica General de los Negocios Terminados y Pendientes de Despacho en la Real Audiencia de Filipinas, 1865, AHN Legajo 2243 (expediente s/n); Estadistica de las Causas Criminales, negocios civiles y espedientes de Gobierno, despacados por la Real Audiencia de Filipinas, 1871–1887, AHN Legajo 2238, Expediente No. 99.

[a] Crimes against public order include crimes against "liberty and security" (such as illegal detention and threats and coercions) and jumping bail.
[b] Crimes committed by public employees in the exercise of their duties.
[c] Mainly crimes of "falsehood" and "against honesty."
[d] Total figures include a tiny fraction of nonconvictions ("hechos que no constituyen delitos").

Table 5.
Crimes against Property Adjudicated in Negros by Type of Offense and Total Cases of Crimes against Property for Filipinas, 1864 and 1870 to 1886

						Totals	
Year	Robbery	Theft	Fraud[a]	Arson[b]	Damages	Negros	Filipinas
1864	4	2	0	0	0	6	n.d.
1870	17	16	5	1	1	40	2,006
1871	12	19	2	2	2	37	1,876
1872	6	27	1	1	0	35	2,049
1873	3	24	0	2	0	29	2,043
1874	10	54	4	0	0	68	2,039
1875	16	45	0	0	0	61	2,188
1876	9	26	2	1	0	38	2,001
1877	4	34	2	2	0	42	1,997
1878	19	22	1	2	0	44	2,305
1879	21	56	2	5	2	86	2,370
1880	32	41	0	8	0	81	1,645
1881	50	58	7	13	2	130	1,869
1882	52	94	17	7	0	170	1,735
1883	24	33	11	2	0	70	1,952
1884	44	50	6	11	0	111	2,471
1885	36	32	4	31	0	103	2,394
1886	18	8	4	6	1	37	1,811

Sources: Estadistica General de los Negocios Terminados y Pendientes de Despacho en la Real Audiencia de Filipinas, 1865, AHN Legajo 2243 (expediente s/n); Estadistica de las Causas Criminales, negocios civiles y espedientes de Gobierno, despachados por la Real Audiencia de Filipinas, 1871–1887, AHN Legajo 2238, Expediente No. 99.
[a]Fraud and other forms of deceit (Estafa y otros engaños).
[b]Arson and other damages (Incendios y otros estragos).

resort to a policy of attraction to coax natives to join the *hacienda* labor force. Transient farm workers were not only offered cash advances and their taxes paid for in their various places of origin but were given rudimentary but free meal rations and living quarters.[18] As a later section shows, returns to labor were also raised. As a further enticement for labor to settle permanently and pacifically on the *haciendas,* sharecroppers and their families were provided with individual native huts, at that time a unique social innovation.[19] Sugar planters endeavored to offer "fair treatment"—no shaming, no insults or abusive words, decent pay, adequate rest—for, as one planter admitted, "bad treatment" of workers created tremendous problems (Stewart 1908, 183).[20]

In addition, the *hacenderos* agreed to organize the work process according to a scheme that suited the *indio*'s preferences, but by no means did this imply that *hacienda* production relations were not exploitative. Although referred to as day laborers receiving a supposed daily wage, the *jornaleros* often eluded the regimentation of industrial time-controlled activities by acquiescing to work only through "the system of contracts" *(pakyaw)*.[21] Under the headship of a *kapatas,* labor gangs were usually contracted to undertake specific tasks ranging from land preparation, planting, weeding, and harvesting of cane to the hauling of the cut cane to the mill. Analogous to the *kabisilya* who headed the labor gangs in the rice tenanted farms of Luzon (first studied in the 1960s, cf. Takahashi 1969; Murray 1970), the *kapatas* in the sugar *haciendas* acted as labor recruiter and foreman, as well as a type of petty exploiter who obtained a percentage share of the contract fee, the remainder being divided as the "wages" earned by members of the work group (cf. Ross 1920, 4). Given the unsettled conditions that prevailed in Negros, it was imperative that planters rely upon the *agsa* and the *kapatas* as the real supervisors of labor, for it was frequently the case that planters could mobilize a work force on their plantations only through the use of such intermediaries. Given the conditions of the period, a wage-labor regime could not be made to arise in Negros. As we shall see in the next section, planter weakness was intensified by lack of support from the colonial state.

State "Neutrality" and *Hacendero* Class Weakness

Although some local officials were individually supportive of the sugar planters, and even tolerant of *hacendero* infringement of colonial statutes, the Spanish administrators at colonial Manila were unwilling to adopt measures that were plainly pro-planter. The seeming neutrality of the colonial state was vividly displayed in the central government's rejection of a proposal formulated in 1883 by the *Gobierno Politico Militar* of Negros to regulate workers—an issue that in other colonial settings had been resolved in favor of capital.[22] The island's governor since 1878, Ramon Estevanez, recommended the adoption of regulations that would, as he claimed, settle the mounting "mutual complaints of the property owners *(propietarios)* and laborers *(braceros)* concerning the non-fulfillment of contracts" and, in so doing, "harmonize deeply injured private interests" in conformity with "equity and justice."[23]

To justify the proposed regulations Estevanez argued that, although "the abuses" were committed on both sides, Negros' agriculture would seriously suffer from a continuation of the system of cash advances and the *indio* practice of flight that left work obligations unfulfilled. The reduction in harvests and sugar production, it was argued, would result in a decline in tariff revenues at the Iloilo customhouse, rendering the colonial state the "main loser" in the planter–worker tussle unless it intervened. Thus the proposed regulations were designed to enhance the leverage of sugar planters vis-à-vis labor. At the same time, there was a visible move to prosecute workers for flight, at least in the first half of the 1880s (Table 4).

Estevanez's recommendations included a set of procedures that would require every worker to *(a)* register within twenty-four hours immediately after being contracted to work on a farm; *(b)* carry an identification booklet that contained his personal characteristics and place of origin, the terms of the work contract, the dates of actual work, and employer "observations" concerning his "conduct"; and *(c)* renew the registration at the beginning of each year.[24] To ensure compliance, Estevanez recommended creating a "rural police force" that would generate its funds from the registration of workers, whose annual fees would be 2 *reales* for day laborers and 1 peso for sharecroppers. If a worker moved to another *hacienda* without the express permission of the previous employer recorded in the booklet, the worker would be made to pay, "without further investigation," a fine of 10 pesos or be imprisoned for fifteen days, and the new employer would also be fined 10 pesos.

Although the proposal contained clauses that would apparently mediate the competition for farmhands among employers and protect workers from *hacendero* abuses, such as barring planters from punishing errant workers,[25] the regulations on the whole were meant to buttress planter strength in the face of a problematic labor force. Indeed, the proposal embodied the quintessential Spanish concern to outlaw "slothfulness." Apart from illness or other legitimate causes, failure to work consecutively for six days made a native liable to incarceration: fifteen days for the first offense, twice that duration for the second, and deportation to a military camp in "the South" of the archipelago in the case of a third. Extra-economic compulsion was deemed essential to compel natives to behave like genuine proletarians.

Notwithstanding the claim of local officials concerning the "equity and justice" of the proposed measure, Spanish Manila was unconvinced about arguments adducing pecuniary benefits from the suggested rules.[26] The withholding of support for the sugar planters would appear to have been motivated by the concern that the colonial state remain "neutral" and hence not bestow undue favor upon any specific segment of the subject population. Spanish Manila's ratiocination centered on upholding the preeminence of the colonial state's own legal traditions and institutional interests. In 1885, a plea addressed by sugar planters to Madrid for "energetic dispositions against vagrancy and agreement on a *Reglamento* which should establish the rights and responsibilities of property owners and laborers" went unheeded,[27] depriving sugar capital of the legal ammunition and ideological apparatus for social control. The colonial power's refusal to act in their favor prompted the planters to devise their own schemes to deal with native labor. Some of these schemes another governor of Negros, Fernando Giralt, would seek to formalize and codify into another proposed set of regulations six years after the Estevanez plan was first propounded. The official response to Giralt's proposal of 1889 unequivocally demonstrated the colonial state's posture of neutrality in the interest of its legal system.

With its purported object of identifying tax evaders among *hacienda* workers, Giralt's proposal stipulated that, after workers were issued official identity papers *(cedulas personales)* and receipts of provincial imposts, their names and other personal information such as their places of origin would be entered in a registry to be kept by the town magistrate, the *gobernadorcillo*.[28] On the plausible pretext that this procedure might lead to the arrest of a fugitive, *hacienda* owners and overseers would be obligated to inform local officials of the presence of undocumented workers so the latter could be accounted for in the registry; failure to make the report would incur a fine of from 5 to 50 pesos. To prevent flight, planters would be allowed to keep the workers' identity papers, although *"papeletas"* would be issued to serve as duplicates or evidence of the sequestered official document.[29] The sugar planters would also be empowered to employ armed guards. In his letter of recommendation to Manila, the governor of the Visayas explained that the suggested procedures would "reconcile" what were perceived to be the conflicting interests of the colonial state and the sugar planters: the latter would obtain the labor they needed, while the former would be able to collect taxes from otherwise undocumented workers.[30]

Unimpressed, the Manila officialdom rejected the proposal and maintained that the retention of workers' identity papers by sugar planters was patently illegal.[31] Since the *cedula,* it was argued, had the character of a *documento de seguridad* (a security document), the power that would devolve upon employers was rightly perceived as lending itself to an infinite number of abuses. Moreover, Spanish Manila justified its action by invoking the European legal concept that demarcated the public from the private spheres, a precept Spanish officials otherwise routinely breached in their colonial practice. The central government declared it could not countenance state intervention "in so direct and specific a manner as to defend the private interests of the *hacenderos*" in a transaction in which "no single governmental authority had any part whatsoever." Spanish officials in Manila emphasized that the employment of plantation workers was "guided solely by private interest or convenience."[32] The proposal that could have given a formal edge to the propertied class of Negros over the laboring mass was disapproved.[33] The only exception to the wholesale rejection of Giralt's proposal was a separate permission for planters to use "one or two shotguns *(escopetas)*" for personal protection.[34]

In general, the sugar *haciendas* evolved under the conditions set by a colonial state whose perceptions of its own institutional interests forbade it to acquiesce to landed interests and rendered it adamant against adopting measures that would signal clear support for the *hacenderos*. Given the contradictions that beset the colonial state, the overall effect of state action at this time was to thwart the planters' project of engineering the control of *indio* labor and solidifying the hegemony of sugar capital. Because the colonial state refused to buttress their position legally, the planters proceeded to deepen their gamble against the state.

Planter Competition and Further Antistate Strategies

Deprived of state support and left to their own devices, the sugar planters embarked on a competitive intraclass struggle by offering increasingly higher cash advances and also by grabbing workers already employed on other farms. Certainly, the planters displayed no qualms about hiring workers who had absconded from other *haciendas*.[35] In the 1890s, Echauz (1894, 162–163) pictured the *cabo* as a henchman extremely adept at accosting workers in the public market or in the cockpit and luring them to his employer's farm. In at least one recorded incident in Negros, a village chief allegedly threatened four workers in the town of Saravia that they would be arrested for being tax delinquents *(retrasado)* unless they agreed to transfer to work on his *hacienda*.[36]

The planters also competed for day laborers by raising wage rates in tandem with the secular rise in the returns to labor coincident with the onset of the export economy. In 1861, Loney's report suggested that, excluding meal rations, nominal wages in Negros increased 2.5 times within about five years after the opening of the Iloilo port to world trade in 1855.[37] As mentioned earlier, cash advances were also raised. In addition, some *hacenderos* apparently engaged in cutthroat competition that jacked up the wage rate, as may be gleaned from the (rather exaggerated) report of Remigio Molto, governor of the Visayas, who averred that Europeans like Ives Leopold de Germain Gaston supposedly paid workers

(4 *reales* per week in 1861) substantially more than what Chinese mestizos such as Lucas Locsin paid theirs (6 *reales* for five weeks of work in 1863).[38] Molto's report would suggest that mestizo stratagems made labor acquisition difficult for non-native planters.

But though they offered a general increment in wage rates, the sugar planters as a group maneuvered to lower the wage bill. Among other such strategies, planters did not pay day laborers for work done on Sundays and feast days, and they reduced the workers' food rations and forced them to accept part of their compensation in kind (rice, clothing, alcohol) valued at higher than prevailing market prices.[39] The workers were particularly irked by the conversion of their pay into consumer goods. They complained that *hacenderos* cheated them by paying part, some say about half (Hord 1910, 14), of their wages in terms of overvalued goods instead of in actual money, which, in specie form, was flexible in use and enjoyment, especially in the gambling den. The resentment became especially acute when coinage became scarce during the harvest and milling months, which ran from May to November.[40] As will be shown later, a more subtle attempt to minimize costs pervaded the *agsa* sharecropping arrangement.

The planters' tactics, it would seem, did not always succeed in producing the required number of farmhands, for some employers went further in contravening the law through the organized recruitment of fugitives.[41] Particularly contentious was the *Guardia Civil*'s discovery in December 1888 that the peninsular Spaniard Rafael Alvarez' roster of workers in Hacienda San Juan in the town of Cabancalan had included forty-five undocumented persons and ten tax delinquents. Of these, seven belonged to two "gangs of evildoers *(malhechores)*" each led by a traditional magical leader, Dios Buhawi and Valentin alias Cadula, whom the Civil Guards had been seeking to capture at the eastern side of Negros.[42] Also found in Alvarez' possession was a "Bulldog" revolver and a rapier *baston*, both without the appropriate license.

The discovery that alleged criminals were being employed as plantation workers was even more controversial, for it happened after the Civil Guards chanced upon another undocumented person, a sacristan who was on his way to deliver to Alvarez a message from the local curate, Juan Perez Zuñiga, concerning the arrival of five more fugitives.[43] Described as a "very frequent" occurrence, friar collusion in the contravention of state policies was deplored by a captain of the Civil Guards who testified that such collusion was "the general rule here where the *Guardia Civil* is not appreciated and respected."[44] As Spanish friars occupied an integral position in the colonial state apparatus, this incident further evidenced the increasing fragmentation of Spanish rule in the late nineteenth century. In a subsequent attempt to patrol the *hacienda* during the New Year festivities, the three Civil Guards on duty claimed to have been "verbally abused" by Alvarez and the parish priest. Alvarez allegedly declared, among other things, that "he did not want any *Guardia Civil* in the *hacienda*."[45] Considering that in most parts of the colony "[e]ven to resist the *Guardia Civil* was so great a crime that the sentence of a court-martial in such a case was penal servitude for life" (Frederic Sawyer 1900, 29; cf. Worcester 1914, 378–789), the *hacendero*–friar alliance in defying the Civil Guards exemplified the flouting of the law in colonial Negros, where the lines separating loyal subject and lawbreaking deviant had become vague and uncertain.

In order to settle the case expeditiously, Negros' Governor Tovar told Alvarez to pay a 25-pesos fine for employing undocumented workers and 50 pesos for the unlicensed firearms.[46] Tovar also lent credulity to the friar's explanation that his participation in the whole affair had no purpose other than to "avoid considerable differences between persons who for some time have been resentful toward each other."[47] Capping the governor's peculiar handling of the case was the release of all the prisoners, to the consternation of the Civil Guards.[48] The Manila central government reacted with a stiff reprimand of the Negros official.[49] To the Recollect Provincial, the governor-general issued a warning that Fray Zuñiga, although "no doubt guided by his good intentions," should not "obstruct and harden the creditable services of the *Guardia Civil*,"[50] apparently the only branch of the colonial state that attempted to interpose the official empire upon the near anarchic society that had evolved in late-nineteenth-century Negros.

On the whole, the *hacenderos* of Negros pursued their sugar interests by systematically undermining the colonial order and engaging in intraclass competition, rendering unwieldy the struggle of sugar capital to dominate labor. The planters' labor recruitment practices, ranging from the fostering of illegal migration to the hiring of outlaws, highlighted the colonial state's disintegration at a time when elsewhere in Southeast Asia strong centralized states were being built. The collective realization by *indio* peasants of their relative strength resulted in serious problems of labor recruitment and control on the plantation. The planters responded by offering an attractive employment package, based on the cash advance, to draw the needed work force and reduce the menace to planter interests. On the part of workers, the growth in the value of cash advances would appear to signal the redistribution of *suwerte* afforded by the dynamic economy of Negros.

It can be surmised that planters were well aware that the system of advances carried the risk that some potential workers would disappear with the money, but the big profits from sugar exports made the risks worth taking. The planters did attempt to minimize their risks by seeking central state patronage, but to no avail. The controverted character of the colonial state–planter relationship and the sugar planters' lack of hegemony as a class resulted in a curious admixture of labor arrangements and practices, but with sharecropping remaining strategically pivotal in the organization of *hacienda* production. Serving as the principal link in the control and supervision of the plantation labor process, the *agsa* sharecropper, as we shall see in the next section, underwent a process of incorporation into the *hacienda* system.

Sharecropping and the Crafting of Paternalism

Most sugar planters relied upon sharecropping agreements, as these were advantageous for a landholding class that knew little about agriculture, whose dominance had not been indubitably established, and for whom the colonial state had become virtually irrelevant. Thus it was convenient and profitable to subdivide the *haciendas* and delegate the actual task of cultivation to tenants. The share tenants, in turn, performed the role of harmonizing the labor process with the *kapatas* and other native workers hired for the different phases of sugar production. At this juncture, rather than acting as a fetter on economic development in Negros, sharecropping represented a progressive means by which planters sought to optimize farm productivity (cf. Byres 1983; Pearce 1983). Under the circumstances, sharecropping proved most appropriate for mobilizing labor and minimizing risks of flight, and for generating and extracting a surplus product. As Echauz (1894, 44) emphasized, the *agsa* system was "the best, the easiest and the most convenient" method to organize agriculture in Negros, as it allowed the *hacienda* "to march forward, earning more each day and imperceptibly creating for the planter a state of fortune under relatively comfortable circumstances."

Echauz also noted that the share-tenancy system attracted a "hard-working" labor force willing to put into sugar production either "some resources" they possessed or the labor power of their "large families" (Echauz 1894, 44). The sharecropper was allocated a piece of land and provided with work animals, farm equipment, and a cash advance, the latter to be repaid with interest in the form of clayed sugar.[51] Acting like middlemen-bankers, the *hacenderos* supplied tenants with funds to pay the day laborers hired on the tenant's sector of the plantation and other farm expenses. If the *agsa* sharecropper owned carabaos to work the land, interest charges on the capital advanced were sometimes waived.[52]

In contrast to the sugarcane tenant in Luzon who was expected to bring to the "partnership" tools and draft animals (Larkin 1972, 81), it would appear that some Negros planters had intended the provision of cattle as an added incentive in soliciting the services of a sharecropper.

The *hacienda* tenant took care of growing and eventually harvesting the sugarcane and providing the mill with cut cane, with all farm expenses being charged to his account. In return, the *agsa* received a portion, generally equivalent to one-half, of the produce; from the tenant's share the amounts advanced by the planter were then deducted.[53] The expenses incurred in the manufacture of clayed sugar were usually borne by the *hacendero,* but the tenant and planter each had to take charge of packing their respective share of the sugar output. In large plantations with iron mills, from 9 to 12 percent was removed from the tenant's share as payment for milling expenses.[54]

Negros Governor Tovar noted that *indio* tenants—whom he described as illiterate and *gente ruda* (crude, dull, stupid)—could not, and did not, question the financial accounting presented to them by the *hacenderos.* The latter's bookkeeping practices were in turn described by Tovar as badly organized.[55] It would appear, then, that the partitioning of expenses between landowner and sharecropper was a grey area the sugar planter could utilize to appropriate more of the surplus product to the detriment of the tenant. It must be noted, however, that contrary to the governor's racially skewed judgment, *indio* tenants would later assert their right to scrutinize *hacienda* accounts, especially in central Luzon where the intricacies of cost accounting constituted, in the first half of the twentieth century, a major bone of contention between planters and sharecroppers.

From Tovar's vantage point, sharecropping in the sugarcane plantations of Negros offered "very few guarantees" for economic advancement. From the point of view of some *indios,* however, the tenant's position was a way of acquiring a foothold in the exciting new world of Negros, where the colonial power could be held at arm's length. Certainly, share tenancy had to be a mutually, though not equally, beneficial arrangement. Mutual benefit as assessed by the actors concerned was important, as the natives retained the option to disregard work contracts and leave the plantation either to work for another employer (who was always in grave need of workers), to make their own clearings and become peasant smallholders near the towns, or to settle in the island's remote and inaccessible interior. Hence, for the many who opted to become sharecroppers, the *agsa* arrangement offered enough incentives to make it acceptable in terms of several viable options.[56]

Having decided to become part of the *hacienda* complex as a share tenant, the *agsa* did not mind it when the planter paid the tenant's official identity paper and illegally sequestered what the colonial state considered an organic document.[57] The planter's precaution to curtail flight became frequently irrelevant to the tenant who had opted to claim a stake in the sugar industry. And developing a stake came within relatively easy grasp because the tenant's effective share of the produce was considerably greater in value when compared with the wages earned by other plantation workers.[58] Moreover, as in the *kasamahan* contract in Luzon, the *agsa* system in Negros conformed to the native's preference for a sufficient degree of autonomy in the work process.[59] To a considerable extent, the tenant controlled key decision-making variables in sugarcane cultivation in his own part of the *hacienda*. But it was precisely the perceived space for sharecropper independence in the production structure which sugar planters skillfully manipulated to bind *(ligar)* the tenants to the plantation (Echauz 1894, 44) and thereby minimize problems of social control.

As in the recruitment of day laborers, the tenant–planter relationship revolved around the issue of cash advances and debt. Both parties would most certainly have been aware that the cash advance was imperative for the land to yield profits. But given the volatile social situation on the island, debt per se was a fragile mechanism of labor control. Therefore debt had to be instrumentalized within the overall context of the *hacienda* production structure. In my view, the fact that the cash advance was to be repaid out of the tenant's share of the final clayed sugar produced by the *hacienda* mill—not in terms of the cane stalks harvested, which would have been the counterpart of unhusked rice in rice tenancy—was crucial to the binding process.

From the mode of repaying the cash advance, the *hacenderos* could extract several advantages. The costs of claying the sugar could be miscalculated, as indicated above, to the detriment of the tenant. Even more tricky was the grading of sugar in a setting where no standard system for grading, hence for determining market price, existed—an issue that Tovar also raised against the merchant houses at Iloilo. But such underhanded tactics, which might have been necessary in the extraction of more surplus from the sharecroppers, were extraneous to the social construction of meaning around debt. A situation of dependence, I believe, was created by the inherent need to transform the tenant's share of cane into a marketable commodity, clayed sugar. Whereas rice was manually pounded in large wooden mortars in the tenant's home, sugarcane had to undergo a slightly more complicated process of claying for which the sharecropper had no alternative but to rely upon the planter's muscovado mill. By virtue of the manufacturing process, the planter thus had a huge leverage over the tenant. The reality of several sharecroppers waiting upon the plantation's crude but indispensable mill was eloquent testimony to the tenant's subordination to the planter.

The supposed independence of the *agsa* as a partner in the sharecropping contract underscored the *hacendero*'s generosity in granting use of the mill, for free or at nominal cost, as well as in extending other forms of timely assistance, such as farm animals and implements lent to the tenant at the opportune moments (Echauz 1894, 44). The supposedly autonomous partner was in fact highly dependent upon the employer, who had greater access to resources. Ultimately, within the local cultural framework of meanings and gift exchange, the use of the mill and various other forms of assistance carried with them the resounding demand for reciprocity—which could hardly be recompensed on a par, for how could the tenant fully return such commercially embedded prestations as "time" and "opportunity"? The intermeshing and trapping of wills in a "debt of the inside," discussed in Chapter 3 in connection with the tenancy arrangement, was therefore even more profoundly operative in the sugar *haciendas* of Negros. Consequently, where the threat of social disorder was relatively high, simultaneously available was a most effective mechanism for the cooptation of labor. The sugar sharecropping arrangement demonstrated that problems of social control and possibilities for social bonding could well coexist. It is thus comprehensible why contemporaries considered sharecropping "the best, the easiest and the most convenient" approach to *hacienda* organization.

Sharecropping also obviated the need for flagrant cruelty on the part of sugar planters. Although local native officials, such as village chiefs, were known to have flogged *indios* for nonpayment of tribute, *hacenderos* deflected responsibility by utilizing native intermediaries who acted on the basis of socially acceptable concepts of discipline and punishment (Worcester 1898, 255–257). During the early American period, the fatal beating of a worker whose negligence reportedly led to the melting of a cartful of clayed sugar was perpetrated by a "foreman" (White 1928, 111–113). Moreover, there were clear boundaries to the punishment that planters could mete out, because workers could strike back with relative impunity, as one *hacendero* family feared even during the 1920s when a relative caused the unintentional death of a tenant. The incident compelled the family to move to Iloilo.[60] As emblematized by the mill and the cash advance, sugar planters in Negros in general would appear to have fostered an atmosphere of "fair treatment," lest they suffer the consequence of either inadequate fieldhands or personal vendetta (Echauz 1894, 84; Worcester 1898, 260).

Unable to use force as they wished, sugar planters turned to the reinvention of paternalism to mitigate their incomplete class dominance and tighten social ties. The pinnacle of such magnanimity was demonstrated in such acts as those of Pedro Pullicar who, in his last will and testament executed in the town of Saravia on 4 September 1891, reduced the debts of his *agsas* and *jornaleros* by a third and fully pardoned the 200 pesos owed him by one faithful tenant.[61] Consideration was also shown in the 1890 sale of Raymundo Alunan's 60-hectare Hacienda Cabiayan in the town of Minuluan to Efigenio Lizares, as the former stipulated in the deed of sale that the buyer should respect the original agreement between the planter-vendor and his sharecroppers concerning the tenants' share of the standing crop.[62]

In the course of time, the tenant developed "fondness and affection" for the farm in which he had formed a stake, where his family had taken root, and where colonial state interference in his life had been mitigated. The sharecropper was transformed into "an integral part of the *hacienda*" and started to willingly assist "the *hacendero* in kind and sometimes with money, gathering firewood, consenting to cut and transport sugarcane, and doing other things" (Echauz 1894, 45). The social requisites of gratitude—in which the gift economy overlapped with the world-capitalist economy—afforded the sugar planters a subtle incursion into the tenant's surplus labor time. Through interest charges on advanced capital and other informal means of surplus expropriation, the sharecropping arrangement was gradually tilted in the planter's favor, although the fifty-fifty division of the sugar product remained as a symbol, albeit an increasingly empty one, of parity.

As partner in the sharecropping contract, the *agsa* felt personally responsible for growing the cane and for reducing uncertainty by appeasing the spirits. Until early in this century it was common practice to make *halad* offerings of food, sweets (like *bayi-bayi* and *alupi*), and drinks (especially *tuba*, the local liquor) to the spirits the night before harvesting the cane. Other beliefs to augment *suwerte* dictated that planting of cane points begin on days other than Tuesdays and Fridays, inauspiciously associated with the "sorrowful mystery" in the Catholic rosary tradition recounting Christ's crucifixion, which, to the local people, signified that malevolent spirits could be roaming the earth, as "God was dead." That cane plants became lean and sparse was believed to be a reprisal from preternatural beings or a case of burglary by *engkantos*. In response to unhealthy cane growth, the *daga* offering was made, which entailed the sprinkling of blood from a sacrificial chicken or pig in three corners of the sugarcane field. To obtain maximum cane juice, banana trees were planted in one corner of the field to coax the sugarcane plants to imitate (*panudlak*) the banana's sappiness.[63]

Individually responsible for negotiating with the spirit-world in the same manner as the prototypical *kasama* tenant, the *agsa* sharecroppers occupied a structural position in Negros that must have been a desirable one, as may be gleaned from the innumerable natives who decided to settle in the *haciendas*. Although sharecroppers could not benefit from the sugar game in the same way that planters did, they nevertheless saw themselves as enjoying a mutually beneficial arrangement. Even if tenancy allowed only a modicum of economic improvement (which would be wiped out at the turn of the century due to the unsettled political situation, rinderpest, marketing, and other problems), life on a Negros *hacienda* was evidently preferable to that in the tenant's place of origin. In Negros, the demands of the colonial state could be alleviated, if not altogether avoided. Indeed, Negros plantations became a refuge of sorts for natives whose traversal of the Guimaras Strait, sometimes under the cover of darkness (cf. Loney 1964, 109), was a palpable act of resistance against colonial authority. Thus, in Negros, workers and planters alike pursued their *suwerte* and jointly gambled against the Spanish colonial state.

Work and Gambling in the Negros *Haciendas*

But flight from other islands to settle in Negros was also an act of resistance in a more mundane sense. The natives were attracted to *hacienda* life for what it offered: the monetary payment. Because the local economy was negligibly commoditized, money was enjoyed not so much through consumer goods (which were scarce and for which the *indio* had few requirements), but primarily through the pleasures of gambling.[64] Given the liberal atmosphere fostered by local Spanish officials, natives with vices appear to have predominated in Negros' agricultural work force. One *hacendero*, described as "the proprietor of one of the best-ordered sugar plantations in the colony," told Worcester (1898, 258, 262) that "in hiring laborers he rather preferred to get men who drank, gambled, or played the *gallera* [cockpit]; for they had more wants than the moral and sober native, and would work more days in the year in order to earn money to satisfy them."[65] Thus, both the act and the object of illegal migration were defined by the gambling mentality. Aware of this, the planters, themselves inveterate gamblers, made life in the *haciendas* culturally delectable by nurturing a gambling milieu. In so doing, they were implementing an insight that emerged in the early nineteenth century as part of the Bourbon-inspired effort to convert Filipinas into a profitable colony.

Because of the intermingling of social categories in the cockpit that implicitly challenged the rigid social classification scheme of the colonial state, gambling was explicitly seen, from the early 1800s, as leading to the "perversion of customs" and the "disruption of respect."[66] However, the need to raise public revenue militated against the prohibition of gambling, particularly cockfighting, and de Comyn's 1810 program to generate funds through cockpit license fees was an early example of gambling's economic utility to the state (de Comyn 1969, 76–78). Moreover, because the dearth of reliable native labor for agricultural enterprises was perceived as a major hindrance to development, the state began to view gambling from a new but contradictory angle: the cockpit was recognized as a powerful magnet to draw labor into *haciendas*. This insight was crystallized by a royal order issued in 1828 that granted *haciendas grandes* the special privilege of establishing tax-exempt cockpits "for workers and their families" who had been connected with those plantations "for years." Several property owners in Pampanga were quick to seize the opportunity, but the provincial franchise holder immediately protested on the ground that there were no grand *haciendas* in Filipinas.[67]

Notwithstanding the revocation of the concession granted by the 1828 decree and the official denunciation of the so-called *juegos prohibidos* (prohibited games), the cockpit as well as card games like *monte* and *panguingui* continued to serve as vital aspects in the slow process of shaping an incipient labor force and in generating wealth for the colonial coffers. Since work *(trabaho)* was associated with obtaining money usable for gambling, which in turn provided income to the (after the 1850s usually Chinese) gambling franchise holders, the colonial state could only regulate, not ban, gambling despite the "perversion of customs."

In 1838, a decree was issued "to contain the pernicious effects" of "all games of luck *(suerte)* and chance *(azar),*" which were compared to a *cancer devorador* that, among other things, "obstructed the sources of public wealth."[68] Artisans and day laborers, in particular, were prohibited from engaging in any form of gambling starting from 6 A.M. to 12 noon and from 2 P.M. to 6 P.M. during working days. In 1847, another decree stated that the playing of prohibited games was rampant and had been "propagated scandalously" to the detriment of "the country's prosperity."[69] However, the order simply extended the gambling ban to cover the period from 10 P.M. onward during working days. The same schedule was adopted by Madrid's *Ultramar* when regulations "to repress" the prohibited games were formulated in 1863.[70]

Notwithstanding the regulations aimed at reversing the counterproductive effects of gambling, there was no attempt totally to extirpate the natives' habit, for the colonial state continued to farm out franchises to gambling houses as a revenue-generating measure. By the late nineteenth century, the government's cockpit monopoly was yielding hundreds of thousands of pesos in annual revenue (with perhaps a considerable proportion illicitly diverted into the pockets of public officials),[71] while the lotteries raised for the royal coffers about one million pesos a year in the 1890s (Sawyer 1900, 416). With its inherent ambivalence, gambling coincided contradictorily but nicely with the colonial state's goal of generating income, but it also lent itself wonderfully to the project of social control.

In Negros, everyone had virtually free rein to gamble even as the island attracted a preponderance of gamblers. Implementing, albeit illegally, the insight of the aborted 1828 decree, sugar planters were able to offer this attraction with little effective interference from the authorities, the erstwhile enforcers of regulations on prohibited games. While officials in many parts of the country were tolerant of gambling, their liberality on this issue in Negros was particularly overflowing. As Jose Marco (1912, 58) noted, "the Guardia Civil was impotent to contain the vice which had been propagated all over the province, even in its remotest corner." In one rare episode in October 1881, when several *monte* cardplayers were apprehended by the Civil Guards in a private house in Silay, the Negros governor adamantly refused to turn over those caught to the judiciary, asserting that the Negros government considered itself "competent" to handle infractions of the regulations on prohibited games.[72] The protective attitude of Negros officials must have been a great boon to gamblers from all social strata in Negros.

Consequently, the *hacenderos* deliberately cultivated gambling on their farms with few, if any, apprehensions about the colonial state. Indeed, unlike other Southeast Asian countries where native elites had been largely co-opted into the colonial state apparatus and where even the underworld had been subordinated to state control,[73] the colonial state in the Spanish Philippines had effectively disintegrated, as local functionaries allied with the Negros *hacenderos* in contravening what were perceived as inapplicable Manila policies. Given the state's virtual irrelevance, gambling was institutionalized as an illicit labor-enticing strategy. The social habit became entrenched: in 1939, a report of the National Sugar Board (Alunan 1939, 38) candidly admitted that, although illegal, "cockpits and other gambling places . . . are maintained in some plantations by the plantation-owners."

Sheltered from the state and the hazards of its official strictures, workers in Negros *haciendas* could gamble unimpeded except by the schedule of the cane fields and the muscovado mill. Given the expansion of the money supply and its accelerated circulation as a result of closer ties to the world economy—reversing the earlier scarcity of metallic currencies[74]—a democratization of sorts in access to gambling transpired. As various forms of gambling flourished, the need for currency to gamble with tied the natives deeper to the source of money, *hacienda* work, and fostered indebtedness in a social process whereby the indulgent gamblers turned out to be the more reliable workers. Despite differences in the structural positions for the enjoyment of gambling, Negros did seem to offer a level playing field for all.

Gambling, Inequality, and the Negros Landscape

Gambling was inscribed upon the geography of Negros. As described brilliantly by "an American resident," presumably a volunteer teacher: "The town population consists of perhaps a dozen wealthy families, who own all the land, and are called *hacienderos* [sic], and some thousands of ordinary folk, who do all the work of fishing, agriculture, and manufacture. The houses of *hacienderos* and laborers are tumbled together in the most democratic fashion" (Thomas 1902, 211). The physical layout of a Negros town where the houses of rich and poor were "tumbled together in the most democratic fashion" was a testimony to the gambling worldview, for just as social categories were intermingled in the cockpit so were they in the unsegregated residential mosaic. Thus, not only did Negros have an interspersion of *haciendas* of extremely diverse sizes, but the town centers had emerged as a "democratic" hodgepodge. The unregulated sprouting up of homes in the newly developing and fast expanding towns echoed the ambivalence of the colonial cockpit.

The evolving morphology of Negros diverged from the rigid pattern of the well-established town *(pueblo)* the colony had known: at the center of town was the plaza, the open square around which tightly and distinctively clustered were the church, the town hall, and the residences of the local elite (cf. Reed 1967, 59–71). Although the nucleus of the old towns in Negros most likely conformed to this pattern, the huge influx of migrants would seem to have created new growth poles that disrupted the extant organization of space. Newer settlements, such as Cadiz Nuevo, added flavor to the island's spatial jumble: by 1888, Negros had ten more new towns (mostly on the west coast facing Iloilo) compared to the thirty-six extant in 1853.[75]

The *haciendas* were most disruptive of the old town pattern. Whereas social prestige was normally measured by proximity to the town plaza complex, in Negros some planters erected their houses on *hacienda* grounds at considerable distances from the town center. Contrary to the customary form, relatively well-built structures scattered in different plantations began to dot the Negros landscape (Table 6). In newer towns like La Carlota and Manapla, more of the better houses were located on the farms than in the town centers. In thriving Minuluan, a considerable

Table 6.
Dwelling Units, by Type of Material, in Selected Towns of Negros Occidental, 1896

Pueblo	Strong Materials	Light Materials	Totals	
			Units	% of Pueblo
Binalbagan	17	779	796	100.0
Poblacion	7	101	108	13.6
Haciendas	8	381	389	48.9
Barrios[a]	2	297	299	37.5
Cabancalan	25	971	996	100.0
Poblacion	11	226	237	23.8
Haciendas	11	384	395	39.6
Barrios	3	361	364	36.5
Ginigaran	8	2,288	2,296	100.0
Poblacion	5	555	560	24.4
Haciendas	3	371	374	16.3
Barrios	0	1,362	1,362	59.3
Granada	6	480	486	100.0
Poblacion	1	35	36	7.4
Haciendas	5	135	140	28.8
Barrios	0	310	310	63.8
Isabela	26	1,701	1,727	100.0
Poblacion	9	581	590	34.2
Haciendas	5	149	154	8.9
Barrios	12	971	983	56.9
Jimamaylan	17	1,308	1,325	100.0
Poblacion	12	245	257	19.4
Haciendas	4	432	436	32.9
Barrios	1	631	632	47.7
La Carlota	27	1,291	1,318	100.0
Poblacion	5	502	507	38.5
Haciendas	22	611	633	48.0
Barrios	0	178	178	13.5

Table 6. (continued)

Pueblo	Strong Materials	Light Materials	Totals Units	% of Pueblo
Manapla	31	1,228	1,259	100.0
Poblacion	11	156	167	13.3
Haciendas	14	806	820	65.1
Barrios	6	266	272	21.6
Minuluan	64	2,213	2,277	100.0
Poblacion	43	800	843	37.0
Haciendas	20	926	946	41.5
Barrios	1	487	488	21.4
Murcia	6	710	716	100.0
Poblacion	6	185	191	26.7
Haciendas	0	125	125	17.4
Barrios	0	400	400	55.9
Pontevedra	12	1,349	1,361	100.0
Poblacion	7	320	327	24.0
Haciendas	3	96	99	7.3
Barrios	2	933	935	68.7
Silay	113	2,522	2,635	100.0
Poblacion	101	755	856	32.5
Haciendas	9	1,328	1,337	50.7
Barrios	3	439	442	16.8
Sumag	6	698	704	100.0
Poblacion	6	200	206	29.3
Haciendas	0	239	239	33.9
Barrios	0	259	259	36.8

Source: Estado urbano-agricola-commercial for respective towns, 1897, PNA *Estadistica, Negros Occidental.*
[a]Includes caserias, sitios, rancherias, etc.

number of the sturdier dwellings were also built on *hacienda* premises. At the same time, "the little town" of Minuluan contained edifices that an American described as "exceedingly substantial," a fact that further destabilized established spatial patterns according to which the provincial capital was supposed to showcase the finest homes (Stewart 1908, 198).

Moreover, the growth of sugar plantations and the expansion of the migrant population led to the proliferation of small settlements, each with its own nucleus as a village or hamlet. By the 1890s, the town centers of Negros accounted for only about 20 to 40 percent of all dwelling units, with the proportion declining to less than 14 percent in a few cases (Table 6). In transgressing the colonial norm of spatial concentration, Negros somehow harked back to the pattern of dispersed settlements in the stateless islands of the precolonial age. As geographic dispersal made colonial state control unwieldy, the morphology of Negros stood for and actually abetted the apparently unbounded opportunities for the exercise of gambling as a game and a worldview.

Table 7.
Haciendas, by Type of Dwelling Units Reported, Selected Towns of Negros Occidental, 1896

Town	With One Strong Unit	All Units of Light Material	Total Reported Haciendas
Binalbagan	6	12	18
Cabancalan	11	10	21
Ginigaran	3	15	18
Granada	3	4	7
Isabela	5	0	5
Jimamaylan	4	17	21
La Carlota	17	16	33
Manapla	14	15	29
Minuluan	18	44	62
Murcia	0	9	9
Pontevedra	3	5	8
Silay	7	10	17
Sumag	0	12	12

Source: Estado urbano-agricola-commercial, for respective towns, 1897, PNA *Estadistica, Negros Occidental.*

Like the cockpit, *hacienda* space articulated a mixed message. Available (but evidently problematic) data for 1896 showing the distribution of houses classified as being made of either "light" or "strong" materials suggest that not every *hacienda* possessed one well-built structure (Table 7). In a discernibly greater number of farms, all dwelling units were made of light materials, presumably bamboo and *nipa* thatch. Although the *hacendero* (or the overseer, in case the planter lived elsewhere) might have built a slightly larger house, in many cases there was no imposing residence that visibly dominated the rest of the native huts. One planter recalls that the lessee of a 150-hectare *hacienda* in Minuluan lived only in "an old *nipa* house that was badly in need of repairs." In 1915, after he had leased the property from the landowner, who was his father, he built a house made "of strong materials," an act that brought him not praise but rebuke: "My critics said that I was just beginning to work but already I was building a palace" (Diaz 1969, 48). Apparently, the configuration of space in most farms served to confirm the parity in the sharecropping system as well as in the broader immigrant gambling milieu.[76] At the same time, the ostentatious residence of some *hacenderos*, "costing from $10,000 to $20,000 each, in Mexican money," and one or two worth an astronomical value, trumpeted the *suwerte* they had firmly attained about which the luckless could do little.[77] During the American period, however, planters would as a rule reside in the town centers of Negros, or in Iloilo or Manila, and build impressive structures on their plantations that for some became mere vacation houses.

Firmly etched in the minds of planters and workers alike and in the Negros landscape, gambling itself became an arena for the production of symbols that lent comprehensibility and legitimacy to the socioeconomic transformations impinging upon the island. In Negros, the view of society as an egalitarian gambling match has provided a cogent justification for the huge disparities in wealth that persist today: the affluent *manggaranon* have *suwerte*, the wretched *imol* none. Gambling has also offered an even more credible legitimatizing device because, in the absence of an ancient aristocratic class, intraclan socioeconomic differences are explicable in terms of *suwerte*: some planter families are less well-off because someone in a line of descent literally gambled away their properties. In Negros there are countless such lores. The conspicuous social differentials intermingled within clans have thus furnished the obfuscatory message that differences between rich and poor are part of the natural order of things and are determined by the skill with which individuals have pursued their given, but changeable, *suwerte*. By no means deprived of human agency, the poor see themselves as among the unlucky players and as pitiful *(kaluluoy)* underdogs relative to the wealthy.

In turn, the middling factions, many of whom have risen from the common mass, have attained their own *suwerte* through a form of gambling-as-cheating that entails attachment to the affluent, who provide them with jobs as farm overseers and foremen. Seen as legitimately pursuing their *suwerte*, the middle groups in *hacienda* society have withdrawn their loyalties from their poor colleagues to work and spy for the planter. Concomitantly, however, they have been wont to pilfer objects from the *hacienda*, pad the payrolls, and demand a "cut" from the workers' wages. In the words of an old *hacienda* hand, "*Tanan nga encargado kawatan kung indi gago.* [All overseers are thieves, if they are not stupid.] That is why they can send their children to school and live in fine houses."[78]

In the 1890s, Echauz (1894, 160–163) caricatured a certain "Cabocabo Mingoy" who, as organizer of work in a small *hacienda*, exploited his position between the sugar planter and the workers. The factotum was said to have known "the rule of addition for himself, and that of subtraction for the laborer." He stole from the *hacienda*'s rice granary, attributed the loss to mice, then asked for a fresh supply of grain from the *hacendero*. He accepted bribes from workers who wanted to put in fewer hours without reduction in pay. He took advantage of the fact that he distributed work assignments and made himself indispensable by aiding anyone to obtain an emergency loan from the *hacendero* in case of sickness or death in the family. He "menaced" workers and sometimes punished them, probably with a stout club that most foremen carried (cf. White 1928, 112). Mingoy's crowning glory as a toughie was that he took care of the planter's cocks destined for the most celebrated fights.

Nurtured under the patronage of the *hacenderos* as a labor-drawing strategy, the working class' penchant for gambling has become a weapon the planters have deployed in the present century as a facile explanation for the workers' poverty. The creative response to colonialism in an earlier period has become the very channel of alienation in a later era. With a ring of truth as well as hypocrisy, *hacenderos* have pontificated: if the laborer is "no vice addict, he can feed and clothe himself and his family well and can even save" on his paltry wages (Runes 1939, 31).[79]

However, like sugar production itself, gambling as a symbolic field produced contradictory impulses as it drew attention to the oppositional relationships surrounding money and money wages. While gambling legitimated social inequalities, the need for gambling money simultaneously highlighted the contentious issue of how much money the workers obtained for their labor. However, despite the squabble over money and the surplus value that leisure represented, the attraction of *hacienda* life was incontestable. Working in combination, the need for gambling money that constituted the peculiar titillation of the new economic age, the use of the muscovado mill, as well as the free housing provided to resident workers arising from the elite's paternalistic veneer, and the emotional attachment to the *hacienda* as the tenants' children were born and reared there, all contributed to the native worker's deeper incorporation into the export economy, which provided a liberation of sorts from Spanish colonial society. On this basis, tenants and their families eventually put down deep roots in the plantations, producing later generations of resident workers called *duma-an*, who would know little about what it meant to be a migrant worker but would nonetheless perpetuate gambling as a social practice.

Like a cockpit, *hacienda* life was a matter of *suwerte* and gambling in a multitude of guises: the skillful manipulation of social relations and the rules of the colonial state; the calculated game of migration; the assertion of *dungan* by planters vis-à-vis their rivals and by workers vis-à-vis their employers; *hacendero* paternalism toward tenants and those who cast their loyal bets on the planter; ambivalent relations with fugitives who became useful farmhands; and an intermingling of races, classes, winners, losers, new and old lifestyles, organized spaces and tumbled geographies, of seeming equality and easily justified inequality. Amid the flux of Negros' new society, the sugar planters struggled to dominate an emerging labor force that could demand its own terms. The planters had few options but to acquiesce to the natives' preferences and to package the *hacienda* as an exciting, even rewarding, alternative to the prosaic life elsewhere in the colony. Although undoubtedly an economically ascendant class, the *hacenderos* did not possess a preponderant hegemony—until they seized upon the conjuncture of events at the turn of the century that radically altered the balance of power in Negros.

6

Toward Mestizo Power:
Masonic Might and the Wagering of Political Destinies

Imperial Disgrace and Clerical Contradictions

As the nineteenth century drew to a close, Spain tried to meet the material and symbolic challenges posed by foreign merchant capitalists by rearticulating its sovereignty in Filipinas. Overturning the liberal reforms of 1869 and 1871, Spain reimposed a protectionist schedule of duties in 1891 (Legarda 1955, 336–349). This change propelled average annual revenues from import duties to soar from 1,664,875 pesos during the 1881–1890 decade to 3,305,281 pesos during the 1891–1895 half-decade (351). Before the final eclipse of Spanish power, the Iberian rulers thus wrested from the formidable foreign capitalists a measure of economic self-respect, leading Madrid to congratulate itself for the "prudent protection" of its national interest (348). But the royal exchequer was only one dimension of the attempt to salvage Spain's shattered glory.

The steady incursion of foreigners into colonial society necessitated that they, as well as the native subjects, be reminded who the true sovereign in the islands was. To assert Spain's so-called historic rights, the past had to be resurrected in the solidity of stone to memorialize, belatedly, the conquest of 1565. On 16 February 1891 the governor-general decided to erect in Manila a monument of the conquistadors Legazpi and Fray Urdaneta.[1] Costing about 20,000 pesos, the monument was to be built through public subscription. In this task the Recollects eagerly participated, two leading members of the order sitting on the project's three-man committee. The Recollects also systematically exhorted their parishioners, especially in Negros, to contribute generously to the endeavor.[2] The fund-raising was completed in May 1894, but the contract to build the monument was not signed until 23 July 1895. As the penultimate embodiment of the dilatory style of Spanish colonial administration, the mute reminder of Spain's suzerainty arose just in time for members of the revolutionary Katipunan to tear their *cedulas* in the Cry of August 1896.

Somewhat earlier in the 1880s, however, the Spaniards had acted to silence the imagined scoffing by foreigners who derided Spanish rule and its inability to suppress the multitude of unsubjugated natives—variously termed *indocumentados, malhechores, remontados, criminales, pulahanes,* and so on—who lurked in the colony's interiors. These marginal elements reduced the colonizers, in their own words, to being "owners only of the littoral."[3] That infidels *(infieles)* attacked the towns and flagrantly violated the law was "certainly humiliating." Officials in Manila and Madrid saw this state of affairs as constituting a veritable stain *(mancha)* on the "national honor,"[4] an insult, a disgrace, and a discredit to "civilization."[5] To erase this ignominy, a new colonization program was launched to subdue the infidels and conquer internal space.[6] Referred to as the *"reducción de infieles,"* the program called for the use of both military means and a newfangled benevolence to coax mountain dwellers to submit to Spanish rule. The January 1881 decree of Governor-General Primo de Rivera stipulated that natives who volunteered to live in new towns to be set up by the state would be exempt from tribute and taxes for ten years; they were at liberty to accept or reject Catholicism, but if they did embrace the colonizer's religion, they would also be exempt from corvee labor for eight years. On the other hand, the decree promised twice-yearly expeditions to destroy the houses and farms of recalcitrant natives who resisted the resettlement program.[7]

In the Visayas, the formation of the *Guardia Civil* in 1882 failed to aid the reconquest of the interior. As indicated by the flourishing of gambling in Negros, the intervention by the Civil Guards failed to deliver a clear signal that the Spanish empire was committed to consolidating its strength. Consequently, a bolder plan was hatched, one that would rely once more upon the magic of Friar Power. Paradoxically, however, the enlistment of the Catholic Church—an institution not without blemish—only crystallized the contradictions of the defense of Spanish honor. In Negros, the collusion of friars and local civilian authorities with sugar planters in infringing the colonial legal system was a potent sign of the fissures in the Spanish apparatus of power. But, with few other means at their disposal, the colonial state managers sought to take advantage of the friars' "art of dominating the *indio* spirit" to advance the plan of internal colonization. The assignment of military commanders to this program, for instance, was based on their "complete conviction of the need to sustain the prestige of the religious orders as an indispensable element in the pacific domination of this country."[8] Not considering for a moment that the friars' usefulness might have been spent, Spanish officials persisted in pushing the so-called religious prestige to advantage.

In Negros, the plan called for the creation of new mission outposts in some twenty-nine locations in the island's interior. Forwarded to Madrid in May 1894, the plan was packaged as the proposal drawn up by the Recollects, but the latter attributed it to the directives of Governor-General Valeriano Weyler.[9] Approved by Madrid six months later, the mission plan was praised by the Recollects as being "for the good of the faith, for the glory of Spain and for the growth of the Philippine Treasury."[10] With their characteristic passion, the Recollects added that the *reducción* plan would procure "peace for the *haciendas* of Negros."[11] The strategy, however, did not bear fruit, as in a few years Spain was to lose its colony. At any event, even before the plan was conceived, the friars had already extended unremitting support to the sugar industry. And, indeed, friar involvement in Negros had become a sign of the contrarieties that pervaded the closing stages of Spain's colonial engagement in Filipinas.

In 1848, the ecclesiastical administration of Negros was turned over to the Recollects, whose coming to the island, as the Spanish historian and Recollect Martinez Cuesta has aptly described, marked the start of "the most brilliant stage in the history of Negros. Their arrival was much more than a simple change of missionaries, for their impact was not limited to the strictly religious. They would soon come to bear on all aspects of life on the island" (Cuesta 1974, 68). Given the reorientation of the curriculum of Spain's mission schools under the *Ultramar*'s direction since the late 1830s, the Recollects were well prepared for massive intervention in Negros. The most distinguished of them was Fernando Cuenca from Zaragoza, Spain, who arrived at his parish of Minuluan in 1850 at age twenty-five, living there until his death in 1902 (Simonena 1974). He is credited for having constructed, in 1873, the first functional waterwheel in Negros and earlier, in 1850, for pioneering the construction of what until the early twentieth century was the island's only stretch of respectable road running parallel to the coastline from Bacolod to Saravia (Echauz 1894, 65–66, 86–88). In the monument built to his honor in 1974 in front of the local church, Cuenca—deftly portrayed as a Negrense through the omission of his birthplace—is proclaimed as the "moving spirit of the socio-economic progress of Negros Occidental."

Besides Cuenca, other Recollect priests also exhibited zeal in the agricultural development of Negros, as they offered technical advice to *hacenderos*, organized parishes to incorporate the migrant worker population, and managed the construction of basic infrastructure projects. The parish priest of Tayasan, the Sevillan Antonio Moreno, "built churches . . . initiated the construction of roads, and organized the barrios . . . peopling them with heretofore scattered dwellers of the wilds. He baptized more than four hundred adults" (Cuesta 1974, 74). Eladio Logroño caused the springs of Barrio Maiti to be channeled to the town plaza, providing the inhabitants of Nueva Valencia with potable drinking water in 1878 (82). In 1858, Agustin Olmedillas "introduced sugar cane to Jimamaylan bringing in seeds from other towns and distributing them among local farmers," while Jose Maria Martinez "took an active part in the organization and administration of a *hacienda* at Ginigaran" (Cuesta 1980, 369).

Although muted in official histories, a number of Recollect priests were more thoroughly immersed in the sugar industry through actual ownership of private *haciendas*, including Fernando Cuenca, whose properties were distributed to his numerous heirs sometime in the 1920s.[12] As Jose Ruiz de Luzuriaga (1900, 419) testified to the Americans in 1899, "There are in Negros at present six women who have been mistresses of friars and they have sons of the friars there. There are friars who have had relations with the women of honorable citizens." The number of women who became, in peninsular language, "devil's mules" would most likely have exceeded Luzuriaga's estimate. As part of the general atmosphere in Negros, dalliance with the *engkanto* was but another strategy of gambling. The offspring of such unions had to be provided for and their future secured, particularly if their mothers were "honorable citizens." And so the Fathers acquired land and, not dissimilarly from other planters, gambled with sugar in a sphere of action pregnant with contradictions.

The *Hacendero* as Mason: Capital Accumulation Mystified

In pursuing their mission of sugar and souls, the Recollects engaged, ironically, in a field of signs opposed to Spain and the Catholic realm. Confirmed by the momentous wedding of George Sturgis, the "Protestant" merchant capitalists' aura of cosmic strength had pervaded the export economy. That Negros' sugar industry, dependent as it was on the foreign commercial houses at Iloilo, derived its vitality from the same anti-Catholic wellspring of power was attested by native explanations of Isidro de la Rama, one of a handful of mestizos who had amassed unprecedented wealth by the end of the nineteenth century.

In Varona's hagiographic version of Negros' history, Isidro de la Rama is enshrined as "a great magnate"—impetuous, aggressive, prescient, original, and shrewd.[13] A consummate gambler, he supposedly began his career when he was eighteen years old as a leaseholder of a small farm in Minuluan, with only 500 pesos as starting capital. He then proceeded to acquire vast tracts of land as well as a fleet of vessels that plied the Iloilo–Bacolod and Iloilo–Manila routes.[14] Driven by a vaulting ambition, de la Rama, after barely seven years as a planter, was able to penetrate the exclusive circle of sugar merchants and warehouse owners at Iloilo. After ten years in the lucrative sugar trade, he moved on to the "most spectacular phase of his financial career" as a large-scale importer of manufactured goods from Europe and North America. He reportedly used his own vessels to transport his imports, which were sold through his "flamboyant shops" in Iloilo and Manila.[15] He traveled twice "around the world" and sent his two sons to study in Europe, one in Paris, the other in London. When Isidro de la Rama died in Manila in 1898, he left a fortune worth more than two million pesos. He must have gloated at the feat of dying rich: some years back he had felt triumphant in knowing that Julian Hernaez, a close competitor, died with just one million to his name.[16]

Varona's anecdotes about Isidro de la Rama suggest that he was possessed of an extremely strong *dungan,* the soul stuff that fortified him in his struggles and granted him enormous success. He astounded not only *indios* and his fellow mestizos but Spaniards as well, including Iloilo's harbor master, who tried to obstruct the movement of de la Rama's goods at the pier. Possessing the temerity to put himself above the law and exact his own form of justice—an attitude common among the flatulent Philippine ruling classes—de la Rama publicly confronted the official with a gun in hand,[17] and got his way. His retort to Spanish abuses against the natives is encapsulated in a supposed quote: "Well, these injustices have never been committed against me, and anyone who does so, I either beat up or kill."[18]

Exuding the bravado of ancient magical men, de la Rama was also undaunted by Friar Power. Like one who had the luck of befriending the *engkantos,* de la Rama mingled freely with the clergy. Varona recounts that, on one occasion, he was the only non-Spaniard among dignitaries invited to a banquet hosted by the Minuluan curate, Fernando Cuenca. One religious, who was new to the place, supposedly demanded in a loud voice why an *indio* dared to impose his presence on that august crowd. With "icy calmness" and without asserting his mestizo origins, de la Rama delivered his riposte: "I am an *indio* and your reverence hold [*sic*] me unworthy of this gathering of Spaniards. Come down with me and, by my honor as a native, I assure you that I will smash your face." Only Cuenca's intervention was said to have prevented the situation from deteriorating further.[19]

In not succumbing to Spanish caste and racial hubris, de la Rama must have taught the fledgling *hacenderos* an object lesson: wealth—money—meant the ability not only to be accepted by the colonial masters but to resist Spanish importunings and the various impositions of late colonial society. (The text of Varona, who belonged to a later generation of that social stratum, exuded such an appreciation of de la Rama.) Animated by linkages to the international economy and by gambles against the colonial state, wealth signified parity with the Iberian colonizers. Wealth made Isidro de la Rama, the individual, simultaneously opposed to and intimate with the Spanish rulers. Manifestly, de la Rama's *suwerte* as planter and merchant capitalist gave him a foreign tinge, but in answering as well to being a native, he personified the possibilities of an ambiguous and subtle anti-Catholic and anti-Spanish ethos.

That Isidro de la Rama was an emblem of opposition to Friar Power is richly captured in folklore. Accounting for more than his awe-inspiring bravery and unsurpassed *dungan,* folklore provides a popular explanation for the enigma of the man who supposedly said, "I [began] selling goods in the streets and today I possess a fortune."[20] To *hacienda* workers, Isidro de la Rama was the quintessential anti-Catholic persona: a Mason with horns *(sungay)* jutting out of his head, as in the image of the devil.[21] To attain his tremendous riches, de la Rama was rumored to have sold his soul *(kalag)* to a supernatural entity known as *Yawa,* the "lord of the forest" of precolonial Visayas (cf. Alcina 1960, chap. 15) but now commonly equated with the devil of European cosmology. In return for allegedly selling his soul, de la Rama acquired magical power *(anting-anting)* and was equipped with the extraordinary ability to produce wealth. All he had to do was scourge with all his might the *Santo Cristo,* a large crucifix, using a whip fashioned out of a sea creature's tail *(ikog sang pagi),* and the icon would vomit strength *(kusog)* and money *(pilak).*

In other versions of this myth, a mysterious person is said to have delivered sacks of money to de la Rama on Tuesdays and Fridays, coinciding with the two days of the Catholic rosary's sorrowful mystery recalling Christ's suffering and crucifixion. In the localized version of a Roman Catholic tradition, the act of reciting the Passion signifies a "real" diminution of power: it means the divine's "death" and incapacity to defeat his foes. Thus, on those days, malevolent spirits are thought to roam the earth freely, and peasants stay away from their fields for fear of coming in contact with those unseen beings. But it was precisely on such days that de la Rama reaped the rewards of his exchange relations with the devil. The legendary tales end by saying that Isidro de la Rama grew a long tail and his whole appearance was transformed into a likeness of *Yawa*. Embarrassed by this unusual development, his family supposedly locked him up in the basement of their Iloilo mansion. Other versions aver that de la Rama actually became *Yawa* and never really died: the casket that was interred was not that of a dead man, and even now he periodically arises and emerges from his tomb.[22]

If these representations can be taken as indicative of how Isidro de la Rama was popularly perceived during his lifetime, it would appear that, for both planters and *hacienda* workers, Isidro de la Rama, the local capitalist, was a symbol of opposition to Spanish colonialism equated specifically with Friar Power. But whereas the elite version focuses exclusively upon de la Rama's character traits as instruments and weapons for financial success that allowed him to penetrate the inner sanctum of colonial power, the workers' mythical version apprehends de la Rama's historical significance more pointedly from the perspective of indigenous spirit beliefs. In the latter, de la Rama was what he was because he sold his being to the devil. To a people for whom long-distance trade antedated the colonial epoch, selling was a familiar idiom. Moreover, in the capitalist epoch when merchant capitalists were seen negatively vis-à-vis Catholic power, selling was a potent act of connection with the anti-Catholic realm. Through the act of exchange with the fearsome *Yawa*, de la Rama gained prowess from an alternative source of cosmic strength.

The relationship to Masonic Capitalism was underscored by the fact that de la Rama's putative act of selling was unimaginable in relation to the Catholic Church. Only the Protestant opponents of Friar Power had the capacity, the resources, and the interest "to buy" adherents to its side. The devil operated through the capitalist market. In contrast, as what used to be the only showcase of wealth in the towns that dwarfed the usually modest municipal buildings (cf. Reed 1967, 65), the Catholic Church was seen not so much as selling but as extracting money, forcibly or by means of compunction. From such a perspective, myth offered a richly textured portrayal of the contradictions in colonial society.

The mystified explanation of de la Rama's success would seem to leave out the contribution of native farm labor to the accumulation of wealth. Folklore also left hardly any complaint that he was a brutish *hacendero* who maltreated his workers. The predominance of sharecropping in his farms probably blunted the contradictions in the planter–tenant relationship.[23] Moreover, the injustice felt more acutely by the natives centered around the colonial state as personified by the friar. Through its religious and civilian bureaucratic tentacles, the Spanish colonial state exemplified the "swallowing up" of money in the form of tribute, taxes, produce, and church offerings and the "swallowing up" of human energy and strength in the form of corvee labor. Money and human energy were then, as suggested in the de la Rama legend, fed to the Catholic deity, supporting the supernatural sphere that enabled the reproduction of the colonial state and all its organs. Friar Power was characterized by gluttony, in marked contrast to the appeasing cash advance and other renumeration offered by export agriculture.

In this context, opposition to the colonial state entailed a reversal of the resource flow that de la Rama achieved by "whipping" the image of Christ, compelling it to "vomit" both money and vitality—in much the same way that natives paid taxes and rendered corvee on pain of the whip. Flogging served as the most potent act to release what had been withheld, as indeed the colonial state extracted what it wanted by having its functionaries perform the very same act on errant natives. At the same time, whipping the cross fully identified de la Rama with the inquisitorial portrait of the heretic, who was believed, in Negros as well as in the Peninsula, to have whipped the cross and assaulted it in other ways (Christian 1981, 190–193). In the de la Rama legend, however, the heretic produced wealth. Capitalist accumulation was thus incidentally fetishized.

Empowered by *Yawa*, the translated antagonist par excellence of Friar Power, Isidro de la Rama was in a unique position to "whip back" at the Spanish colonial state. Utilizing the individualized attributes granted by cosmic forces, he recovered those basic elements of native existence which had been gobbled up by appendages of the colonial state. But the vomited wealth was only for himself: de la Rama was not expected to be a social liberator. Consistent with the indigenous worldview, a colonial-era man of prowess had arisen from among the natives, his mestizo ethnicity glossed over. Folklore nonetheless took note of the fact that Isidro de la Rama left the historical stage before the final demise of the Spanish regime by stressing that Friar Power managed to inflict retribution. According to various legendary sources, while alive, de la Rama was merely like *Yawa;* in death he was believed to have been transformed into *Yawa*. The "failure" to die a human death meant eternal suffering for one who, according to Philippine Catholic expectations, ought to have had his soul undergo purification in Purgatory in the initial stage of the afterlife. So in death Friar Power exacted its ultimate vengeance.

The tales spun around de la Rama's persona showed that elite opposition to Spanish colonialism was expressible as a negation of the Catholic Church. But like the subtle transactions with environmental spirits engaged in by locals even now, de la Rama's opposition to Friar Power entailed the concomitant befriending of friars. With Spain's purported enemies residing in one man, de la Rama's sign, akin to that of Sturgis, was that of a cosmic strength capable of engaging in skillful and cunning combat with the dominant power. Isidro de la Rama represented the apogee of mestizo contests with the state. As we have seen in previous chapters, the lesser *hacenderos,* beneath their appearance as loyal subjects, were also carving out an enclave of resistance to the colonial regime through patent gambles against the state. Embedded in Masonic Capitalism, the emergent planter class was in symbolic opposition to Friar Power. The planters' fascination with unloading their sugar stocks on the market during the Holy Week commemoration of Christ's crucifixion with the expectation of earning more money than at other periods of the year hinted at the overall Masonic quality of their class.[24] Thus, McCoy's (1982b, 173) contention about a "predictable symmetry of secular and spiritual allies" in which "the planters, both Spanish and Visayan, guided by the temporal and spiritual powers of the Holy Roman Catholic and Apostolic Church," placed themselves on "the side of Spain and order" may be seen as failing to grasp the planters' various antistate strategies and the sugar economy's undercurrents of meaning.

Isidro de la Rama conveyed the surrealistic anti-Catholicism that reverberated through the *hacendero* class at the turn of the century and beyond, in Negros and in other parts of the country. The martyrdom of Jose Rizal, which accelerated the revolution against Spain, for instance, was first memorialized in the shape of a Masonic-inspired obelisk on which were written the titles of his two great novels, a monument erected in the throes of the revolution.[25] Thereafter, many prominent members of the landed class in Negros and elsewhere in the Philippines came to be widely believed by ordinary people as practicing Masons.[26] The continuity in the Masonic label of the rich suggests, at least in Negros, a sustained mystification of capital accumulation. Today's Masons may not be perceived as possessing the spectacular prowess attributed to Isidro de la Rama, but the belief among many *hacienda* workers has persisted that the purported object of Masonry is to desecrate the cross—a legacy of the past that present-day Masons would protest as black propaganda against their movement.[27] At any event, some people still believe that the soles of Masons' feet are said to bear images of the cross profaned in the very act of stepping on the ground. In their houses, Masons are said to have affixed crosses on their floor tiles and on the steps of stairwells to ensure a daily dose of profanity against the Catholic deity. A leading Mason has purportedly continued the practice of flogging the cross. From Isidro de la Rama onward, a pervasive anti-Catholic aura has suffused the sugar capitalist class of Negros.

In pursuing their patriotic commitment to sugar, the Recollect priests unwittingly participated in a sphere that symbolically represented defiance against Spain and the Catholic Church. To the extent that the Recollects also decided to be on the side of the *hacenderos* in contravening colonial state strictures, the friars in Negros became key actors in painting a fading picture of Spanish hegemony. Buffeted by Masonic Capitalism, the friars no longer seemed to know their place, and their *dungan* began to wander. It was as though the friars had stayed too long in the cockpit, oblivious to the fact that they were betting against their own cocks.

The Breakdown of Friar Power and the *Babaylan* Challenge

The crisscrossing of signs and the confounded directions of social alliances were inescapable signs that the Spanish colonial regime was breaking down and Friar Power waning. The malaise of colonial society did not escape the acute discernment of the shamans, generations of whom had sustained the subversive imitation of elements in the friars' magical ensemble. In the second half of the nineteenth century, *babaylan* subversion thus progressed from the use of *orasyon*, the ritual formularies appropriated since the seventeenth century from Latin Catholic prayers, to a proliferation of armed challenge to the colonial state. From its individuated privacy, magic reemerged with a collective public face.

The disintegration of Friar Power was keenly felt on Panay Island, cradle of the *babaylan* tradition and sacred center of the universe where the "four pillars" supporting the world are believed to be located, linked by subterranean passages to Negros' fabled mountain, Kanlaon (Magos 1992, 53–55). In 1864, the Augustinians recorded an unusual event in which the native shamans of the town of San Joaquin in Iloilo Province handed over amulets, books, gongs, medical concoctions, and other sacred instruments to the local parish priest.[28] The Augustinians interpreted this episode as meaning a surrender and admission of religious error. Conducting a meticulous examination of the *babaylan* materials, the friars learned about the veneration of ancestors and the gathering of herbs and other objects for medicinal purposes in a nearby sacred mountain on the Tuesdays and Fridays of Holy Week; about the priesthood of the male and female shamans under the leadership of one whom the friars referred to as "their Pope"; and about *babaylan* rites and ceremonies and the grand celebration held every seven years during Holy Week in the mountains of Tubungan.[29] The detailed knowledge thus acquired by the friars, however, proved of no avail in weakening the *babaylan* challenge.

Perhaps the shamans' intentional self-exposure indicated a renewed confidence that, no matter what the Catholic clergy knew, they themselves had learned to be invulnerable vis-à-vis Friar Power. Or perhaps it was a daring provocation that presaged a later display of cosmic strength. For instance, in 1874, more than a thousand people scaled Tubungan for the great assembly of shamans in a communal ceremony called *samba*. An Augustinian friar attended the event and was killed for disrupting the solemn occasion (Echauz 1894, 141; Magos 1992, 35). The colonial state was so perturbed by this incident that, seven years later, soldiers were posted at Tubungan to prevent the expected meeting from taking place. The governor of the Visayas claimed to have prevented the festival of 1881 from being held,[30] but the Augustinians were forthright in reporting that the assembly was moved to the mountains of Miag-ao.

A more spectacular demonstration of the resurgence of *babaylan* strength transpired in San Joaquin. Oral tradition among shamans of Panay recount "three years" of drought and famine that ravaged this town and left people dying of starvation and thirst, as all the rivers and springs had dried up (Magos 1992, 35). This was probably the same famine, mentioned in the Augustinian record, that befell San Joaquin in 1877 and 1878, when corpses were literally strewn around the town.[31] According to the lore, people sought help from the parish priest, but he failed to induce rain. Desperate in his inability to alleviate the disaster, the curate advised the town leaders to call upon a *babaylan* known as Estrella Bangotbanwa, who ordered that seven black pigs be butchered, shaved, and covered with black cloth. She then took a black pig from the convent to the plaza, where she pressed its mouth to the ground until it gave a loud squeak. Suddenly, the sky turned dark and a heavy downpour followed. The butchered pigs and the sea crustaceans at the offering table sprang back to life. Bangotbanwa had brought back rain and life to San Joaquin.[32]

The popular belief in Estrella Bangotbanwa's salvational role demonstrated that, in a crucial contest of cosmic strength, the *babaylan* emerged triumphant over Friar Power, whose efficacy had been so diminished as to be utterly helpless in the face of a natural disaster. (It would seem that even the mestizo *gobernadorcillo* of Jaro, Eugenio Lopez, fared better, at least in the kin group's memory, than the friar because he distributed "both rice and money to the countless famine victims who flocked to their homes for help"; O. Lopez 1982, xlix.) Estrella Bangotbanwa's success over the Catholic Church has been immortalized through her elevation to the status of an ancestor *(papu)* with mythical prowess, and at present she is revered as the matriarch and "founder" of a group of shamans in Antique Province (Magos 1992, 33–37). In the same vein, oral tradition has remembered the magical impotence of the Spanish priest.

The disgrace cast upon Friar Power—further exacerbated by Spanish inability to control the massive cholera epidemic that struck the western

Visayas in 1882—must have inspired the shamans to mount direct challenges to a disintegrating colonial state, converting the whole of the Visayas into a theater of resistance by the late 1880s. Various groups led by different individuals, all of whom were animated by links to the spirit-world, waged minirebellions against Spain. Among the leaders in Panay were "Dios Gregorio" (God Gregory) [33] and Clara Tarrosa, an eighty-year-old woman from Tigbauan, Iloilo, who claimed to be the "Virgin Mary."[34] Although resistance took the familiar form of a retreat to the interior, this time they did so in larger numbers, and many groups developed an armed corps to defend their mountain holdouts and even assault local town centers. The Civil Guards in the Visayas decided that the shamans had ceased to be a mere nuisance, as they were now engaged in the spreading of "subversive words and threats against the Nation, the authorities, the police forces and in particular against the Spaniards."[35] Superstition was more than a religious matter to be extirpated, for the claim to be the Virgin Mary had become an overtly political act. The authorities in Manila, however, mistakenly viewed the various shamanic groups as mere brigands, or as "disturbances to public order" having no political significance.[36]

Similar movements emerged in Negros at about the same time as in neighboring Panay Island. In the late 1880s, in the southeastern town of Zamboanguita, the effeminate *babaylan* Ponciano Elopre, better known as "Dios Buhawi" (God Waterspout), attracted a considerable following who trooped to his isolated village of Nahandig for religious ceremonies and revelries every Saturday (Hart 1967, 373–378). Heading a movement that was as much political as it was magical, Buhawi's reputed powers included the ability to summon rain, as his sobriquet indicated. He could also make coins appear from a fresh squash or from leaves afloat on a river, the money obtained in this manner being said to have come "from America," by then the major importer of Negros sugar. Tacitly, Buhawi expressed a link with Masonic Capitalism even as he preached that people desist from paying taxes to the colonial state. Buhawi and his band became hunted outlaws, but he was an elusive target, and a raid organized by the local priest and the Civil Guards failed to capture him. His followers subsequently became active in stealing cattle and carabaos, to which the Civil Guards responded with a string of murders of innocent civilians.

Around 1889, a government force of more than a hundred men attacked Buhawi's group before they could mount their planned assault on the town of Siaton. Buhawi was killed, but legend insists that he is still alive, although since then he has not been seen in Negros Oriental (Hart 1967, 382–389). His brother-in-law, Valentin Tubigan nicknamed

"Kachila" (Castilian), a name deriving supposedly from his long black beard, assumed the movement's leadership. Kachila bore the title "Dios unico Rey de la tierra" (God Only King of the Land), a status supposedly acknowledged by another famous outlaw called Kamartin.[37] In October 1888, the Civil Guards reported Kachila's death in Siaton.[38] Three months later, eight members of Buhawi's and Kachila's movement were apprehended on Hacienda San Juan in Cabancalan, having been hired with the collusion of a friar, as narrated in the previous chapter.

Also in October 1888, Kamartin's group robbed a house, killed three men, and kidnapped several people from Tolong, ten of whom were rescued by the Civil Guards.[39] By March 1889 Kamartin's "band of criminals" was still at large.[40] To the colonial state, Kamartin was a bandit and fugitive from justice, but to the people he—like many other outlaws in the colonial history of the islands—was credited with the magical *anting-anting* that gave him invincibility and other awesome powers (Worcester 1898, 269–270). Not only could he escape gunfire, but Kamartin was believed to possess a charm that enabled him to take a giant step from one mountaintop to another or to precipitate a sudden torrent of water to drown his pursuers—feats that stupefied even the "intelligent mestizos" of Negros (272–273).

By 1897 various shamanic groups in Negros were in open rebellion, armed with spears, daggers, machetes, and bows and arrows. They came to be known, among other names, as *Civil-Civil* because many wore striped clothing patterned after the uniform of the Guardia Civil.[41] From the *babaylans'* emulation of the friars at the onset of colonial rule in the sixteenth century, they had graduated to imitation of the colonial army by the close of the nineteenth. Intrinsically political, shamanic mimicry had become both religious and military and had spread across the islands. Copying Friar Power, some *babaylans* also donned priestlike robes adorned with red belts (red being the magical color of courage) and proscribed the eating of meat on Wednesdays, a reworking of the Catholic abstention on Fridays (Hart 1954, 235–236). The *babaylans* punished natives who wore their shirts outside their trousers, a symbol of subservience to the dress code imposed by the colonial state following its categories of race and status (235).[42] The shamans had formed the conviction that natives, as a mark of equality, had as much right to tuck in their shirts as Spaniards did. Thus the *Civil-Civil* were dressed in the same manner as the Civil Guards.

On the western side of Negros Island, *babaylan* resistance was led by a former *hacienda* worker popularly known as "Papa [Pope] Isio," whose supernatural powers made him the epitome of "the good man" *(maayo nga lalaki)*. Said by some to have been a migrant from Panay, Isio's flight to the mountains of Negros supposedly followed an incident in which he gravely wounded a Spaniard in a quarrel (Cullamar 1986, 24). Isio's core group, numbering some fifty followers who camped in the village of Alabhid in the central mountain district of Isabela-Jimamaylan, was bent on "killing friars and Spaniards," according to a Recollect.[43] By late 1896, the Civil Guards in the town of Ilog received reports that people were "absent everyday from the towns and plantations."[44] After a relative of Isio, known as "San Juan" (St. John), was killed in an encounter with the Civil Guards in December 1896, the group transferred to the southern vicinity of Cabancalan, "where the band acquired greater importance."[45] In January 1897, the Civil Guards learned that men, women, and children had disappeared from Cabancalan town to join Isio's group in celebrating a *misa de campaña* (campaign mass), and immediately made plans to mount an assault against Isio's forces.[46]

After an arduous trek to the summit of the mountain, the Civil Guards reached a plateau where they encountered more than a thousand men in battle formation. As the natives began to circle the government forces, the latter gave shouts of "Long Live Spain!" "Long Live the Queen Regent!" "Long Live the Captain General!" and "Long Live the 22nd Regiment of the Civil Guards!"—to which Isio's group countered with "Long Live Rizal!" "Long Live Filipinas and its Liberty!" and "Death to Spaniards!"[47] Trusting in the protection of their amulets, many of Isio's men drew within ten meters of the barrels of the Civil Guard's guns. The fierce fighting left some forty dead and countless injured among Isio's followers. Despite this setback, Isio's movement continued to control the central mountains of Negros until around February 1899, when they launched the second phase of their struggle, this time against U.S. colonialism.

The various discourses from Isidro de la Rama to Papa Isio disclose that, toward the end of the nineteenth century, both mestizo *hacenderos* and *indio* workers shared a common sentiment of opposition to Spain and Friar Power. Planters and shamans, in particular, claimed equality with the Iberian rulers, but their rejection of colonial subjugation found diverse expressions, which varied according to different modes of identification with the colonizer. Planters asserted themselves through wealth and Masonic Capitalism, essentially the negation of Friar Power, whereas shamans penetrated and adopted Friar Power's magical ensemble, such appurtenances as the Spanish dress code and the naming system, including those of divine entities and the *Cachila* alias of one who fought the children of Castile.

Initially overwhelmed by Spain's "spiritual conquest," the native had learned to coexist with the conquering *engkanto* and other alien spirits that, by the late nineteenth century, had become fully integral to the local spirit-world. Not only were these Hispanic beings domiciled in the islands and the desired communion with the *engkantos* fulfilled, but they had also been transformed into local sources of cosmic power the natives could tap through Catholic formularies, rituals, and medallions like the Agnus Dei.[48] As one group of shamans in the eastern town of Majuyod claimed, their invulnerability stemmed from a collection of Catholic prayers contained in a booklet called *bugna*, which an angel had handed over to a priest who was saying mass; the cryptic invocations were believed to carry the Roman pontiff's blessings.[49] No longer just the friars', this secret power had become the shamans' own to behold and to use. Having found rapport with the spiritual fount of colonialism, the shamans could claim to personify the key figures of the Spanish spirit-world: God, pope, King, Saint John, or the Virgin Mary. Imitative magic defined the possibility of the shaman's liberationist project.

Although springing from different sources with different meanings, the aspirations of *indios* and mestizos, planters and shamans, to see an end to colonial oppression and humiliation somehow converged, signaled broadly by the battle cry "Long Live Rizal!" Seemingly transcending class divisions, a shared historical experience had been forged among the subjugated natives, with *engkantos*, friars, Civil Guards, and restrictions to geographic mobility and prohibited games becoming common foci of struggle. But because those struggles were spatially circumscribed (say, the illegal migrations), and because elite interests were focused on specific regional enclaves—in contrast to the more encompassing secular pilgrimages afforded by colonial bureaucracies elsewhere (Anderson 1991, chap. 4)—mestizo and *indio* nationalisms assumed a decentered, if not multicentered, form. At the same time, because there now seemed to be a chance to resolve the clash of spirits in the cosmological space in which the *indio* lived, the shamanic *indio*'s anticolonialism actively sought to reclaim not one or the other but both of the competing worlds that had initially prompted the *dungan* to wander. As we shall see, *indio* nationalism also imitated Friar Power's inquisitorial mentality and its implied stance against "foreigners." Appropriating the colonial legacy while affirming indigenous spirituality, the *babaylans*' belief in the sacred center of Panay over secular Manila did not dissuade them from raising the battle cry "*Viva Filipinas Libre!*" However, it was the *hacendero* living the life of the consummate gambler who emerged the ultimate victor.

The *Hacendero* as Revolutionary: Masonic Might Triumphant

While the shamans mounted their uncoordinated but serious challenges against the colonial state, the symbolic opposition posed by Masonic Capitalism also started to materialize. The colonial authorities began to be alarmed by reported acts of prominent individuals in Negros, among them the celebrated Teodoro Benedicto, adversary of the peninsulars who denounced him for land grabbing in the 1870s, who in 1884 was suspected of spreading separatist propaganda. The provincial governors of both Iloilo and Negros, in an unprecedented display of cooperation, organized a joint secret police unit to undertake surveillance of Benedicto, who was codenamed "Z."[50] Despite covert planter opposition to colonial rule, however, the Negros elite generally avoided direct political provocation. One exception was an 1891 incident in which the *gobernadorcillo* of Dumaguete allegedly insulted and threatened to kill the Spaniards in attendance at one public gathering.[51] This episode aside, *babaylan* groups by comparison were more openly confrontational. Nevertheless, the *hacenderos* of Negros eventually took the upper hand in finally subduing Friar Power in one swift dramatic turn of events in 1898.

The idea of revolution was finally brought home to the Negros elite as the direct consequence of Spanish paranoia in the wake of the uprising that erupted in Luzon in August 1896. Writing in the newspaper *La Republica Filipina,* Jose Ner (1898) explained that more than twenty *hacenderos* in Negros were arrested by the Civil Guards for supposedly being in league with the revolutionaries of Luzon. The arrests were instigated by defamatory letters written by Recollect friars who, terrified by the prospect of the final blot on Spanish honor, began to denounce the planters for alleged treachery. Among those incarcerated were Juan Araneta, Carlos Zamora, Sabina viuda de Higgins, and Ventura Magalona, each of whom carried a price of several thousands of pesos for their release. In Zamora's case, for instance, 15,000 pesos were demanded, although after some haggling the amount was reduced to 8,000 pesos. Many languished in jail, some intentionally, including Juan Araneta, who remained in prison for ten months from January to October 1897 (Cuesta 1980, 467 n. 72). The Civil Guards also reportedly scoured the south of Negros Island, killing some 150 persons on the slightist suspicion of being a rebel (Ner 1898). In an anonymously written newspaper report, a Spaniard confirmed that suspected insurrectionists in Negros were tortured (Anon. 1898). The atrocities and indignities suffered by planters must have convinced them of the need to shake off "the heavy Spanish yoke" in alliance with the movement on the "sister" island of Luzon (Ner 1898).

However, in Luzon the trajectory of the revolution against Spain was disrupted by the outbreak of the Spanish-American War, which ramified in the immediate capture of Manila by U.S. forces on 13 August 1898. With the loss of Manila to the Americans, the Spaniards tried to maintain their hold over the Visayas and Mindanao by moving the seat of government to Iloilo. They also offered to implement the autonomy reforms earlier demanded by leading mestizo reformers, but this gesture was spurned. In Panay, an armed uprising was already being organized by planters and other elites at the time of the U.S. invasion of Luzon. In October, news reached Negros concerning the outbreak of hostilities in Iloilo and in the neighboring islands of Cebu, Bohol, and Leyte, where the insurrectionists were scoring successes. The Negros *hacenderos* were finally emboldened to launch an armed confrontation with the small Spanish force on their island.[52] They organized a provincial revolutionary committee with links to the Central Revolutionary Government of the Visayas, solicited funds to buy firearms, and supplied *hacienda* workers with machetes and shotguns. Advising every planter to consider himself as "the military superior of his men" (Fuentes 1919, 42), the committee decided to simultaneously launch the offensive in the different towns of Negros Occidental on 5 November 1898. At day's end, only Bacolod, the capital, and Jimamaylan were in Spanish hands, the latter subsequently surrendering three days later.

In Bacolod the Spanish forces were undermanned, having only about seventy-five Civil Guards, fifty soldiers, and a handful of peninsular volunteers as against some eight thousand insurgents from the north led by Aniceto Lacson of Silay and nearly two thousand from the south captained by Juan Araneta of Bago. Beginning at midday of 5 November, Araneta's corps of fighters steadily advanced toward the capital, where pandemonium had broken loose at the prospect of heavy fighting. From the church tower, the Spanish signal corps reportedly noticed the movement of a huge column of men equipped with Japanese *murata* rifles strategically poised to attack Bacolod. The rumor spread rapidly that a Japanese vessel had landed near Pontevedra town loaded with thousands of arms for the rebels, a tale that caught the Spaniards by surprise. Faced with what seemed like impossible odds, the governor of Negros Occidental commissioned the Spanish mestizo planter Jose Ruiz de Luzuriaga to negotiate with the rebels. Believing the Spaniards would lose, the emissary persuaded the governor that an "honorable surrender" was better than a shameful defeat. Late the following day, 6 November, an act of capitulation was signed. Friar Power was finally vanquished in Negros.

With the Spanish forces disarmed and quartered in the convent, Araneta's troops made their triumphal entry into Bacolod. Only then did it become known that, save for a number of shotguns, the troops actually had little in the way of arms or ammunition. What had appeared from a distance to be Japanese rifles with bayonets attached to their muzzles turned out to be poles fashioned from *nipa* palms uniformly trimmed and cut to size, with polished blades of knives affixed to one end. A clever ploy that destroyed the morale of the Spanish troops, the idea was reportedly concocted by Juan Araneta.

The day after the capitulation, Negros' Provisional Revolutionary Government was formed, with Lacson—an extremely affluent planter who owned Negros' second most expensive house, valued at $80,000 (Mexican)—as the President, and Araneta—another exceedingly wealthy man who was "looked upon as the greatest agriculturist in the island"—as the Delegate of War.[53] A struggle for superiority between the two men was resolved when Araneta invoked the state of war existing on the island, which justified his assumption of full state powers. Araneta was later designated brigadier general as well as interim governor of Negros Island by Emilio Aguinaldo, who headed the national revolutionary government based in Luzon.[54] As one of his first official acts, Araneta moved thirty-nine friar-prisoners to the *La Granja* (model farm) in La Carlota town, where they might, in Araneta's terms, finally fulfill the "divine precept which they had been satisfied to preach: 'By the sweat of your brow shall you eat your bread'" (Fuentes 1919, 72).

The story about Araneta's ingenuity in engineering the Spanish downfall in Negros was widely believed by planters and workers alike. Relying upon its elite version, an American attributed the Spanish surrender to "a series of diplomatic bluffing" by Araneta (Stewart 1908, 182). The historical veracity of Araneta's role is not at issue here, but rather the consequences of popular interpretations of that role. The event which Araneta (1898) rather self-servingly celebrated in a newspaper article as a "brilliant national episode" that "brought honor to the land" transformed him into a mythical hero and earned him a unique place in the history of Negros, for where the *babaylans* failed he succeeded marvelously.

According to various legendary fragments,[55] Araneta was a man of exceptionally strong *dungan*. In the league of the traditional *maayo nga lalaki*, he possessed a commanding voice and was full of knowledge and wisdom. His reputed powers included the ability to heal, to take giant steps and to see from afar, to vanish before one's very eyes and reappear in another place instantaneously, to produce objects from nowhere, and to fly on his white magical horse. His body was invincible to bullets. Considered a kindhearted man *(buot nga tawo)* who treated his workers well, he used his incomparable talents for "the good of the people," particularly in driving away the Spaniards. Harking back faithfully to the image of the preconquest *datu*, Araneta was a man of prowess who received guidance from and enjoyed a special friendship with the spirit-world.

His spiritual rapport was no ordinary one, for he had a privileged relationship with Sota (or Suta), the preternatural being that displaced the precolonial female divinity Laon, who once had dominion over the magical mountain of Kanlaon. As Araneta's spirit-guide, the Sota was a composite power source, reputedly Spanish-looking but with a body that was half black and half white, a feature described in Ilonggo as *kambang*.[56] Some informants suggested that Sota was Araneta's spouse, or that Araneta himself was a type of Sota. Others claimed that Sota was a spirit-being of small stature who regularly visited Araneta at his residence in Bago. Most certainly linked to Araneta's mystique was a "white monkey" he claimed to have captured, such creature being, in the opinion of an American writer, "the only one found in the islands" (Stewart 1908, 198).

The uncertainly gendered Sota stood for the combined forces of Hispanic and native wellsprings of otherworldly power the mestizo Araneta was enabled to tap into in order to perform an unprecedented role in the island's history. To reciprocate Sota's friendship and assistance—indicative of the elite's anti-Catholic aura—Araneta supposedly would scale the slopes of Kanlaon during the annual Holy Week in order to, like the *babaylan,* augment his powers and presumably commune with Sota. He was said to have collected medicinal herbs, roots, barks, and other objects that were later mixed with oil made from a peculiar coconut *(bugtong nga lubi).* Consistent with the present-day practice of shamans in Panay and Negros Islands, Araneta allegedly gathered Kanlaon's magical objects during Good Friday, when Christ was believed dead and thus unable to exact divine retribution. At Kanlaon's summit, Araneta was said to have planted a white flag to announce his presence on the fabled mountain.

To immortalize his triumph over Friar Power, legend has elevated Juan Araneta among the venerable ancestors who died of "old age." Blessed with longevity relative to the average life expectancy of the period, Araneta's life span extended into the first half of American colonial rule. He was seventy-two when he passed away in 1924. In contrast to the tales about Isidro de la Rama, Araneta's death from "natural cause" became the ultimate guarantee of the victory of Mestizo over Friar Power. In the end, the latter was unable to avenge the utter humiliation suffered by the friars in Negros.

During his lifetime, Araneta amassed some 7,000 hectares of agricultural land divided into several *haciendas.*[57] He purchased land parcels, as vouched for by one of his descendants, from several small owners and aggregated them into larger plantations, in the same way that attestations to claims of legitimate title were made during the Spanish period. To *hacienda* workers, to try to legitimate Araneta's landholdings in legalistic terms would be superfluous, as the Juan Araneta of mythical fame they knew did not err in deploying his powers to acquire land. Even before the end of Spanish rule, Araneta already owned substantial properties, including his two favorite *haciendas,* which he had named Louisiana and California after being much impressed by the agricultural conditions he had seen in those two states during his visit to the United States in 1893 (Stewart 1908, 191). But the turbulent turn of the century was also a splendid time when "certain planters" were reportedly "enlarging their plantations very rapidly" (Brill 1901, 554). Whether Araneta was implicated in the illicit acquisition of property cannot be ascertained.[58] What is telling is that, in the fragments of the Araneta legend, the social setting is invariably one in which only the mystical hero had ownership of land.

To acquire the land he desired, Araneta would simply swing his magical kris-like sword and all occupants within its radius would vacate the land. Described literally as merely "moving a bit" *(idog)* to make way for Araneta, the displacement of peasants was not portrayed as unjust or burdensome. Exercising his exceptionally strong *dungan,* Araneta had authority, as signified by the sword, to issue directives that people, in fear and awe, followed. As in the preconquest age, the early-twentieth-century peasants could be deemed as reciprocating the great service Araneta had rendered in expelling the Spaniards. He was thus depicted as being perfectly justified in using his spiritual endowments in order to accumulate land. Indeed, within the indigenous spirit-belief system, there was no question about Araneta's legitimacy in amassing land and, thereby, wealth. The resulting structure of economic inequality could not have found a better justification that would have impressed the average person at the time as valid and credible.

Reportedly already having a reputation among his own workers "of having his orders obeyed" even before the downfall of Spain (Stewart 1908, 182), Araneta catapulted himself onto the center stage as a mythical figure through the single decisive act of dismantling Friar Power and, in so doing, fulfilling the desire of the repressed colonial subjects. He thus personified the historical assertion of the *hacenderos'* Mestizo Power. Although reportedly he was kind toward his workers, and thus behaved appropriately within the paternalistic framework that became a marked feature of Negros *haciendas* in the late nineteenth century, Araneta was at the same time pictured as the feared planter whose supernatural powers, the existence of which was confirmed by the Spanish surrender, made him absolute master and overlord. His whips *(ikog sang pagi)* were reputedly capable of lashing at offending workers even if they hid under water. Despite the entreaties of his workers, Araneta meted out punishment for offenses, such as theft, which violated planter property rights. Araneta thus fit the pattern of ancient men of prowess who, depending upon the circumstance, were both paternalistic and ruthless. The fear he generated and the spiritually sanctioned authority he wielded placed Araneta in a cosmological position of dominance, yet equally positioned him to profit from the capitalist age.

As a modern-day man of prowess, was Araneta conscious of the magical powers attributed to him? That he was aware of the fantastic tales about him could be inferred from the recollection of a granddaughter, who remembered hearing other household members tell such stories with her grandfather within hearing distance.[59] That he believed in the spirit-world could not be established, however. But for him to have believed was not implausible, as many mestizos in Negros, although Catholics, not only believed but were said by their fellows to have rapport with *engkantos* and other spirit-beings. In any event, to the extent that he was aware of the magic attributed to him, Araneta's exercise of Mestizo Power was a superb act of structural opportunism that finally won for the planter class undisputed symbolic capital. In turn, his manipulation of indigenous beliefs might have been possible since, reminiscent of the classic cultural ambivalence and vacillation of mestizos, Araneta could objectify the native belief system even as he himself probably believed in its various elements. Although the exploitative use of native culture cannot be established in the case of Juan Araneta, the popular lore about him was a veritable lesson for other planters who, during the American period, would extend and expand Mestizo Power in their own *haciendas*.

Wagering National Destiny:
The Planters "Invite" the United States

In articulating their demands to Luzuriaga, who acted as emissary of the Spanish camp, the Negros planter-insurrectionists emphasized, as Araneta (1898) wrote in a newspaper item, the determination to lay down their lives for the sake of freedom: *"A este albur jugamos nuestras vidas ques estamos decididos a perder."* The choice of words was distinctively *hacendero*: the decision to throw off the Spanish yoke, when it did occur, issued from a deeply felt anti-Spanish feeling, but it was also—or, probably, above all—a calculated risk *(albur)*, a decisive seizure of the chance *(albur)* to alter the course of history for which they were willing to stake *(jugar)* their lives. In its intransitive form, *jugar* means to gamble, a signification similarly embedded in Araneta's statement. Steeped in gambling, the planters approached the turnabout against Spanish colonial rule as a high-stakes political gamble. They were lucky. As recalled by a then youthful participant (Diaz 1969, 14), "Our bravura paid off and luck smiled on us"; he added that, "if the Spaniards had put up

FIGURE 3. The Jack *(Sota)* of the Spanish Playing cards

a stiff resistance we would not really have been able to successfully fight back."

Contemporary legends would appear to concur in the gambling aspect of the elite's revolutionary feat, for the Sota that gave Araneta the incomparable *suwerte* of defeating Friar Power referred literally to the jack of the Spanish playing cards (Figure 3). Just as the king and queen were real entities of power whom the native subjects never saw except as portrayed in the cards, in like manner the Sota was conceivable as a real powerful being. Just as the king and queen seemed eternal, in the same way that the card game was timeless, so the Sota was given the same attribute in being pictured as a spirit-being. And just as indigenous spirits participated in the cockpit, so Hispanic spirits were conceivably involved in the card games introduced by the Spaniards. In the gamble of November 1898, therefore, the Sota appeared as a Hispanic spirit that used its powers to intervene on behalf of a mestizo in waging a cosmic battle against Spain. In this political game of chance, the Sota rewarded Araneta with spectacular success. The seeming betrayal of Spain by a Hispanic being was highly significant, for it proved an alien entity's sympathy could lie with the native. The native, in return, localized the Sota, attributing to it the *kambang* attribute. Also considered the wealthiest of the unseen spirits, the Sota blessed Araneta with vast landholdings. The Sota is thus doubly emblematic of the spirit-world's relationship to *hacendero* political economy and of the efficacy of gambling.

After the Spanish surrender of Bacolod, the mestizo leaders tried to put in place a provisional government that was meant to last for a limited transitional phase, as indicated in the terms of the capitulation act, which were to remain in force "until the conclusion of the Peace Treaty between the United States of America and Spain" (Fuentes 1919, 60–61). Meanwhile, the Provisional Government, under instruction from the Revolutionary Committee of the Visayas, sought to secularize the state by claiming jurisdiction over cemeteries, by rendering free the use of church altars, and by having church collections entered in the municipal books of account (112–114). No longer would non-Catholics be deprived of a decent burial.

To deliver social services, provide public works, and maintain the island's large security contingents, the Provisional Government generated revenue by imposing—following defunct Spanish colonial state policies—a tax on gambling: 1 peso daily for each game table of *panguingue*, 20 pesos daily for a table of *burro* card games, and 200 pesos daily for a table of *monte* (Fuentes 1919, 157). The levies on these card games proved to be a lucrative source of funds. As recounted by Cornelio Fuentes, the people were "totally dedicated to gambling vices" during the transition period (162). As the most popular card game, *monte* "was so widespread in Negros it was a rare town indeed which could not boast of from four to five gambling houses where *monte* was played" (161). With little else to do as farm work was at a standstill, the people of Negros simply gambled away the interregnum.

Viewing Negros as part of a grand *Republica Filipina*, the planters declared on 27 November 1898 a "federal cantonal" government with the unwieldy name of *Gobierno Republicano Federal de Canton de Ysla de Negros*, composed of the island's two provinces.[60] Although the formation of this government disappointed revolutionary leaders in Iloilo and Luzon (Cullamar 1986, 40–41; Fuentes 1919, 142), the drive toward an autonomous Negros within a national polity was not deemed contradictory, as Negros' relationship to the former Spanish colonial state was conducted along exactly the same lines. In pursuing their sugar interests, the Negros planters learned the mechanics of formal allegiance to a central authority even as, individually and as a group, they found ways to make themselves paramount over the rules of this very central authority.

The *hacenderos* took an even more idiosyncratic action. Two weeks before the Cantonal Government was declared, its leaders dispatched a message to the captain of the U.S. cruiser *Charleston* anchored at Iloilo. The message of 12 November 1898 carried an unprecedented request. Seeking to avert Spanish reprisal, which might spoil their original gamble, the Provisional Government of Negros voluntarily placed itself under "the Protectorate of the Great Republic of the United States of America," heretofore a noncolonial power. The Negros elite sought U.S. protection to safeguard lives and properties from "foreign aggression"—as though the United States were one grand *datu* with whom they could have a compact. The reverberation of this ancient notion was discernible in that the invitation to the United States to impose its sovereignty was extended on the condition that Americans recognize the island's "internal independence" and that "limitations" were later to be agreed upon by both sides, as though they were equal negotiating parties.[61] The request for protectorate status from the United States reeked of naiveté, but it also bore the marks of a calculated move by a daring and intrepid gambler.

When representatives of the United States and Spain met in Paris starting on 1 October 1898 to negotiate a conclusion to the war, the American demand that Spain relinquish Cuba, Puerto Rico, and Guam was already well enunciated. Regarding Filipinas, the United States continued to deliberate whether islands other than Luzon should be acquired, and for how much indemnification. For their part, the Spaniards were trying to hold on to the Visayas and Mindanao. Apart from Luzon, the Americans deferred their decision on other parts of the archipelago until an inquiry on the conditions of the prospective colony could be completed. With the eruption of open hostilities in the Visayan islands, it became evident that Spain's claim of sovereignty was no longer tenable. On 31 October, the United States finally demanded the cession of the whole archipelago, a demand the Spaniards rejected and for which, in Europe, the Americans were seen as brutal aggressors (Le Roy 1914, 354–377).

Perceiving the uncertain fate of Negros, the planter-revolutionaries decided to gamble with their destiny. In a move born of the conviction that Spain was no longer a credible colonial master, and with the thought that they had been excluded from treaty privileges with the lucrative

U.S. market, the *hacenderos* decided to cast their lot. The Negros elite's request for protectorate status was meant to be their intervention in the power brokerage in Paris. The message it meant to deliver to Europe was that Spain had been resoundingly repudiated and that the United States, contrary to being an aggressor, was irrefutably welcome. Thus, despite possible objections from other Philippine elites, the Negros planters acted recalcitrantly on behalf of their island, whose interests they nonetheless recognized were inextricably linked to the Filipinas they had known.[62] A calculated move, the invitation to U.S. power over Negros was, in effect, an invitation to the entire breadth of the archipelago.

On 21 November, the Americans offered 20 million dollars and equal commercial rights with Spain for ten years in exchange for all of the islands, an offer the Spanish monarch, "moved by lofty reasons of patriotism and humanity" to spare Spain "all the horrors of war," accepted on 28 November 1898 (Le Roy 1914, 368–369, 373). The *hacenderos* believed in the magic of their message. Juan Araneta, however, was reportedly angry that Spain, having lost its "rights" to the archipelago, particularly Negros, could not "sell" its former colony to the United States (Fuentes 1919, 164). When the peace treaty was finally signed in Paris on 10 December 1898, the Negros planters relished the idea that their calculated gamble had won them a major stake. A decade after the imposition of American colonial rule, Negros planters began to reap their reward when initiatives pushing for the duty-free entry of Philippine sugar to the U.S. market finally came to fruition in 1909.

Babaylan Defeat and the Unleashing of Mestizo Power

Having outflanked the shamans in ousting the Spaniards, the planters immediately sought to exercise Mestizo Power by making Papa Isio's group toe the line of the elite victors. On 27 November 1898, Araneta reported that he had notified Isio to report to the Cantonal Government (Fuentes, 1919, 122). Four days later, Araneta issued a circular to his military chiefs and town presidents enjoining them to follow a "policy of attraction" to entice Isio's followers to "return to their occupations" (149). In a communiqué jointly signed by Araneta and Isio in Bacolod on 19 December 1898, the *babaylan* leader was designated "commandant" of the mountain district between Isabela and Jimamaylan and required to order those in his movement with outstanding debts to leave camp and return to the *haciendas* to fulfill their work obligations.[63] Although the sugar economy was virtually at a standstill, the planters sought to coopt Isio to contain the festering problem of labor control that had plagued the *haciendas* since the 1870s. An officer of the U.S.-organized Philippine Constabulary later reported that Isio visited Bacolod like a "visiting potentate," was "wined and dined" by the cantonal president, and was commissioned as "military chief of La Castellana," after which "his power" grew immense (Rivers 1908, 311). This evidently apocryphal story, which could only have been peddled by an angry elite, arose presumably because Isio reneged on his supposed agreement with Araneta.

Toward the end of February 1899—after the *hacenderos* had voluntarily raised the American flag in Bacolod—Isio's movement embarked on the second phase of its struggle. In repudiation of General Araneta, the *babaylans* organized a roster of military officers, with Isio as captain-general and two other leaders as generals. Isio's followers differentiated themselves from the planters by claiming to be the genuine revolutionaries, as they were the "first to shed their blood in the Revolution against the Spaniards" and that "when the Civil Guards and the Spaniards effected a blockade in the mountains, they were the first military leaders who rose in arms against the forces of the Spanish Government."[64] Although the planters had outwitted them in ousting the Spaniards, Isio and his followers were not cowed by Mestizo Power. And because they knew that the Americans were the superior force, they aimed their fight against the new colonial power.

In waging this battle, the *babaylans* adopted a stance of almost total identification with the Catholic Church, with Isio's group positioning itself to fill the vacuum created by the downfall of Friar Power. Deploying the lexicon of the Inquisition to resist U.S. colonialism, Isio's movement denounced the Americans as "Protestants," "heretics" *(ereges nga mga tao)*, and "enemy Jews" *(judious enemigos)*.[65] As had the defunct Spanish colonial state, they labeled the Americans *"Extrangeros"*[66] whose arrival disturbed the native quest for liberty. The American occupation was portrayed as a "temptation" to be resisted lest it result in the "loss of our body and soul."[67] The *babylans* had appropriated the Spanish equation of politics and religion for their own.

Consequently, the struggle against American rule was viewed as the defense of "the holy faith." Isio launched the anti-American resistance from the same mountain site he rechristened "Paradise" *(Paraizo)*.[68] For a fee of 50 centavos, Isio's followers pledged allegiance in a ritual in which, kneeling before a "holy cross," they submitted themselves to Isio's "seven commandments." They then took an oath, swearing by their version of the Philippine flag to defend the "holy fatherland" *(santa patria)* and serve Papa Isio as long as they lived.[69] To a holy war and loyalty to a charismatic leader Isio's followers committed themselves for life. In the same vein, Isio's group solicited contributions of 30 to 50 pesos from planters on pain of being considered "spies of the heretics," but looked kindly upon those who chose to unite "with the one holy faith."[70]

Scarce provisions, limited arms, and intensified operations by colonial forces locked Isio's group into a very tight situation. In early 1899, to protect planters and quash the *babaylans,* the California regiment was sent to Negros (Schurman et al. 1900, 180). In August 1901 a more organized force to defend the planters became available with the creation of the Philippine Constabulary, a native police force officered largely by Americans under the direct supervision of Manila's civilian government. By mid-1902, the constabulary in Negros Occidental had 8 inspectors and 253 enlisted men, in addition to Negros Oriental's 5 inspectors and 123 enlisted men (Taylor 1903, 217–218). Not only was Negros Occidental's the largest force in the constabulary's Third District, but it also possessed the largest number of arms, with 200 shotguns, 112 revolvers, and 290 rifles totaling 602 weapons (218–219). This large arsenal of state weaponry signified a level of military support never before experienced by the planters. Nevertheless, Isio defied capture until, on 6 August 1907, bowing to what he called the "mandate of God,"[71] he surrendered in Isabela and was brought to Luzon, where he died in prison in 1911.

Papa Isio had envisioned a theocratic form of government that essentially would have replicated Spanish rule, in contrast to the planters' move to secularize the state. In July 1900, for instance, Isio ordered the baptism of all infants for a fee of 12 *reales* to be paid by the godparents, "although the administration of the Sacrament should not have all the ceremonies."[72] Conjugal unions were to be solemnized and "matrimonial contracts" executed "in order that in this way the contracting parties are saved from mortal sin." Because of his theocracy and other reasons, Isio did not neatly fit the labels of "anti-foreign, anti-elite" subsequently given him by Renato Constantino, and neither did Isio's group suit Constantino's (1975, 272) celebration of the *babaylan* movement as "a civil war that was virtually a class war." Constantino's interpretations have found resonance, however, in the works of Fast and Richardson (1979, chap. 13), McCoy (1982b, 173–174; 1992, 120–122), Cullamar (1986, 71–72), and Larkin (1993, 137–140). Oddly, academic portraits of Papa Isio as "socialistic" have been based on Negros mestizo elite and American military reports that painted such a picture of him, in the process dignifying the earlier fears of the sugar-planter class and repackaging them as acceptable late-twentieth-century radical representations.

Although Isio was staunchly anti-American, he showed a remarkable respect for property, as is indicated by his instructions of July 1900. In fact, the edict opened with a strict injunction to the town presidents that they "must harmonize with the rich and powerful *(pudientes)* residents who can assist them in their work."[73] Isio ordered that a record of ownership documents be maintained to help settle property disputes, and that local authorities set a day when the town elite could gather to discuss the people's welfare. In the case of American supporters or *"Americanistas,"* Isio ordered the seizure of "all their money, fabrics and other possessions" but an inventory was to be conducted immediately, and any soldier who filched would immediately be killed and their "fall to hell" assured.[74]

Isio was absolutely against American rule, for he saw himself as the rightful successor to Spain. The government he would establish was to be the vanguard in defending a sacrosanct faith the natives had learned to call their own. Isio envisaged Negros and the Filipinas he would inherit, together with "brothers" from Luzon unshackled from colonial domination, in order that the people might be free to be like the Spaniards, and Isio to rule according to the Catholic ideals the friars had failed to uphold. In pursuing that substantively hybrid but emotionally pure dream, Isio thought there were those among the elite who shared his vision. In particular, he referred to Jose de la Viña, a Spanish mestizo *hacendero* who was initially a supporter of the anti-American cause, as his blood brother: *"¡Oh hermano mio en la sangre!"* Isio exclaimed in a letter.[75] Overall, Isio hardly conformed to images of him as a socialist waging a class war or as a nativistic leader longing for a return to an age of undiluted blood, antedating Spain, *haciendas,* and sugar. A product of his times, Isio was, however, committed to a noble cause, something that cannot be said about his "blood brother."

Jose de la Viña proved to be the cunning player. As the later turn of events showed, de la Viña was the complete *hacendero,* a master structural opportunist, the dreaded enforcer and exemplar of Mestizo Power. Jose was the eldest child of Diego de la Viña, "perhaps the most powerful *cacique* of Negros Oriental" (Romero 1974, 105).[76] A Spanish mestizo born in Binondo, Manila, but educated in the Peninsula, the older de la Viña was lured by the sugar industry to settle in east Negros in 1881. During the revolutionary period, Diego was Araneta's counterpart as Delegate of War in Negros Oriental. His son, Jose, also became a key figure owing to his imputed role in defending the eastern coast of the island.

On 23 November 1898, the Spaniards in Dumaguete boarded a warship and abandoned the provincial capital without a fight (Rodriguez 1983, 88–93). But fearing a possible counterattack by Spanish forces, Jose de la Viña led in guarding Negros' east coast. In one incident, a small vessel owned by a planter from the town of Bais was on its way to Dumaguete to deliver provisions when it was fired upon by a Spanish gunboat, forcing the crew to abandon the vessel and swim to shore. In what would be the province's only encounter with the Spanish navy, de la

Viña and his men engaged the gunboat in a brief volley of shots, after which the Spaniards withdrew. As in the fall of Spanish Bacolod, the successful defense was credited to a *hacendero* ploy. De la Viña and his volunteers were said to have mounted banana and coconut trunks along the shore to simulate cannons and held branches to look like rifles from a distance, a trick that was said to have caused the Spaniards to flee (112).

Jose de la Viña was subsequently imprisoned by the Provisional Government—according to him, without justifiable cause.[77] He later joined the anti-American struggle, and from him even Papa Isio's group sought assistance.[78] By 1902, however, a reversal had occurred such that the local interpreter of insurgent documents for the Americans could affirm that Diego de la Viña and "all his family prided themselves" as *Americanistas*.[79] In all likelihood, Jose de la Viña, like most other *hacenderos* in Negros, realized that his material interests were better served by cooperating with the Americans, whose acquisition of the former Spanish colony had become an ineradicable fact. Indeed, under American rule, de la Viña consolidated his landholdings and acquired a total of about 5,000 hectares in the mountainous expanse between La Castellana in the west and Vallehermoso in the east.[80] Desiring to retire from business in 1926, "due to old age," he decided to sell his 912-hectare Hacienda San Jose to the Spanish firm Ynchausti and Company (*Sugar News* 1926a). Dubbed as one of the best sugar plantations in Negros due to its location and fertile soil, the *hacienda* provided employment to 550 workers and produced 40,000 piculs of sugar annually. Involving half a million pesos, the sale of the plantation was "believed to be the biggest and most important business transaction" that year (ibid.).

No doubt fueled by his supposed anti-Spanish exploits, Jose acquired a tremendous spiritual reputation in the tradition of ancient magical individuals. He was allegedly capable of knowing a person's whereabouts even if the latter were to go into hiding, his body was invincible to bullets, he could reduce a person to dust as he could boulders, and he knew how to cure illnesses using herbs and potions. Immersed in the same anti-Catholic signs and intentionalities as Isidro de la Rama and Juan Araneta, Jose de la Viña reportedly scraped pieces off a crucifix during Good Friday for use as *anting-anting*. Like Araneta and the shamans, he climbed Kanlaon during Holy Week to augment his powers, commune with spirit-beings who were his friends, and collect medicinal objects. The inhabitants were stupefied and overwhelmed by the unspeakable powers of de la Viña. To be sure, there is some record that Jose, along with his father, Diego, and an explorer, Juan Mencarini, accompanied by a huge retinue, climbed the summit of Kanlaon in February 1896 (Marco 1912, 47). His later ascents to the fabled mountain, however, were part of a contrived ritual knowingly undertaken to cultivate an image of himself as a man of prowess. As one informant intimated, because Jose "knew the mentality of the people," he decided to "definitely fool" them.[81] De la Viña made an incisive mestizo reading of the *indio* mind and then proceeded to manipulate local beliefs in his rise to power.

The ascendancy of this brand of Mestizo Power occurred at the same time that the American colonial state gave planters legal access to firearms through a system of registration introduced on 25 March 1903. Three years later, there were in Negros Occidental as many as 1,019 firearms issued legally to individuals, including 624 revolvers, 187 shotguns, and 208 rifles (Philippine Commission 1907, 250). The commanding officer of the Third District was not exaggerating when he reported in 1904 that most planters in Negros Occidental were "well supplied with arms, some of them having 6 or 8 rifles," an estimate raised the following year to "ten or a dozen" rifles and shotguns (Philippine Commission 1906b, 88; Taylor 1905, 93). By the early 1900s, the planters of Negros possessed arms that collectively surpassed the strength of the official military establishment in the region. Never had they been deemed so trustworthy as possessors of firearms. From its foundation, the American colonial state appeared willing to forego its monopoly on violence to enhance the security of their native elite wards.

With unwavering support from the colonial state, someone like de la Viña could rule his estate as a petty tyrant. He reportedly kept seven kinds of whips for each day of the week, which a paid flogger applied, usually twenty-five times, to any errant worker guilty of such a misdemeanor as filching, after which the wounds were customarily doused with vinegar. Following Spanish machismo, de la Viña, who was no different from many other *hacenderos,* had a surfeit of women. In the gambling culture of Negros, some fathers played the social game by actually offering their young daughters to de la Viña, for which they received a carabao or a sum of money. Many, however, obliged out of fear, for whoever opposed the magical planter risked a dreadful punishment. An informant reported that some workers were summarily executed, then buried in an illicit cemetery inside the plantation.[82]

Jose de la Viña probably represented the extreme version of Mestizo Power, but the spiritual ascendancy and cultural hegemony of Negros sugar capitalists were widespread phenomena characterized by the general attribution of Masonry. In outmaneuvering the *babaylans* in the struggle against Friar Power, the planters conjured up the aura of the preconquest *datus* with their tantalizing magical endowments validated by the spirit-world led by Sota. The turn of the century served as the culmination of the planters' engagement in the realm of Masonic Capitalism, whose citadel of power had become the mestizos' own source of cosmic strength. Against such power Isio could not endure. Initially encoded in Isidro de la Rama, the Negros elite conveyed their Masonic prowess in the dramatic climax represented by Juan Araneta. It was, however, Jose de la Viña who epitomized the planters' adroit manipulation of indigenous culture as a brilliant stratagem of class power.

Brashly tempting the limits of fate and exuding the gambler's love of self-mastery and shrewd calculation, de la Viña and peers who followed his example fabricated situations to demonstrate their supposed magical abilities. To this end, some planters even resorted to asinine tactics like prying under the huts of workers to eavesdrop on private conversations, later using the information thus obtained to tantalize the workers.[83] Other planters relied on technologies unknown to the workers, such as binoculars and dynamite, to perform feats for which the natives had no explanation save *anting-anting*. The planter elite's spurious claim to spiritual wonders was given credibility by their generally aggressive *dungan*, their ruthlessness and the fear they evoked, their bravado and gambling, and the accumulated riches that were assumed to be a sign of commendation by the spirits. Mestizo Power was exercised with panache even by planters who did not engage in deliberate play with culture. Aware of the efficacy of the mirage that had been constructed around their class, many planters behaved in the mold of the strong *dungan* spiced with paternalism, in the classic manner of Juan Araneta. Contrary to the conventional understanding of the Gramscian concept of hegemony, in which forceful and violent forms of domination are resorted to when ideological and cultural techniques fail, a framework applied to cases of Southeast Asian peasant struggles (Turton and Tanabe 1984), the specific moment of the assertion of planter class hegemony entailed force and violence as a concomitant of the ideological claim to mythic prowess, buttressed by support from the American colonial state. It may even be said that forceful domination within the *haciendas* became an acceptable and legitimate possibility only after the mystical superiority of the mestizos had been securely set in place.

Not inconceivably, the *hacenderos* could have been cognizant of the tactic of manipulating spirit-beliefs deployed by foreigners like the American Worcester (1898, 263), who wrote of the following experience on a visit to Negros in the late 1890s: "when we boldly entered a country supposed to be dangerous, and went quietly about our business, the superstitious natives were, as a rule, not long in concluding that we had *anting-anting*, i.e. charms, which protected us from harm. After once getting this idea into their heads, they treated us with great consideration." The *hacenderos* would also have been acquainted with the Frenchman Paul de la Gironiere's book, which appeared in the 1850s and was considered "the best seller among books about the Philippines in the 19th century" (Legarda 1962, xv). In running his Hacienda Jalajala in the 1830s, de la Gironiere (1962, 48–49, 53) learned that natives "dislike cowards, but willingly attach themselves to the man who is brave enough to face danger," the lesson being "the imperious necessity of showing myself not only equal but superior in the struggle" for bravery. In what would later be echoed by Isidro de la Rama, de la Gironiere sought "to become master of myself" (49). His project succeeded, for the natives "always believed I was possessed of this secret [*anting-anting*], as well as of many others" (99–100). He stressed, however, that to befriend the bandits in Jalajala he "resolved" that "I must go amongst them, not like a sordid and exacting landlord but like a father" (43). The inroad of foreign merchant capitalists in the nineteenth century (and de la Gironiere was a friend of the firm Russell and Sturgis; Legarda 1962) was thus based on a simulation of the so-called attitude of power that reached its high point in Sturgis' Protestant wedding ceremony at Manila Bay. As outsiders, the foreigners were able to exploit and manipulate native culture.

The mestizo *hacenderos,* for their part, could objectify local beliefs because their cultural ambivalence had been further amplified by modern education and identification with "Protestant" merchant capitalism. They could also draw upon the Masonry movement, which reached its zenith during the revolutionary fin de siècle. But many mestizos, at the same time, had been socialized into the spirit-world, a belief acquired from received native culture. In learned derision of specific local beliefs as superstition while fearing the realm of spirits, the mestizo planters stood on the threshold of two worlds and two cultures, with one foot in the local culture and the other in the European. Mestizo cultural ambidexterity was reflected in their two styles of eating: elegant European-style dinner parties served with exquisite napery and imported silverware when with guests, and eating in the native fashion, using their bare hands, the older ones squatting on the floor chewing red cuds of betel, when by themselves (Stewart 1908, 41; White 1928, 118–119). Living in two overlapping worlds and knowing when to deploy the elements of their cultural armories, the mestizos found themselves in the social position to instrumentalize by playing and gambling with the cosmology of *hacienda* workers, often with marvelous success. The resulting accomplishments of Mestizo Power dwarfed the import of Sturgis' earlier contest with Friar Power.

The sugar workers were bewildered, stunned, and immobilized, even as their bonds of dependency on the sugar economy became even more difficult to untie. Some workers later attempted to retaliate against the oppressive conditions in the *haciendas,* but were overwhelmed on all fronts. As we shall see in the next chapter, the deepening commoditization of the economy, the shift to *duma-an* wage-based labor relations from around the 1930s, and the strategic sponsorship of sugar elite interests by the American colonial state left workers with few options but to tolerate and defer to Mestizo Power. In the sphere of the spirit-world, the favors rested with the planter class, whose mystified accumulation practices and mystique no magicoreligious leader managed to undermine. Individually, the recourse to sorcery known as *hiwit* proved inefficacious as well.[84] By wearing a diamond ring, the planters are said to possess an impermeable shield and powerful antidote to *hiwit.* All that remained of the workers' weaponry was their Masonic discourse and other forms of everyday resistance. As they had done in migrating to Negros Island, the workers took their own chances with life, though resigned to the inimitable relations between elites and spirits. For their part, the Negros sugar planters—with their cultural hegemony and newfound political power, the substantial arsenal of state violence behind them, and the economic bounty opened up under the aegis of America's "benevolent assimilation"—marched on to consolidate their domination as a class.

The American Colonial State:
Pampering Sugar into an Agricultural Revolution

The American Colonial Corporatist State

The advent of American colonialism at the turn of the century made possible the forging of a totally different kind of relationship between the sugar planters of Negros and the colonial state. Eager to pacify their new colonial subjects and subdue various sources of resistance, the neophyte American imperialists pursued this goal by relying upon indigenous elites who were provided with ample room to participate in colonial governance. Unlike the Javanese *priyayi* and the Malayan sultans, the Philippine elites were incorporated by the colonial hegemon primarily as politicians—publicly elected officials and national legislators—rather than as bureaucrats. Their power and privileges were rather peculiar in early-twentieth-century Southeast Asia, allowing them to bypass bureaucratic rationalization for a more direct plunge into the politics of state. Discarding the Spanish legal distinction between mestizo and *indio* and substituting the homogenizing category of "Filipino," these elites included as one of their most formidable segments the ascendant sugar capitalist class of Negros, whose members, at a very early stage of the U.S. engagement, acquired the belief that they played a strategic role in national affairs.

With the assistance of Colonel James Smith, a lawyer from San Francisco who commanded the U.S. forces in the island, a constitution written exclusively for Negros was drafted by a committee of planters. In transmitting the proposed charter to the U.S. president in May 1899,

Aniceto Lacson, titular head of Negros' Cantonal Government, promised on behalf of his fellow *hacenderos* to "endeavor to be worthy citizens of [the great American] Republic which is the model of morality and justice."[1] The military governor of the Philippine Islands, Major General Elwell Otis, was not convinced, however, that the people of Negros would be able "to maintain the character" of a republican government.[2] Addressing this concern, the American civilian governing council known as the Philippine Commission, in its desire to co-opt local elites, asserted that a civilian government for Negros Island would be "promotive of peace and quietness"[3] as long as an American was placed "in full control" (Schurman et al. 1900, 180).

The commission's view prevailed, and Otis forwarded the proposed scheme to Washington, D.C., with the explanation that the "people of Negros deserve great consideration," for in freely welcoming the United States they served as "the wedge by which the American Government has been enabled to split open" the resistance of the Filipino insurgents.[4] The Philippine Commission underscored the necessity "to give to the people of Negros as great a show of self-government as was possible" because it was "desirable to conciliate them."[5] The show went on in November 1899, when the planters of Negros elected a civil governor. However, contrary to the elite's naive expectations that this elected local official would have genuine authority, the civil governor played, at best, an advisory role to Smith, who was promoted to the rank of general and appointed as the island's military governor.[6]

With the pacification of the new colony, local civilian governments were established throughout the archipelago. Organized in May 1901, the provincial governments of Negros Oriental and Occidental, as in other provinces, featured an elected governor, a Filipino, and an appointed treasurer and supervisor, both of whom were Americans. The latter two formed the majority of the governing board to ensure that "ultimate control," especially in the matter of funds, contracts, and public properties, was in "American hands" (Schurman 1902, 59). The scope of native participation in electoral politics was expanded in January 1907 with the holding of the contest for seats in the first national Philippine Assembly. Totally unprecedented in Southeast Asia, the holding of national elections at the dawn of the twentieth century paved the way for the institutionalization of the congressional system in the Philippines.

The devolution of political power fulfilled the recommendations formulated by the first Philippine Commission chaired by Jacob Gould Schurman, the original architect of the policy of "the Philippines for the Filipinos." To the objection that even educated natives did not possess the American concepts of civil liberty and official responsibility, Schurman (1902, 99) countered that "the way men are trained in government" was to "get them in harness quickly and let them tug and sweat under the burden of national affairs." He suggested that Filipinos take control of their affairs even if they "should make mistakes, even if the officials sometimes squander, or even embezzle, local funds" (59–60, 99). After all, as political aptitudes are primarily "the gift of nature" and hence cannot "be donated," the native's capacity for government did not "necessarily mean capacity like ours . . . but merely capacity of some sort" (98–99). Schurman, who was also president of Cornell University, articulated a policy that may be designated as "racist corporatism," which embraced native planters and other elites within the American colonial state, in stark contrast to the exclusivist method of Spanish colonial administration.

No one had anticipated the complications that would ensue from Schurman's corporatist strategy. For example, even after Isio surrendered in 1907 and the last of the anti-American insurgents were vanquished, the Negros planters would not countenance any curtailment of their use of firearms. In 1908, the newly created Philippine Assembly passed a law prescribing procedures for the licensing and distribution of guns to farm hands in *haciendas*. Immediately, the planters of Talisay decried "the hindrances and difficulties" imposed by the law. McCreary, the provincial treasurer, proposed that the Provincial Board of Negros Occidental support the Talisay complaint, stating that the procedures had encumbered planters and disenfranchised them of their arms, which were their "private property." The board approved the motion and sent the protest to the Philippine Assembly, the Commission, and the governor-general.[7] Americans like McCreary demonstrated what Schurman's vision of American control meant in practice: unremitting support of the local gentry.

Accordingly, the Americans helped entrench a structure of private violence that started to become evident in the first half of the 1920s. At a time when, as we shall see later in this chapter, the sugar industry was modernizing and major stakes were up for grabs, electoral ambitions and personal animosities—cloaked in the guise of competing "mutual help societies"—erupted in what Larkin calls "Chicago-style gang warfare." In many towns the *Kusug Sang Imol* (The Strength of the Wretched) and the *Mainawa-on* (Merciful) had "masoniclike" local chapters that attracted a heterogeneous clientele reminiscent of the *datu*'s protectorate groupings that engaged in internecine feuding (Larkin 1993, 190–192).[8] By the 1930s, Negros was secure in its reputation for being "revolver-crazy," where "even mere '*cabos*' sport guns wherever they go" (*The Commoner* 1935c).

The case of gambling reveals in a graphic manner the dynamics of the new colonial state. Despite pressure groups spearheaded by Protestant ministers who sought to curb gambling, the history of gambling in the Philippines persisted as a virtually unbroken chain from the Spanish to the American colonial period. Eager to gain the support of their new subject population, and in connivance with some parasitic capitalists, American colonial officials even tolerated, at least initially, the proliferation of gambling. Changes in official gambling policy were introduced later. However, the enshrinement of electoral politics in conjunction with the congressional system, with Sota's approval of Mestizo Power in the background, augured the rise of the gambling worldview to hegemonic heights. This achievement has been emblematized in the peculiar Philippine English word for candidates, who are known as "bets."[9]

In the matter of horse races, for example, what used to be restricted under Spanish rule to two four-day meets every year had exploded, by 1906, into a splendid annual package of 220 days (60 percent of the year) of race-track gambling offered by the Pasay Jockey Club on Wednesdays and Thursdays and the Manila Club on Saturdays and Sundays.[10] From 7 to 8 million pesos were wagered annually at these clubs, which garnered at least 750,000 pesos from the 10 percent commission on bets, not to mention gains derived from rigged races. When the Chinese and the American Chambers of Commerce at Manila protested the negative effects on labor productivity occasioned by incessant horse racing, the Philippine Commission decided to restrict horse races to the first Sunday of each month, legal holidays, and three days before Lent, which totaled twenty-one race-track gambling days per year,[11] a substantially reduced package, which nevertheless provided more room to wager on horses than what the Spaniards had previously allowed.

In 1908 the city of Manila awarded a cockpit concession for the annual carnival. Notwithstanding protests by Protestant leaders, who even sought the intervention of the U.S. president, the carnival cockpit continued, as both Filipino and American officials had no objections to the "sport" and were equally in favor of the concession.[12] In the eloquent words of an American who wrote to the White House to counter the church ministers' appeal for a ban, "<u>Let the cocks fight</u>. A chance with the spur is surely preferable to the axe of the chef."[13] The following year the American colonial state devolved upon municipal authorities the regulation of cockfighting.[14] This course of action was admittedly "as much as could wisely be done in a matter in which practically the entire people of the Islands were so much interested it being, so to speak, a national sport."[15] Hence, instead of outright prohibition, which the Americans from their brief Cuban experience learnt could not be implemented, "the gradual educating of a people out of what is regarded as a reprehensible custom"[16] was hoped eventually to achieve what some desired: the eradication of cockfighting. On hindsight, it was too sanguine an expectation. By the 1930s, Philippine cocking was undergoing modernization, with the introduction of "Texas cocks" flown in by American airmen stationed in Central Luzon, the conduit for the modern breeding of elite cocks (Guggenheim 1982, 9).

Moreover, while such games of chance as *monte* and *jueteng* and lotteries were subjected to a blanket prohibition after 1907,[17] relegating the enforcement of gambling laws to local officials and functionaries was tantamount to tolerating them. It also meant that the American colonial state was being eroded at its foundation, as its legal system was easily flouted by local gambling interests in connivance with state actors. By 1910, officials of Negros Occidental, for example, noticed that the newly created municipal police forces were not implementing the antigambling laws, "either out of negligence or passivity."[18] The central government in Manila also observed that "some municipal treasurers gamble more or less habitually."[19] Since municipal authorities were deemed unreliable, the governor-general decided in 1912 to transfer the implementation of gambling laws to the American-officered Philippine Constabulary.[20] In a few years official files of the colonial state began to record the evident connection of local authorities to the gambling world. As an example, the extant provincial records of Negros Occidental for 1923 show that the police chiefs of Ilog, Cauayan, and Escalante were suspended for involvement in gambling, some as ordinary players, others in profitable collusion with gambling houses under police protection.[21] In the same year, the municipal vice president of Isabela was suspended for playing the novel game of poker, which was speedily classified as a "prohibited game."[22]

However, the success of the Philippine Constabulary was temporary; its discipline was unable to withstand the onslaught of the gambling mentality. By 1935 Bacolod was under the rule of someone cryptically named in the local press as the "*jueteng* emperor" and obliquely identified as a Chinese (*The Commoner* 1935a, 1935d, 1953e, 1935f). Despite requests by Bacolod's mayor for assistance from the constabulary commander, the "*jueteng* emperor" proved too elusive for the law enforcers. As *The Commoner* (1935e) insinuated, "some of our public officials are in league with this underworld character making his capture very difficult if not practically impossible." Seeking not to undermine the legitimacy of the electoral system, the hallmark of U.S. political tutelage, the American colonial state effectively turned a blind eye and did not prosecute any of these public officials, who, in any case, possessed robust connections with other Filipino elites staffing the various levels of the judiciary. The

Ilog chief of police mentioned above, for instance, was exonerated by the local Court of First Instance in the same year he was caught playing *monte* in the house of a municipal councillor, who also went scot-free.[23] By 1936, Filipino legislators were arguing that the eradication of gambling was "an impossible thing, and that it would be a waste of time, money, and effort to suppress that which is so deeply rooted in us" (*The Commoner* 1936d).[24] With the state apparatus as just one more arena for life's games of chance, the elite wards would gamble with the state to advance their interests, particularly in the sugar industry, which American colonial officials selected as a special focus for developmental efforts.

Statist Magic and the Revival of the Sugar Industry

Immediately upon its establishment, the American colonial state proceeded to find solutions to the ills of the moribund sugar industry, which was reeling under severe problems brought about by rinderpest and locust plagues, the dearth of agricultural credit, the absence of a profitable market, and the industry's generally antiquated techniques of production. The search for solutions to the expressed needs of sugar planters became an important preoccupation of the new Manila government, which established the Bureau of Agriculture in October 1901 to bring the weight of American scientific expertise to bear upon local sugar culture. In the same year, the Spanish model farm in La Carlota was converted into a college of agriculture built around an experimental station (Hayne 1904).

In the early 1900s, rinderpest decimated 75 percent of Negros' working animals and forced about two-thirds of the formerly cultivated lands to be left idle (L. Locsin 1904, 694). A lucrative black market emerged, brokered by local officials, but it could provide only a limited number of stolen carabaos from Iloilo, and at prices most planters could ill afford.[25] The drastic drop in production—from a peak 288,456 tons exported in 1893 to a measly 71,860 tons in 1900 (see Table 2)—pointed to a foreseeable collapse of the sugar industry. Considering "the problem of restocking these islands with draft animals" as "most serious," the colonial state in 1903 decided to import 10,000 inoculated carabaos from China, which were resold to planters at a subsidized price (Philippine Commission 1904, 23–25, 51). In 1910 thousands of carabaos were bought from Saigon "and eagerly bought up by the planters" (Diogenes 1910, 27). Although the epidemic abated, only in 1923 was Negros declared "free of rinderpest in the sugar cane districts" (*Sugar News* 1923e). Seeking another avenue to solve the problem of draft animals, the colonial state introduced tractors in 1919. The superiority of "power farming," as it was called, became immediately apparent. With some tractor dealer companies offering oil and fuel at cost to speed up the adoption of the new technology, a partial but steady shift to tractorization transpired in the cane fields of Negros (*Sugar News* 1921a, 1922g).

The colonial state also tackled the problem of locusts, which had besieged the *hacenderos* since the nineteenth century. From 1906 onward, the Provincial Government adopted measures that enjoined the population to aid in exterminating locusts, but inadequate funds hampered their implementation (M. Lopez 1908, 412). In 1922, along with several other provinces, Negros Occidental was hit by a locust plague that required 6,000 persons under the Bureau of Agriculture's direction to quell. The following year, the colonial state systematized the annual campaign of locust extermination by deploying a battalion of the Constabulary "trained in hopper fighting" (*Sugar News* 1922c, 1922e, 1923a).[26]

As the colonial state mobilized its resources to eliminate the blights to Negros' agriculture, sugar cultivation became a less risky gamble. More importantly, the power of deities to send locusts and plagues was curtailed by a seemingly superior realm, the colonial state. Although secular, the state apparatus exuded a spiritual prowess that allowed Mestizo Power to flourish in ways unimaginable under the Spanish dispensation. Planter dalliance with the Sota of Kanlaon, it would seem, was bringing to fruition a radically different world. Progress was magic.

To enhance geographic mobility within the island, the Provincial Government of Negros Occidental embarked on a road-building program with supplementary financing from the central colonial government. By 1913, the province had 102.25 kilometers of roads designated as "first class"; ten years later, the figure stood at 285.5 kilometers.[27] Although many rivers still had to be traversed on rafts, by the early 1920s there remained but approximately forty kilometers to connect the roadway of the west coast of Negros with that of the east (*Sugar News* 1923c). At the same time, more than 1,000 kilometers of railway had been built as an adjunct to the modern mills (*Sugar News* 1922b). To permit Negros to export sugar directly and bypass Iloilo, the Pulupandan port was constructed and further upgraded in 1926 (*Sugar News* 1922f, 1926e).[28] Under the aegis of American rule, the crude vessels that plied Negros' waterways as the only reliable means to convey sugar became a thing of the past.

An even more insuperable obstacle than the natural and infrastructural constraints to sugar production was the tight credit situation and the near bankruptcy of the *hacendero* class, which resulted in the number of cultivated farms declining from some 678 *haciendas* in 1905 to only 326 in 1907 (Yulo 1909, 388). Compounding the problem, many planters "took advantage of their creditors" who, as a result, became reluctant to advance money "except at exorbitant rates of interest and upon undoubted security."[29] Resurrecting an idea broached in the late Spanish era, the planters appealed for cheap credit to be made available through the establishment of an agricultural bank, what a local governor referred to as "the radical remedy for the parent cause of all our present ills" (L. Locsin 1904, 695). In 1907, the U.S. Congress authorized the formation of such a bank patterned after the British scheme in Egypt, the bank's principal to be guaranteed by the Philippine government with payments not exceeding $200,000 per annum. With private U.S. capital unwilling to invest in such a venture, the Philippine Assembly allocated one million pesos of insular government funds to capitalize the bank the following year (Nagano 1997; Philippine Commission 1906a, 22; Villanueva 1909, 391).[30]

The Agricultural Bank was engaged basically in a bailing-out operation, as most loans were used primarily to redeem the indebted planters, reclaim their mortgaged *haciendas,* buy work animals, and recultivate the abandoned farms. In the first half of 1910, the Provincial Board of Negros Occidental recommended approval of three loan applications totaling 46,200 pesos. The biggest loan was extended to Juan Araneta, who borrowed 25,000 pesos payable in eight years: 4,000 pesos to purchase carabaos; 5,000 pesos to redeem mortgaged property; and 16,000 to acquire mill equipment, trolleys, and wagons.[31] In 1913, the board recommended an aggregate 1,834,300 pesos to be conceded to 112 loan applicants, with the mean at nearly 16,400 pesos and the maximum individual loan placed at 35,000 pesos.[32] Negros Occidental accounted for under a quarter of all loans approved, but obtained nearly half the value of all loans granted throughout the country (Nagano 1997, 319). As the monetary lubricant freely flowed into Negros in 1913 at the low interest rate of from 8 to 10 percent, and with duty-free access to the U.S. market, the *haciendas* began to come to life again. The reestablishment of the sugar industry's viability was evinced by the marked increase in land values, which jumped six times from 1911 to 1913.[33] By provisioning the planters with capital, the American colonial state sponsored the great twentieth-century revival of the sugar economy of Negros.

More funds were needed, however, but the Agricultural Bank's originally authorized capitalization had already been exceeded by about twice the amount of loans drawn. In June 1913, Governor-General Cameron Forbes ordered that a further 1.5 million pesos of state funds be deposited in the privately owned Bank of the Philippine Islands (BPI) "to be used exclusively . . . for the specific purpose of making loans on growing crops to tide the Hacenderos of Negros and Panay over the present crisis."[34] Whereas the Agricultural Bank required real estate as collateral, the BPI was instructed to issue loans on future crops.[35] This served to inaugurate the system of crop advances dispensed by the Philippine state, which took over the role hitherto played by private merchant houses.

In 1921, the BPI crop loan system was instituted at the Philippine National Bank (PNB), which was formed in 1916 on the ruins of the Agricultural Bank. Given the little provision for inspection, the system was described by a critic as "not conducive of best results." Even under the most favorable growing conditions, crop loans were not liquidated, because "planters hopelessly in debt" tended to "overdeclare areas placed under cultivation and not to apply crop loans fully for the purpose for which they were granted" (Keller 1926). The Insular Auditor similarly observed that planters converted "demand" loans into "personal" loans, which the PNB nonetheless liberally dispensed.[36]

In bringing the game of sugar under state auspices, Forbes justified his 1913 directive as a stimulant for the industry: "The fact that the government was loaning money restored confidence, and money lenders began to loosen up. The low rate of interest . . . set the pace for the other owners of money to lend, and the whole thing resulted in bringing out the necessary money to see the sugar crop in Negros through."[37] Ten years later, the BPI was loaning from three to four million pesos in crop advances to planters from Negros (Zafra 1926, 747). Although the secretary of war objected to the government's involvement in private business,[38] the financing policy espoused by the American colonial administrators of the Philippines gradually but irretrievably led the state to relinquish any semblance of neutrality and laissez-faire in order to prop up the sugar-planter class.

To upgrade the industry, the Bureau of Agriculture suggested in 1902 that modern sugar factories capable of producing centrifugal sugar needed to be built to replace the obsolete muscovado mills, which extracted only about half the cane's sugar content (Welborn 1908, 804). Centrifugal sugar also fetched a higher price in the world market because it was demanded by manufacturers of white refined sugar (Pitt 1909, 12). However, until the late 1910s and even with the passage of the Tariff Act of 1909, which allowed 300,000 tons of Philippine sugar to enter the U.S. duty-free, the country continued to export muscovado sugar to China, in part due to the war in Europe.[39] Despite widespread reluctance among planters to abandon the muscovado mills, mounting pressure to produce centrifugal sugar was felt by 1912, when the Iloilo brokers stipulated the quality of sugars they were willing to handle, with the lower grades consigned to China "or remain in godown and beg its market" (*Philippines Monthly* 1912).

As the planters did not have the requisite funds to industrialize, the bureau suggested they try "to interest outside capital in an enterprise that, more than any other, will overcome the present regrettable conditions" (Lyon 1903, 597). Observing the fractiousness of the planter class, the bureau also recommended that "the initiative and the organizing ability must largely come from the outside" for the *hacenderos*—whose farms were individually too small to each house a large modern factory—were "generally suspicious of each other" and had little inclination "to organize and cooperate in a way to make central mills a success" (Welborn 1908, 806). The idea of inviting outside capital to catalyze the modernization of the sugar industry obtained the assent of the high colonial administration. Some planters cautiously welcomed the idea of attracting American capital if it would restore Negros' agriculture to its "former prosperity" (Jayme 1906, 367). However, the contradictions that beset American colonial policy toward the Philippines stalled the march of foreign capital, even as local planters succeeded in diverting more resources of the colonial state to further promote private interests in the sugar industry.

Skillful Play and the Advent of Centrifugal Milling

The principal disincentive to foreign capital was the restriction on the amount of land that could be acquired from the public domain, a legal impediment that favored American beet-sugar producers who sought to forestall potential competition from Philippine cane sugar.[40] When the Philippine Organic Law of 1902 was under deliberation in the U.S. Congress, Governor-General Howard Taft recommended a maximum allowable land limit of 25,000 hectares. But with the debate couched in protectionist and antitrust terms, Congress agreed to lower the land ceiling to 16 hectares for individuals and 1,024 hectares for corporations. No similar restrictions were imposed in Cuba, Puerto Rico, or Hawai'i, places where American sugar capital predominated. The Philippine Commission repeatedly tried to secure from the U.S. Congress a modification of the public land law, but to no avail.

In 1916 the U.S. Congress granted the Philippine Legislature authority to enact land laws, but the 1902 restriction was not altered by the Filipino elite, whose basically mestizo economic nationalism coincided nicely with American protectionism.[41] Subsequently, the restriction on land acquisition by foreign companies was expanded as a result of the establishment on Mindoro Island of a sugar mill-plantation complex by a group associated with the U.S. Sugar Trust, which tried to cash in on the limited duty-free entry of Philippine sugar to the U.S. market. With the approval of Washington's Bureau of Insular Affairs in circumventing a legal technicality,[42] the Mindoro Sugar Company acquired the 22,484-hectare Recollect friar estate in San Jose, Mindoro, where in 1910 they built a centrifugal mill with a rated daily capacity of 1,100 tons. To prevent foreign capital from making a similar land purchase, the Philippine Legislature in 1916 expanded the landownership restriction to include former friar estates. The more leeway the Filipino politicians enjoyed, the more they acted like the old Spanish colonial state.

The Mindoro venture turned out to be a colossal failure. In 1929 the Mindoro plantation had to be administered by the Catholic Church–owned Philippine Trust Company, to which it owed about 10 million pesos (*Sugar News* 1929). In addition to triggering protectionist sentiment, however, the Mindoro mill also served as a showcase of modern technology for the Negros planters to emulate,[43] spurring local capital to build modern, though diminutive, mills with capacities of from 100 to 500 tons per day only. By 1914 five such factories had been constructed in Negros, three of them owned by the de la Ramas.[44]

In 1913, the U.S. Congress allowed unrestricted duty-free admission of Philippine sugar. The following year, American-Hawaiian capital established a foothold in the colony by setting up the first successful large modern mill (with a rated capacity of 1,200 tons per day) in San Carlos, Negros Occidental.[45] Built as an entity separate from cane growing, the San Carlos mill became the centralized manufacturer of sugar from canes supplied by the surrounding tributary farms, whence about twenty muscovado mills had been displaced (Henry 1928, 729). The prototype of a sugar *"sentral,"* as centrifugal factories came to be known, the San Carlos mill entered into thirty-year contracts with *haciendas* that promised to provide ample cane to ensure full-capacity operation during the milling season. The contract stipulated the equal division of the sugar output between the mill and the planters. Following the San Carlos pattern, a 1,600-ton-per-day mill in Canlubang, Laguna Province, on Luzon Island also became operational in 1914. Four years elapsed before another American business concern became involved in a similar mill-*hacienda* tandem. Nonetheless, the massive inflow of American capital was not forthcoming, especially as the retrenchment of European capitals to finance the war at home, inter alia, made Latin America a more attractive ground for U.S. investments compared to the Philippines (Calleo and Rowland 1973; Krenn 1990). The American colonial state was thus compelled to assume the role of industrializer in addition to its money-lending function.

Earlier in 1911, in his address to the Philippine Legislature, Forbes had recommended that, as governor-general, he be "authorized to purchase bonds of companies organized by proprietors of sugar estates in provinces where there are considerable numbers of sugar haciendas, for the purpose of assisting the owners thereof in building sugar centrals, under terms and conditions which will amply secure the government in the matter."[46] In a show of unflinching support to the planter class, Forbes articulated the possibility that massive state resources could be utilized to establish modern sugar factories.

Coincidentally, in 1912, the colonial state was forced to render a legal opinion on various raffles and lotteries known as *rifa, kalog-kalog, beto-beto,* and an assortment of parlor games that proliferated in fairs, churchyards, corner stores, and so on in evasion of the gambling ban. In a remarkable continuity with legal opinion during the late Spanish period (Bankoff 1991, 276–278), the American colonial state decided on what was thought to be a fairly clear-cut and implementable legal distinction: *(a)* if the result of any game depended "wholly or chiefly upon chance or hazard . . . there would appear to be little doubt that it would come under the prohibition of the law as a gambling game"; but *(b)* in cases when "the element of skill may enter to a sufficient degree" and the "ability of the players" was employed, it would not fall within the prohibition.[47] Unknown to the American colonial bureaucrats, the indigenous concept of gambling had always involved "ability" and "skill" in the shrewd negotiations with nature, destiny, history, and social structure. In that sense, state-defined gambling would not be gambling at all. The same cunning was in fact deployed in beating the gambling prohibition discussed above and, more significantly, in pursuing the concept adumbrated by Forbes in his 1911 address.

Philippine sugar interests conjured the urgent need to protect the planters allegedly from domination by foreign capital—the same fear once expressed by late Spanish officials like Remigio Molto. Vociferous demands that large milling facilities be built using state resources were made. As Negros' Hermenegildo Villanueva, in introducing the bill that created the Sugar Central Board, unctuously orated in 1914:

> undoubtedly, we will with pain see [Filipinos] resort to the last and suicidal course of transferring their property to alien hands which, to our sorrow, control now with their capital our commercial and agricultural field. . . . Yesterday Philippine sugar growers were struggling . . . against the encroachments of [foreign] capital . . . so as not to see their holdings pass into other hands, hands not calloused by the labor of cultivating the lands. . . . But now . . . our producers have no remedy except to succumb. . . . But before this happens, I believe it the duty of the Philippine Legislature to forestall the direful consequences entailed by the critical

condition of our sugar production, through the erection of sugar mills in the different points of the archipelago. . . .[48]

The creation of the Sugar Central Board in February 1915 served as the first step in diverting colonial state funds to erect privately owned centrifugal mills. The board, however, was allocated 2 million pesos only for investment purposes, which was not enough to guarantee in notes or securities the full cost of more than one sugar factory.[49] By March 1916, no mill had been constructed, and the planters, seeing the war-induced hike in sugar prices in Europe pass them by, denounced what they charged were the government's empty promises.[50] Convinced that the issue of the sugar mills was "one of general political concern," American officials circumvented the board's financial constraint by establishing the Philippine National Bank (PNB) in July 1916.[51]

The Sugar Central Board nonetheless had laid down the policy that government financing was applicable only in cases where "a majority of the landholders" contracted to supply cane to the sugar factory also constituted the stockholders in the corporation that owned the mill. This so-called cooperative principle along the lines proposed by Forbes was meant to ensure compatibility of interests between *hacienda* owners and the mill. The board also launched a system by which the governor-general's office and the Bureau of Insular Affairs intervened to secure for Philippine sugar the best possible price in the U.S. market, later systematized with the founding of a PNB branch in New York City to handle sugar marketing for the bank-financed mills.[52] The entanglement of the colonial state in private commercial enterprises deepened in a manner not anticipated by its original intent of assisting planters who could not afford to build large sugar factories.

Between 1914 and 1920, three small mills were added to Negros' four Filipino-controlled sugar factories.[53] During the 1920s, two more privately financed mills were erected in Negros.[54] These moves were hailed as "the great awakening of the Filipino capitalists," and the local elite was panegyrized in the Manila press as "taking hold of these large propositions and investing their money to improve the country and . . . retain the control of the industries themselves" (*Manila Daily Bulletin* 1918). The mills had become elite nationalist symbols. However, the recourse to economic nationalism became redundant as the much vaunted encroachment by American capital did not come to pass. Since the board's creation, only two large mills had been constructed by American capital.[55] At the end of the day, Negros had only two American-Hawaiian–owned mills, one in San Carlos, the other in Silay. But the nationalist rhetoric persisted, for it was strategic to the elite's skillful play. With the

Philippine National Bank under the Board of Control composed of the American governor-general and two Filipinos (the senate president and the House Speaker), the PNB became the instrument to put foreign capital at bay.

By 1922, aside from providing loans to upgrade four existing factories in Negros,[56] the PNB had sunk a staggering 40 million pesos, about $20 million, in investments to finance the construction of five large "Bank Centrals" in Negros and one in Pampanga.[57] In the PNB's contractual scheme, the planters who supplied cane were considered mill owners and were required to use one-fourth of their net profits to purchase stocks in the corporation that controlled the mill. The repayment of the loans, however, was not forthcoming and, by December 1922, interest on the principal of the six Bank Centrals had accrued to some 3 million pesos, excluding another 5 million pesos owed to other parties.[58]

Under Venancio Concepcion, a general in the revolution against Spain, the PNB made these astronomical loans "knowingly . . . in violation of the law," as American officials admitted.[59] Since the PNB charter limited the agricultural loans to half the bank's capital stock and surplus, "The records are clear that these amounts were knowingly exceeded and . . . that this fact was known to every one connected with the Bank."[60] As the single largest investment of public funds that used up the country's gold reserve, the PNB loans were "improvident," for their lending "overshadowed the question of a return on its investment."[61] Highlighting the gambling aspect of the loans was the bank's deliberate neglect to safeguard colonial state monies from possible loss. The PNB accepted the planters' already heavily mortgaged properties as collateral by employing a deceptive formula. The value of sugar lands with centrifugal mills in Cuba and Hawai'i, three times the value of local sugar lands, was used to reestimate the net worth of local *haciendas*, the inflated value then becoming the basis for the loans extended by the PNB.[62] Not surprisingly, Concepcion's policies were praised by planters as "daring" and "heroic," his "fighting spirit" and "cold determination" considered worthy of emulation, especially during those "days of passionate and triumphant nationalism."[63]

The PNB floundered: 11 million pesos were written off as bad debts, but the remaining indebtedness still exceeded the actual value of the properties mortgaged to the bank, which was in no position to enforce a lien.[64] By 1926 the debts of the Bank Centrals had escalated to 48 million pesos, with total advances reaching 76.6 million pesos as of April 1927.[65] It was their good fortune that, with the surge in sugar prices in 1927, the mills could begin to pay off their debts.[66] And the sugar personalities were far from apologetic. To underscore their point, some planters organized a "bridge party" for the PNB president, who visited Negros in 1935, "to give bridge fans a chance to collect something from him and to make him realize the debtor's point of view when it comes to collection of accounts" (*The Commoner* 1935b).

In investigating the PNB fiasco, the Americans felt "powerless" in the face of "a carefully worked out and extensive system of graft all of which has its center or head in Concepcion," who oversaw the organization of companies that secured, through unfair advantage, the contracts for mill construction and supplies acquisition.[67] The estimated graft at each Bank Central amounted to not less than 100,000 pesos per year. Except for Concepcion, who was incarcerated, no one else was held accountable. This pattern of skillful gambling within the state apparatus was replicated in the postcolonial period through even more elaborate and sophisticated systems of corruption. With the aid of numerology and secret Swiss accounts, state gambling reached its apogee during Ferdinand Marcos' incumbency. In the immortal words of a Filipino politician, "What are we in power for?"

Concepcion's defense was to blame the bank's financial straits on force majeure, the effect of the global crisis after cessation of the war in Europe (Concepcion 1927, 17–20). He argued that the gargantuan amounts lent to planters "on the verge of ruin" were absolutely necessary if "giant commercial progress" were to be achieved by Filipinos (15–16). The law creating the PNB, he argued, was designed to assist "our business, our industries, and especially our agriculture—and by using 'our' I mean businesses capitalized, directed and dominated by Filipinos" (4). Calling the Philippines "a country of paradoxes," Concepcion bewailed why "the National Bank, whose primordial purpose for its creation was to give impetus to the economic activities of Filipinos, should be subjected to financial canons completely alien" to local business (5). Domestic capital wanted to play by its own rules; after all, it was "their" country.

The local elite also sought to rid mill management of those foreigners whose presence served to hamper illicit activities. In 1921 an attempt was made "to eliminate American management from all the centrals," but it did not prosper.[68] To minimize graft and improve mill efficiency in the wake of the Concepcion scandal, the PNB examiner confidentially recommended that "a competent American [manager] who will be in absolute control of operations" be installed in the Bank Centrals, along with "a competent auditor."[69] Unable to implement this suggestion, a new PNB management decided in 1921 to create the Philippine Sugar Centrals Agency to supervise the operations of the Bank Centrals.[70] Nonetheless, foreign personnel were eased out. In early 1924, apparently to stave off an investigation, the "stockholders," aided by armed thugs of the *Kusug Sang Imol,* raided the Binalbagan Central and dismissed the American and European staff (Larkin 1993, 157–158, 192). The *Sugar News* (1924d) reacted sternly, stressing that "individuals and capital have certain definite rights which cannot be tampered with so long as these Islands are under American sovereignty."[71] By 1927 there had been an "almost complete elimination of American personnel."[72] Unmindful of the costly consequences of mismanagement, domestic capital adamantly insisted on its own way. About seven years after their construction, the state-sponsored mills were in such a "deplorably run down condition" that repairs to bring them up to par were estimated to require 2.5 million pesos.[73]

Resisting "Exploitation": The Triumph of Filipino Capital

With the Republicans recapturing the White House and a new governor-general, Leonard Wood, appointed in 1920, the policy of extricating the colonial state from the Bank Centrals was announced. The unpopular Wood justified his policy by pointing out that "too much of the public money has been put into these sugar centrals and that the large investments in the two or three sugar provinces have curtailed the necessary aid to other provinces and limited our extension work in public education, public health, public works, etc., etc."[74] Wood's argument resonated in certain sections of the Manila press, which reported that even the 1 million pesos deposited by the Postal Savings Bank with the PNB were used for "rich merchants borrowers" [*sic*], depriving "farmers who are more badly in need" of financial aid (*Manila Daily Bulletin* 1925). Certainly, the Philippine population as a whole was made to subsidize the industrialization of sugar, and Wood's policy conceivably might redress the injustice. In response to the policy, two serious offers to acquire the Bank Centrals were made: the first in 1923 by the consortium of Hayden, Stone and Co., New York bankers, and E. Atkins and Co., sugar operators in Cuba; the second in 1927 by the group of Hallgarten and Co. and the J. Henry Schroder Banking Corporation. Both attempts failed to transfer the Bank Centrals to private American capital.

In the first place, Wood's policy never received the support of Washington, D.C.'s Bureau of Insular Affairs, headed by Frank McIntyre. In October 1922, before any serious offers had been made, McIntyre admonished Wood not to be "too critical of the method in which the investment was made" because, "if these centrals did not exist, you would be forced, as have all of your predecessors, to try and build them."[75] The bureau chief stressed that "The building of these sugar centrals has been one of the most successful acts of the Philippine government since we have been in the Islands."[76] Wood replied, "Do not be unduly concerned," adding that "things are [not] being hurried in the way of liquidation of the bank or disposition of the sugar centrals."[77]

Although the Bank Centrals' total indebtedness was rising, McIntyre firmly believed the $20 million investment would be recouped eventually. For him, there was no urgency to sell, especially because the offers were made on a 50 percent basis of book value (replacement value was estimated at half the original construction cost, which was extraordinarily high due to wartime inflation). To accept a loss of $10 million would be "unfair," "unwise," "embarrassing to the government," and would unduly antagonize "public opinion."[78] Even formally to consider a losing proposition, McIntyre envisaged, might prompt planters to evade paying their debts or, worse, lead them to make identical offers the Philippine government might not be able to turn down.[79] McIntyre's optimism was confirmed by the highly profitable global sugar market in 1927, which improved debt repayments to the PNB. In the same year, the secretary of the treasury decided that the Hallgarten offer was prejudicial to the PNB's interests and advised against accepting the offer.[80]

Another obstacle to the liquidation of the Bank Centrals was resistance on the part of the planters, whom Wood referred to as "the sole beneficiaries of the great investments the Government has made in Negros, which, by the way, are limited to a small number of families."[81] As legal owners despite their minuscule equities, the planters' goodwill, especially those of the few dominant families, was necessary, for, as Wood admitted, "nothing can be done in the way of sale which has not the approval of the hacenderos."[82] But such an approval was not forthcoming. In February 1923, Filipino legislators passed a resolution emphatically declaring that any measure to dispose of the Bank Centrals "would mean transfer of control to foreign hands," which would lead to the "denationalization of Philippine industry" (*Sugar News* 1923d).

In alliance with Filipino planters was a small group of Americans entrenched in the Philippine sugar industry who reacted to Wood's divestment policy by playing on nationalist fears that "it would only be a short time before their plantations would be absorbed by the 'foreigner.'"[83] (Note that, unlike the Spanish, American colonials saw their fellow Americans as "foreigners" in the Philippines.) Even the American official who succeeded Concepcion at PNB was concerned to "insure the ownership of these centrals by the planters, and that in becoming the owners, they would not be exploited."[84] Another American lamented that it was "too bad some plan cannot be worked out to organize a Philippine corporation and let the planters themselves get the benefit of it, without turning them over for exploitation to the same group that dominates the Cuban situation."[85] Surely, these expressions of support by sympathetic Americans warmed the planters' bosoms.

The American–Filipino planter alliance became strategic in blocking Wood's attempt to change the PNB's board of directors that would have carried out the liquidation policy. In a fait accompli, the ailing governor-general was outmaneuvered by "high sources pretending loyalty, but really determined to defeat his plans."[86] A new board was formed and immediately decided against the Hallgarten offer. The Insular Auditor bewailed, "we are drifting into an impossible situation—politics—politics and more politics," adding that "the National Bank has been the most conspicuous failure, and, further, it has been, and inevitably will continue to be, a most effective instrument for continuing and maintaining in power the few political leaders who dominate the government."[87] The coup de grâce to Wood's policy came in August 1927, when the Philippine Legislature passed a resolution that no sale of "government properties" could be made without its prior approval.[88] Finally, in December 1927, another offer was made, one that did not "contemplate taking these Centrals away from the planters"; the promise was also made not to involve anyone "interested in the production of Cuban or other sugars."[89] But the offer did not prosper, as it came at a time when American sugar capitalists in Cuba and on the mainland were making moves to restrict the entry of Philippine sugar to the United States. In March 1928, a joint resolution was introduced in the U.S. Congress that the annual duty-free admission of Philippine sugar be limited to 500,000 tons.

However, the unrestricted entry of Philippine sugar lasted until 1934, allowing two decades from the date of the free admission act of 1913 for the American colonial state to oversee the sugar industry's modernization. By 1929 the Philippines was producing 667,263 tons of sugar, 85 percent in centrifugal form (Fairchild 1929, 18). With Negros accounting for half the total number of modern sugar factories in the Philippines, the island produced 69 percent of the country's centrifugal sugar. A veritable agricultural revolution in Negros had displaced some 568 muscovado mills extant in 1912, replacing them with 19 modern centrifugal factories by the early 1930s. Of these, only 2 mills were controlled by American interests and only 3 small and 2 large modern mills were owned by Spaniards. Negros' modern mills were predominantly in Filipino hands.

By the early 1930s, the land area devoted to sugar cultivation in Negros had expanded more than two and a half times from the area cultivated before the 1920s (see Table 9), and Negros itself accounted for nearly half the total hectareage planted with sugarcane in the country

(Gordon 1933, 609–610). Before the effects of the Great Depression began to be felt in the colony, therefore, the American colonial state had ushered Negros Island and the Philippine sugar industry into the modern era. The mill-*hacienda* arrangement, however, obstructed economies of scale, and the smaller farms failed to optimize sugar yields per hectare (Ilag 1964; Schul 1967; Wernstedt 1953, 159–160). Average sugar yields in the Philippines fell below world standards (Larkin 1993, 155). Market distortions and the dispersed control over land restricted the industry's aggregate productivity and overall efficiency, but this less-than-desirable state of affairs seemed to suit the individualistic preferences of the sugar planters.

On the whole, the strategic participation of Filipino interests in the colonial state prevented the dominance of American capital. By 1935, although the largest American investments had been made in sugar (equivalent to twice the amount invested in the coconut industry), American capital accounted for a small 10.5 percent and Spanish capital only 9 percent of total investments in the sugar industry (Table 8). Filipino capital controlled 79 percent of all sugar investments. Filipinos held most of the sugar lands. Except for some inconspicuous *haciendas* owned by a few former U.S. Army soldiers in the remote southern part of Ne-

Table 8.

Capital Investment in Selected Philippine Industries, Proportion Invested by Citizenship, 1935 (in pesos)

Industry/ Citizenship	Land and Improvements	Mills, Refineries, etc.	Totals
Sugar	362,640,000	168,100,000	530,740,000
Philippine	94.0%	47.4%	79.2%
American	3.0%	26.7%	10.5%
Spanish	2.0%	23.8%	8.9%
Others	1.0%	2.1%	1.4%
Coconuts	418,640,000	23,790,000	442,430,000
Philippine	93.0%	7.6%	88.4%
American	4.0%	46.6%	6.3%
Spanish	2.0%	4.4%	2.1%
British	—	29.4%	1.6%
Others	1.0%	12.0%	1.6%

Table 8. (continued)

Industry/ Citizenship	Land and Improvements	Mills, Refineries, etc.	Totals
Fibers	374,300,000	15,630,000	390,130,000
Philippine	94.2%	12.9%	90.9%
American	2.9%	50.2%	4.8%
Japanese	1.9%	9.6%	2.3%
British	—	18.0%	0.7%
Others	1.0%	9.2%	1.3%
Tobacco	41,990,000	18,500,000	60,490,000
Philippine	97.0%	1.0%	67.6%
Spanish	2.0%	65.3%	21.4%
Others	1.0%	33.7%	11.0%

Source: Philippine Statistical Review 1935a.

gros Island, American capital owned hardly any sugar lands.[90] In sugar manufacturing, Filipino investments reached a 47 percent plurality, while American capital accounted for 27 percent and Spanish capital 24 percent (Table 8).[91] In terms of actual production, Filipino-controlled mills produced 51 percent of aggregate centrifugal sugar, American-controlled mills accounted for nearly 28 percent, and Spanish mills for 20 percent (Fairchild 1929, 8). Toward the end of the decade, as access of Philippine sugar to the U.S. market was narrowed, a number of American interests divested and their equities were taken over by domestic capital.

The local sugar capitalists could allay their fears of American domination, a theme that later resurfaced in the work of Constantino (1975, 300–301), whose bold strokes dramatized "the triumph of American business," which he claimed was characterized by the "prompt and strong" response of American traders and investors "to the lure of profits in the new colony." Contrary to that scenario, sugar—the most heavily capitalized industry—presented the paradox that, because of the contradictions of American colonialism, American capital had no hegemonic position in the Philippine colonial economy, particularly in Negros. Mestizo Power had reason to be exultant over sugar.

Preparing for Independence: The Miller–Planter Struggle

The American colonial state's intervention on behalf of elite sugar interests proved deleterious to the avowed high mission of political tutelage preparatory to the full independence of the Philippines. Policies ranging from firearms to technical change, from gambling to crop loans, from lobbying for a lucrative market in the United States to underwriting modernization, taken together, schooled the elite in the intractable tradition of gambling within the state. Corroborated by the PNB experience, the laxity, graft, and pliability of the legal system—justified passionately in the nationalist rhetoric—perpetrated the Spanish legacy of corruption and cultivated the idea of the state apparatus as totally manipulable. As in the stratagems used to stave off American capital from the Bank Centrals, nationalist defiance came in handy as a tool for the promotion of ruling-class interests.

However, with the U.S. Congress under the sway of American agribusiness, political independence became a distinct possibility and, with it, the loss of the U.S. market, prompting the Filipino elite to amend their nationalist ploy. The first half of the 1930s was spent in legislative wrangling in the United States and in the Philippines, which was ultimately settled by bartering political status for a fixed sugar quota (equivalent to 982,000 short tons raw value) and a graduated increase in the export tax, with full duty to be imposed upon independence—a deal first rejected by the Filipino elite in 1933 and then forlornly accepted in 1934 as the inexorable death knell of the sugar industry.[92] Because the U.S. tariff had shielded Philippine sugar from the rigors of international competition, it was believed the industry would not survive without U.S. support. There were forebodings about the massive closure of sugar mills[93] and apocalyptic visions of disaster striking "within a few hours after independence is granted.[94] The end was imminent, the days of easy living slipping out of grasp. But before it came, there was a mad scramble for the remaining surplus to be gained, with the agitation centered on the mill–*hacienda* relationship. The sugar capitalists applied the tactics used in resisting foreign capital to their intraclass clashes, in what an American official aptly termed as "squabbling over the profits to be made before independence" (Graham 1939, 371).

Whereas the concept of the centrifugal mill as a centralized factory for the processing of canes grown by the adjoining farms was officially described as a "cooperative" venture, the mill–*hacienda* relationship, from the outset, was seen in oppositional terms. Planters resisted the passing of the muscovado era, as the modern production complex meant a loss of autonomy for the *haciendas,* where cultivation and milling used to be one integrated affair. A rather profitable market for low-grade sugar also existed, and several planters refused to enter into contracts with the new mills. Thus, until the late 1930s, Negros had sixty-four muscovado mills that produced 147,039 piculs, or 18 percent, of total low-grade sugar production in the Philippines (*Sugar News* 1939c). A visitor to the island noted that "The influence of the muscovado days is still felt and large *camarins* and muscovado mills meet the eye at every turn" (Barnshaw 1924, 661). However, the displacement of the open-kettle contrivances was a foregone conclusion considering that, out of the same stand of cane, a muscovado mill could produce 41 piculs per hectare only while a centrifugal mill could extract 96 piculs, a higher yield that also commanded the top price in the raw sugar market (*Sugar News* 1921c, 8).

Along with "a rather strenuous re-arrangement of conditions" that occurred in the early 1920s (Pitcairn 1922, 471), most planters decided to affiliate with a modern sugar factory. The planters felt, however, that they had been reduced to the role of mere suppliers of cane to the centrifugal mills, which became "the important timing agent" (Wernstedt 1953, 149). In contrast to the muscovado mills, which ground protractedly from six to seven months simultaneous with cane growth, the sugar mills operated at full capacity for four months, during which time intensive work in the fields also had to be completed (Henry 1929c, 19). Detailed scheduling of specific activities ranging from the cutting of cane tops, the replanting of cane points, and the harvesting of canes for milling had to be coordinated with other farms as well as with the mill, to avoid costly shutdowns due to inadequate cane supply, and to preserve the cane's sucrose and maintain sugar quality (Pitcairn 1922, 476). Many *haciendas* were thus subdivided into blocks, each with a specific cultivation date harmonized with milling periods, although in many farms harvest dates and the topping of canes were not synchronized, resulting in losses in sugar yield and quality (Pritchett 1926, 459–460; Wernstedt 1953, 145–46). Carlos Locsin (1921, 466) tried to impress upon his fellow planters that they must adjust to the "passing from a period of individual responsibility to one of collective responsibility."

In the ancient mold of *dungan* contests, planter rivalries and jealousies persisted, as well as what the *Sugar News* editorialized as "useless and impotent competition."[95] But in the face of the perceived uneven relations with the *sentrals,* the planters were compelled for the first time to form associational groupings: by the late 1920s there were seven in Negros and three in Luzon (Henry 1929a, 486). With funds raised by assessing 1 percent of the sugar produced by members, these planters' associations hired their own technical staff to organize and regulate cane cultivation, allocate vehicles to transport the canes, and handle the complaints of planters, most of whom knew next to nothing about the technical aspects of sugar production (Barnshaw 1924, 663; cf. Asociacion de Hacenderos de Silay-Saravia n.d.). Reared as every *hacendero* was on the sharecropping arrangements of the past (although this time they were positionally akin to the tenant), the planters feared possible machinations in the crop-division process and so were deeply suspicious of the new terms of the sugar game dictated by the mills. As the visiting chief agriculturist of French Indochina Yves Henry (1929a, 485–486) observed, planters were skeptical of "the modern methods of analysis and calculations" and anxious that they were not receiving their full share of the sugar. Suspicions that they were being cheated were eased when the sugar factories allowed the technical employees of the planters' associations to observe and scrutinize mill operations. Nonetheless, that the mills were getting the better end of the deal bothered planters who did not own any stocks in the mill corporations.

A struggle over the division of profits between mill owners and planters ensued, especially with independence looming. The planters were not happy with the original *sentral-hacienda* contracts, which, on average, allocated to each planter in Luzon 50 percent and in Negros 55 percent of the processed sugar, prompting them to demand a larger share of the final output. In April 1930, the planters' associations of Isabela and Binalbagan initiated the battle by passing separate resolutions calling for a "readjustment" that would entitle them to a 60 percent share.[96] The PNB was asked to "take the necessary steps to intervene or intercede to induce" the mills to heed the planters' demands. In response, the so-called *sentralistas*, or mill partisans, put up a staunch resistance. In the words of one miller, "If you can assure me that the American flag will continue unfurled over these Islands, I will cheerfully accede to your petition for a 40–60 division of the sugar output," but with independence "the machinery in the centrals will only be useless pieces of steel."[97]

In September 1930 a delegation of planters called on the governor-general to seek assistance in modifying the milling contracts. The governor responded that the government "could not interfere in private contracts between citizens by bringing pressure to bear on either party" (*Sugar News* 1930a). In 1932, Philippine senate president Manuel Quezon raised the issue again with the acting governor-general, arguing that since the Bank Centrals were "in the hands" of the PNB, government could intervene on behalf of the planters who were allegedly "unable to make ends meet" with the present division of the crop (*Sugar News* 1932c).

The planters, however, were about to begin a splurge of speculation in securities and particularly in mining shares, which boomed in 1933 and then again in 1936, occasioned by the U.S. government's policy of buying gold. Because of fat profits, the sugar capitalists had the chits to gamble outside of sugar. It appeared that ties with powerful beings sustained, even expanded, the planters' *suwerte*. Not long after the start of the mining boom, the sugar elite became the recipient of more windfall income when the New Deal phase of the "AAA" (Agricultural Adjustment Administration) to limit sugar production, stabilize prices, and alleviate America's farm crisis was applied to the Philippine sugar industry in late 1934.[98] By early 1935 the U.S. Treasury had advanced $14 million to the Philippine government upon which to draw the so-called benefit payments, which were fixed at the rate of 2.40 pesos per picul foregone (*Sugar News* 1935c), although not till the end of the year did the actual payments begin to be issued. Many sank their benefit payments in mining shares, while some invested in the construction of modern buildings in Manila (Hartendorp 1958, 55–57; *Sugar News* 1933a). With cash in hand, Negros gamblers "went en masse to Manila to play real big games," and *monte* dens in Bacolod were temporarily left empty (*The Commoner* 1936a). The planters were also buying cars as a new way to flaunt their wealth. From 1934 to 1935, 251 new automobiles were registered in Negros Occidental, which brought the total number of private vehicles in the province to 2,448, the largest concentration of automobiles outside Manila (Philippine Statistical Review 1935b, 1936). The planters themselves belied the claim that they could not "make ends meet."

Meanwhile, the PNB had "conditionally" granted the request of the Isabela planters in February 1932. The incremental 5 percent share was called a "bonus" that could be revoked anytime and not "a modification of existing milling contracts" (*Sugar News* 1932b). The evasion technique was recommended by a special bank committee composed of two Americans and one Filipino. The PNB decision was explained as an egalitarian measure to redistribute the benefits from the Isabela Central beyond the "small group of people who have invested little or nothing and who have assumed very little risk" and to include the planters who "own very little if any stock in the central" the bank had capitalized (171). The PNB action was referred to by a Manila paper as "a harvest of complications"; and the legalistic trick goaded a few Negros mills to follow suit in conceding "bonuses," despite the bank's disavowal that "its action shall [not] be used as a precedent for compelling action by other centrals" (172–173).[99]

Seeking to make the new ratio mandatory, planters attempted legislative subterfuge by inserting a section on the 40–60 crop division in a bill aimed at limiting sugar production. An even more disturbing clause sought to legislate that "any party to a contract of lease or to a milling contract may have the same modified or cancelled and terminated, without payment of damages, upon satisfactory proof" of adverse effects that would "make it unprofitable for him to continue with such contracts" (*Sugar News* 1934f). In explaining his veto of December 1933, Governor-General Murphy pointed to the crop-share section as patently "extraneous to its subject and in contravention of the well recognized principle of inviolability of legitimate contracts." In their shortsighted wrangling over profits, the sugar planters were prepared to throw out the contract as a social institution, in much the same way that Concepcion derided accounting principles as alien and unfavorable to Filipinos.

Accentuating the miller–planter conflict was the failure to achieve the original intent concerning the cooperative ownership of the Bank Centrals. One clear obstacle was the numerical increase in the category of sugar growers who leased *haciendas* and who could not possess shares in the mill corporation. By the late 1930s, this category had become preponderant in Negros, where three-fifths of *hacenderos* were lessee-planters in contradistinction to owner-planters.[100] Although leaseholding was already practiced in late-nineteenth-century Negros, the massive expansion in the category of lessee-planters suggested the formation of a new rentier stratum that lived off the technologically enhanced profits from ground rent. In Negros, the average leasehold rates varied from 8.3 to 18.3 percent of gross sugar output (Alunan 1939, 82), the net effect being the reduction in the lessee's crop share to from 36.7 to 46.7 percent, instead of the full 55 percent. In the characteristic pattern of multiple and overlapping circuits of capitalist exploitative relations in the Philippines (Aguilar 1989, 59–61), the lessee–rentier oppositional relationship added fuel to the miller–planter struggle over the surplus product.

The cooperative ideal was also thwarted by the floating of mill corporation shares in the Manila stock market, which was established in 1927. The profitability of the sugar factories spawned insider trading, and shares were controlled by the most influential buyers (Larkin 1993, 168). Awash with cash and benefit payments, certain groups founded on marital-familial alliances monopolized the equities of mill corporations in a stratagem dubbed as the "*rigodon* of millions," exemplified by the Lizares-Mapa consortium that obtained control of the PNB-financed Talisay-Silay mill.[101] Having purchased the Danao Central in June 1930, the Lizares group in 1936 was able to buy out, for 2 million pesos, the Alunan family from the Bacolod-Murcia mill, also a Bank Central.[102] Contrary to the ideal of cooperative ownership, control of mill equities was concentrated in only a few hands.

The sharpening of the divide between farms and mills created two fractions of the sugar ruling class—one in agriculture, the other in manufacturing—which continued to fight over sugar profits. With the Aranetas and the Lopezes among the most conspicuous, the wealthiest of mill interests began to diversify their portfolios, gradually displacing the foreign sugar merchants and acquiring interests in key industries located in Manila and elsewhere (Larkin 1993, 168, 211–212; McCoy 1992, 127–128, 135). On the other hand, the planters continued to press their demands, using the welfare of workers to justify their claim to a bigger share.[103] The struggle did not abate, particularly as sugar prices plummeted in the late 1930s, although the outbreak of World War II in 1939 initially brought higher prices to Philippine exports (Hartendorp 1958, 47–48).

Because of the mill–*hacienda* conflict, the lingering racial tension between Spanish and Filipino *hacenderos* was overshadowed. The war did serve as a catharsis for the Filipino planter-guerrillas, who oversaw the burning and looting of *haciendas* and small *sentrals* owned by Spaniards who generally supported the Japanese invasion (de Uriarte 1962, 19, 30–38, 54–58; Zaragoza 1982, 85; cf. Hart 1964). After the war, many Spaniards left Negros for Europe.[104] On the whole, the war period saw a temporary hiatus in the miller–planter contest. To survive, one segment of the sugar capitalist class collaborated with the Japanese, while another led the pro-American guerilla movement (whose members punctuated the monotony of their mountain redoubts by holding dances and engaging in cockfighting and gambling).[105] The elite's hedging of bets by alternately supporting the two opposing war powers paid off during the postwar reconstruction effort.

Of the country's forty-two mills in operation before the start of the war, only three were left intact after the war (Mirasol 1950, 146). With the devastation of the sugar industry, the U.S. Congress again rescued their elite wards. In view of the country's formal independence in July 1946, rehabilitation funds and preferential market access were made available in exchange for two constitutional amendments: "parity rights" that placed American businessmen on an equal footing with nationals of the Philippines; and territorial grants for U.S. military bases and installations. Thus, rather than the complete termination of duty-free entry after independence, Philippine sugar (with a quota of 982,000 short tons) was granted eight years of tariff exemption, followed by a graduated annual increase of duty until the full duty was reached in 1974.[106] The state apparatus and old colonial ties were again crucial in reinfusing life into the sugar industry.

In Negros, twelve sugar districts were formed around the reconstructed sugar factories, which replaced eighteen prewar mill districts (Wernstedt 1953, 153–154). Despite the retention of its structural inefficiencies, the industry was highly profitable because U.S. sugar prices were far in excess of world market levels. Not surprisingly, the fight between mills and *haciendas* resumed and became a major preoccupation of the Philippine Congress. As in the preindependence period, elections in Negros were fought bitterly along the *hacendero-sentralista* divide (Larkin 1993, 168–169, 190–192, 217). With more *hacienda*-backed "bets" winning national and local electoral seats, a hiatus in the miller–planter struggle occurred with the passage of the Sugar Act of 1952, which enlarged the planters' share to from 60 to 70 percent contingent upon actual production (Marcelo 1953). In the struggle over the surplus product, members of the sugar capitalist class fought in the primeval *dungan*-versus-*dungan* manner. The sugar workers were not oblivious of these contests and, at this time, they explained the seeming stalemate in the clash of spirits and the immense wealth the sugar factories continued to create by ascribing mythical prowess to selected mill owners and officials.

The Demise of Sharecropping: Wage Labor in Negros *Haciendas*

As a result of the profound societal transformations occurring in the American colony, sharecropping was discontinued in Negros plantations, as it had become an impediment to production and the new frenetic struggle between mill and farm interests. Because the circuits of exploitation were a tangled web of overlapping relations, the planters decided to buttress their advantageous position at the *hacienda* level while they waged a more arduous and highly politicized battle with mill interests at the national level. Being in a clearly dominant position over the plantation work force, planters asserted their superiority by reorganizing production relations. Consequently, while centrifugal milling eroded the *hacienda*'s autonomy relative to the *sentral,* planter control within the farm was tightened through a drastic overhaul of labor processes.

The imperatives of the new production complex required a more precise and systematic cultivation scheme not realizable under share tenancy in which there was "absolutely no system or control" over tenants, whose planting style one agricultural scientist caricatured as "one of the most haphazard ways of growing cane ... [which], when harvested, is not cut in the same order as planted" (Anon. 1922, 119). The *Sugar News* (1924a) remarked that "The tenant does not lend himself to intensive methods." Another American noted that "Under old methods [sharecropping] may have been a good system, but under modern methods, such as tractor plowing, deeper tillage and better methods of cultivation, with which the [tenant] will not conform, as it materially increases his expenses without benefiting him in return, a new method must be found" (Ross 1920, 3).

Sharecroppers might have been willing to alter their cultivation techniques but were not encouraged to do so because, as suggested by the extract just cited, the sugar planters decided to pass on the costs of modernity to tenants. Fertilizers, for instance, had become a vital ingredient of the new sugar culture, but sharecroppers were made to shoulder from 50 to 75 percent of the cost of fertilizer as well as the total cost of fertilizer application (C. Locsin 1923, 75). The tenant, who labored under the illusion of a partnership, was also made to assume the costs of draft animals and tractors, and to bear "the loss of the animals through death, or loss of equipment, paying for the complete ownership of the same in annual instalments" (Pitcairn 1925, 31–32; cf. Henry 1929b, 124).[107] The costs of technical change were thus added to the usual expenses incurred in land preparation and cultivation and to the interest charges upon the money advanced by the planter. As though the sharecropper was not sufficiently exploited, some planters were wont to confound the tenant with the mill's complicated formulae for computing cane quantity and quality, the calculations being spelled out in documents that many planters themselves did not fully comprehend but used to deceive the tenant. The odds of the sugar gamble were more than tilted against the tenant.

In the struggle at the farm level, sugar planters sought to maximize their gains over their tenants, especially because a larger surplus product was at stake. Aggregate statistics suggest that the colonial state-sponsored technical changes increased sugar yields from 25 to 30 piculs in 1898 to about 75 piculs per hectare in 1925 (Fairchild 1929, 11). In Negros, the centrifugal sugar yield rose to nearly 100 piculs per hectare in the 1932–1933 crop season in contrast to the corresponding figure of only 66 piculs in Luzon (Gordon 1933, 609). If the sharecropping arrangement had not been altered, the tenant's profit could have been truly substantial, but the planters were quick to realize the cost of inaction.

Negros planters who initially retained the tenancy system strove to squeeze the surplus out of the share tenants by heaping the burden of production costs on the cultivator's side while formally reducing the tenant's crop share. In the early 1920s, Carlos Locsin (1923, 74) reported that the *agsa* assigned to about four hectares of cane with his own work animals received as "compensation a share of from 20 to 25 per cent of the sugar produced." On the assumption that the statement referred to the total processed sugar and that the mill–*hacienda* sugar partition was 45–55, Locsin's data meant a 40–60 to 45–55 crop division between

Table 9.
Population and Selected Agricultural Data on Negros, 1918 and 1939

	Negros Occidental	Negros Oriental[a]	Negros Island
Population			
1918	396,636	215,750	612,386
1939	824,858	335,173	1,160,031
Area planted with sugarcane (hectares)			
1918	37,069	3,231	40,300
1939	97,281	6,145	103,426
Total number of farms under share tenancy (% of all farms under share tenancy)			
1918	1,950	2,927	4,877
	(14%)	(9%)	(11%)
1939	23,371	20,665	44,036
	(65%)	(44%)	(53%)
Total number of sugar farms under share tenancy (% of all sugar farms)			
1939	1,380	1,135	2,515
	(38%)	(44%)	(40%)
Total area in hectares of sugar farms under share tenancy (% of total sugar area)			
1939	12,715	1,034	13,749
	(13%)	(17%)	(13%)
Reported share of sugarcane crop received by share tenant, 1939 (% of total reported)			
<50%	446	71	517
	(75%)	(30%)	(62%)
50%	49	110	159
	(8%)	(46%)	(19%)
>50%	103	57	160
	(17%)	(24%)	(19%)
Not reported	314	8	322

Sources: United States 1921, 80 and 342; 1940, 4–7; 1941, 993, 1061, 1073, 1272–1276.
[a]Excluding Siquijor Island.

tenant and *hacendero* (cf. Henry 1929b). By 1939, available census data suggest that at least 75 percent of the sugarcane share tenants in Negros Occidental Province received less than half of the sugar output (Table 9). In fact, of those reporting their shares (and many did not), 61 percent said it was less than 20 percent, which could have meant a crop division of at most 35–65 in favor of the planter. Data problems notwithstanding, it is evident that the sharecropping agreements that remained in Negros in the late 1930s had clearly departed from the 50–50 parity of earlier times.

As an intermediate step between sharecropping and outright wage work, many Negros planters instituted a system of contract work called *pinico* or *linacsa,* in which a plot of land was assigned to a worker whose remuneration was calculated at a certain rate per picul produced or per *lacsa* planted, the latter being a measure of 10,000 cane points (C. Locsin 1923, 73–74). In the early 1920s, the usual rates were 1.00 to 1.50 pesos per picul and 30 to 40 pesos per *lacsa*. If the laborer, called *manogpinico,* had his own work animals, the rates were increased to from 2.50 to 3.00 pesos per picul and 50 to 60 pesos per *lacsa*. In general, farm expenses were charged to the *manogpinico,* although the latter, unlike the tenant, was not made to share the fertilizer cost (Henry 1929b, 125; C. Locsin 1925, 133). Phenomenally closer to the wage-labor form, contract/piece work by picul or *lacsa* was admittedly "a cheap way of working" the land, particularly in low-yield areas (C. Locsin 1923, 74).

To maximize returns from the modern sugar complex, however, most planters in Negros shifted to a variant of the wage-labor form. In the early 1920s, Carlos Locsin (1923, 74) reported that "The *'aparceria'* or share-work system is very much in vogue in the *haciendas* of Negros." By 1939, with the spread of centrifugal milling, only about 38 percent of all sugar farms in Negros, or a meager 14 percent of total sugar hectareage, was reportedly operated on a share tenancy basis (Table 9). A definite trend away from sharecropping transpired sometime in the late 1920s and early to mid-1930s, culminating at about the time planters were frolicking in the mining boom, the windfall income from benefit payments, and a momentary surge in sugar prices in 1936 and early 1937. While in part an issue of production efficiency, the large-scale shift away from sharecropping was preeminently a gambler's strategy to optimize gains in what many perceived as the closing stages in the game of sugar. The *agsa*'s share was "too much for the planters to forgo," especially in the mad race "to make the best out of the remaining period" before the dreaded arrival of Philippine independence and the loss of the U.S. market (*The Commoner* 1936c, 13).

The planters did not encounter stiff resistance in restructuring *ha-*

cienda production, given the class dominance, political power, and symbolic capital they had consolidated with the aid of the American colonial state. Farm reorganization was accompanied by a drastic alteration in the cultural prescriptions surrounding the roles of *hacendero* and worker. Around this time, as the previous chapter's discussion has shown, some planters began to overlay their contrived paternalism with cruel and brutish tactics to intimidate recalcitrant workers. Most of the laboring class succumbed to planter hegemony, as they were left with few viable options.

The demise of sharecropping initially resulted in the formation of "a wandering class of farmers" who possessed carabaos and agricultural implements but with no land to till (*The Commoner* 1936c, 8). With the dislocation of the tenantry, many *agsas* opted to become *duma-an*, the local term for the wage-dependent laborer who resided on the plantation. As a result of past planter strategies, the tenant-turned-*duma-an* had become deeply entrenched in the *hacienda* mode of life. The very curtailment of the cognitive availability of choice of alternative livelihoods possible in a wider labor market was the principal force behind the *duma-an*'s "unfree" character, what Tom Brass and Henry Bernstein (1992) have theorized as "deproletarianized" plantation labor. Subsequent generations of *duma-an* would be born in the same *hacienda*, where they were socialized into a life of poverty, dependency, and subservience. Many a *duma-an*'s identity was fused with the farm where he or she learned to call the planter *toto* (brother) or *inday* (sister) as though the plantation constituted one big familial community—the ease with which the ruling class allowed themselves to be referred to, not as aunts and uncles, but as siblings of poor workers serving as a reminder of the invention of the family construct during the Spanish colonial epoch. *Duma-an* parents also learned to depend upon the planter for names to christen their children. Fully supported by the state, the planter class succeeded in producing its requisite mass of compliant workers through the cultural entrapment of labor.

But there were other constraints on labor mobility in the early part of the twentieth century. Workers could no longer flee to the hinterland, as this possibility was restricted by the virtual disappearance of the frontier, what with most arable lands having been encroached upon by the expanding sugar industry. The only remaining area for settlement was the island's distant southwest, where the 15,000-hectare Tablas Valley absorbed from one to two thousand settlers embarking on a spontaneous colonization project in the 1930s—in the process weakening the impact of "radical propagandists and agitators" in Negros (*The Commoner* 1937a, 1936c, 1936e). However, because of rapid increases in internal rates of population growth, the population of Negros Island soared from over half a million in 1918 to over 1.1 million by 1939, an 89 percent increase over two decades (Table 9). In Negros Occidental, the population rose by more than 100 percent, markedly increasing population density, leaving little fertile land for occupation, and creating unemployment in the towns (*The Commoner* 1937b, 1940a). Labor in Negros had lost the space for maneuvering it had enjoyed in the late nineteenth century.

Overall wage rates were also depressed: in the early 1920s, while the wage rate of the day laborer in Pampanga was 1 peso without food (or 70 to 80 centavos plus meals), Negros' rate was a low 60 centavos without meals (Stower 1924, 28).[108] Women and children, who were hired to pull weeds by hand and to scrape away the earth that had fallen on young shoots of cane after plowing, were paid minuscule wages of from 10 to 30 centavos a day in Negros (Costenoble 1923, 112). The centrifugal mills, which offered only a limited number of job placements, paid somewhat higher wage rates, 1 peso being the usual minimum at the time (Stower 1924, 88). The pay varied according to task, an electrician, for example, receiving 2.00 to 2.50 pesos per day. The sugar factories also attracted workers, as the mill management—under the initial influence of American personnel who were concerned about "Bolshevism"—generally provided schools for workers' children and amenities like light, water, dispensaries, hospitals, playgrounds, and recreation halls (Runes 1939, 27). It was reported that "Men for all these posts in the mill are plentiful, as the average native likes mechanical work, in preference to any outside labor, and we do not foresee any shortage for many years to come" (Stower and Urquijo 1924, 89). In 1926, the oversupply of mill labor resulted in real unemployment in the vicinity of the factories (Walker 1926).

In the sugar-growing areas of central Luzon, share tenancy remained because labor conditions were not as repressive as in Negros. Attempts by Luzon planters to reorganize their *haciendas* met with stiff resistance. Reports indicated that those "who have operated under direct administration have taken great risks and have often incurred greater liabilities than the planter who farms his land by *aparceros*" (Barnshaw 1924, 662). The ferment among workers and the influence of labor unions that proliferated in the early 1920s, which spread from rice to sugar *haciendas* (Larkin 1993, 193–196), made any sweeping change in sugar production a thorny issue. Moreover, unlike Negros, Luzon labor retained a bargaining space because the urbanization of nearby Manila raised the demand for industrial and "informal sector" workers (Doeppers 1984). The earlier competition among Negros planters was now seen operating in Luzon, where agents pirated contract workers: "The *scavenger* who waits for the chance to take labor from one *hacienda* to another always has glowing tales to tell of the better conditions existing at the next place...." (Milne 1925, 40; my italics). In such a context, the sugarcane tenant *(casamac)* of Pampanga was said to be "in a peculiar situation" since, despite work instructions from the planter or plantation staff, he nevertheless continued to reserve for himself "a certain degree of independence" (Onrubia 1925, 22). Even in introducing changes in cultivation methods, tact was needed to convince the tenant to adopt the desired techniques.

Because Negros had become the enclave and preserve of the sugar capitalist class, the tenant's autonomous space was fast obliterated to expose the commoditized and alienated condition of deproletarianized plantation labor. In contrast, the relations of sugar cultivation in central Luzon were also beset by the contradictions inherent in the oppositional relationship between capital and commoditized labor, but the latter did not take the conventional form of wage work. The Luzon sugar tenant was encouraged to increase production by the promise of additional compensation for every ton of sugar raised in excess of the standard fixed by the *hacienda* management (Onrubia 1925, 24). The tenancy contract also stipulated that the decision as to when to sell the sugar and for what price was the planter's exclusive purview. However, the planter and the tenant had to agree on the price per ton of cane that was to be used in calculating the sharecropper's compensation. Still, Luzon sugar planters strove to reduce the tenant's returns to labor by manipulating mill documents on sugar quantity and quality (23), rendering the struggle over the surplus product in the planter–tenant relationship more marked. Tensions began to fester, as planters, apprehensive about limited access to the U.S. market and other grim scenarios after independence, drove harder bargains in tenancy relationships.

The perturbation among the Luzon tenantry persisted and the colonial state was compelled to enact a sugar tenancy law in December 1933, which required planters to furnish their tenants with a clear and complete accounting of farm expenses and of the sugar obtained based on receipts issued by the sugar mill (*Sugar News* 1934d). But, like previous pieces of legislation, this law was difficult to implement. Beginning with disputes over the benefit payments as Luzon tenants rightfully demanded a share of the U.S. bonanza, the late 1930s witnessed a period of radicalism among the sugar tenantry (Larkin 1993, 205, 219–235). Still seemingly oblivious of the depth of the agrarian unrest seething in central Luzon—which exploded in the late 1940s and 1950s into the Huk movement (Kerkvliet 1977, Lachica 1971)—the sugar planters had to be exhorted in 1948 by the new republic's secretary of labor to comply with the legal requirement "in such a manner that the tenant will feel nothing is being hidden from him" (Magsalin 1948, 2).

Meanwhile, Negros *hacenderos* were spared the agitations of plantation labor by their having seized the decisive juncture of the 1920s and the 1930s to veer away from sharecropping while concomitantly instituting a tight grip over labor. Although a few sugar factories in Negros became sites of labor unrest in 1931 and in the late 1930s, these sporadic episodes were easily contained due to the existence of a reserve army of labor (including scab labor from other islands), the inherent weaknesses of Negros' labor union movement, and the unequivocal support of government troops, particularly the Constabulary (*The Commoner* 1939a, 1940b; Larkin 1993, 192–193, 222; McCoy 1982a, 342–343). Subsequently, unlike the central Luzon planters and landlords who fled the countryside, Negros planters strengthened their dominant position in Negrense society by assuming leadership of the guerilla movement during the Second World War.

In Negros, problems in the recruitment of transient cane cutters, today's *sacadas*, had also become mere "annoyances" by the 1920s, despite the greater need for such workers; a procedure was also established to "afford the greatest security for capital."[109] With funds lent by the large sugar mills, planters employed labor agents who scouted for workers from Capiz and Antique Provinces (Stower and Urquijo 1924, 87). Paid on a commission basis, the agents arranged the hours of work, the rate of pay, and the work tasks of the temporary laborers at the time they received their cash advance, which clearly had become part of an enforceable labor contract. The situation was a far cry from the nineteenth century, when planters had to offer "appeasement" and "beg for labor."

Without the internal passport system of the Spanish period, labor recruitment was also no longer encumbered by legal and administrative obstacles. Workers, especially from more distant Antique Province, were fetched by "barges towed by small tugs" and brought directly to Negros (Stower and Urquijo 1924, 87). Furthermore, grease money converted local officials into touts who facilitated the recruitment process. As one American industry man in Negros reported:

> we always go to the *Presidente* of the town where the laborers are recruited and see that he gets his bit. This insures the *Presidente*'s cooperation in case the men run away without completing their contract. Last year about 80 men ran away.... We went to the [Justice of the Peace] of the town where the men were recruited and he made every one go back to work. (Stower 1924, 29)

In addition to the agents' knowledge of the people and the locality where workers were obtained, planter control of labor was further enhanced by obligating the agents to repay the advance money should any temporary worker abscond. The planters' jubilant report that "[c]omparatively few laborers in Negros fail to complete their contracts" could be made by the 1920s (ibid.). The dictates of sugar capital had become preponderant over labor.

Labor, Sugar Capital, and the Phantoms of Capitalism

The fundamental reorganization of sugar production in Negros during the American colonial period was apprehended by mill and farm workers from the perspective of indigenous culture, leaving behind a rich body of folklore that depicts the hegemony of the sugar capitalist class as well as the contradictions of Negros' agrarian capitalism. Local culture thus offers simultaneously a critique and a legitimation of *hacienda* life.

A popular story, with variants told by both planters and workers, concerns the son of Isidro de la Rama, Esteban.[110] Every year, he was said to travel "around the world." Whenever he was abroad, instead of using toilet paper, Esteban de la Rama allegedly used money so that people would know he was wealthy; bellboys in hotels would even retrieve the money from the receptacle *(urinola)*. In a remarkable continuity with the legends about Isidro, the Esteban de la Rama lore exuded nationalist pride in one wealthy mestizo-qua-Filipino's ability to make "foreigners" stoop down for the sake of the money a Negrense could afford to dispose of like toilet paper. Just as Isidro had not been intimidated by Friar Power because of the status his wealth had brought him, so did Esteban transcend cultural ambivalence through money.

Like Jose de la Viña, Esteban de la Rama was reputed to have had several mistresses *(kerida)*, at least one in each plantation. Showcasing the strength of Mestizo Power, Esteban reportedly contracted venereal disease but nonetheless lived a long life; he was about eighty-one years old when he died "of old age" in 1947. The fortune Esteban inherited was believed to have originated from his father's ties with *Yawa,* but Esteban himself was also portrayed in mythical terms: a Mason supposedly with horns and *anting-anting,* his abilities catapulted him into the position of a magnate in the sugar, maritime, and mining industries. *Hacienda* workers explained his wealth in Masonic terms, although the folkloric fragments were not as embellished as in the Isidro de la Rama legends. Esteban's "natural death" demonstrated Friar Power's inability to exact reprisal and, ultimately, its dissolution in the face of twentieth-century individuals of prowess.

Mestizo Power seemed naturally to flow in the veins of the Negros ruling class, its continuity passed on from generation to generation. A daughter of Esteban de la Rama is purported to have continued the practice of Masonry. The mark of a Mason, horns are said to have grown on her head, for which reason she would normally sport a bouffant hairstyle. Her Masonic powers, it is believed, have allowed her to produce money unendingly, even though some of her farms have gone bankrupt since the crisis of the late 1970s. With the lingering mesmeric effect of Mestizo Power, wealth and capital accumulation continue to be mystified and fetishized.

During the American colonial period the impregnability of Mestizo Power was evinced by its ability to withstand the fleeting challenge posed by Flor Intrencherado, who claimed to be an emperor and attracted a large following in the mid-1920s in both Iloilo and Negros as well as in other Visayan provinces.[111] In the manner of the *babaylan*, Intrencherado reputedly had a spirit-friend and power to control the elements; he also claimed to receive enlightenment from the mestizo-Catholic conglomerate of Padre Burgos, Jose Rizal, and the Holy Ghost. Warning of a deluge that would begin with a predicted eruption of Kanlaon on 4 February 1929, Intrencherado preached salvation from the impending cataclysm, with the survivors reclaiming the whole province after the floodwaters had subsided. He gained the allegiance of *hacienda* workers, who donated money in exchange for the emperor's photograph. He promised to abolish taxes and accused the Chinese of being interlopers who should be repatriated. Displaying the mestizo worldview concerning the exchangeability of money and sovereignty, he expressed willingness "to pay" the United States for Philippine independence. In late 1926, Intrencherado was investigated by the Constabulary, temporarily jailed, then tried by the Iloilo Court of First Instance and declared insane. Because his lawyers appealed his case, he was temporarily released, a fact interpreted by his followers as proof of his omnipotence.

The Supreme Court upheld the lower court's decision to commit Intrencherado to a mental hospital in Manila, and his departure by boat from Iloilo was scheduled on the night of 13 May 1927. Intrencherado's departure, however, was delayed by an uprising staged by his followers on the morning of the thirteenth, the day the rivers in Negros were said to have flowed with ash from a restless Kanlaon. In Bago, La Carlota, La Castellana, Silay, and Victorias, the emperor's followers, led by a defeated candidate for municipal mayor in the 1926 elections, took over municipal buildings, killed a policeman in Victorias, seriously wounded the chief of police in La Carlota, and placed at least three Spanish planters in La Castellana under house arrest. The Negros elite did not foresee this "rebellion," and even the local constabulary was unprepared, engrossed as everyone was in the political rivalries and bloody quarrels between the *Kusug Sang Imol* and *Mainawa-on*. But the aura of the sugar capitalist class emerged unscathed from this incident. On the contrary, Intrencherado's followers became disheartened when the prophesied arrival of their emperor in Negros did not materialize.

Intrencherado proved to be a disappointment to the workers, who had clung to the hope of a historic intervention by a *babaylan* from the masses. His fall merely served as an affirmation of Mestizo Power. (Folklore is replete with stories of magical planters, but there is hardly a surviving legend about Intrencherado. Recollections by a few workers who remember him are that he was a con-man, no different from the memory of elderly *hacenderos*.) Workers have since learned to desist from following theocratic liberators, and magic has once more receded to the individuated realm. The failure of Kanlaon to erupt since that fateful day in May 1927 was a reminder of the emperor's hollow utterances. The *hacienda* workers soon found a more convincing explanation of Kanlaon, one attuned to the intense social transformation of Negros.

Kanlaon, many workers believe, is not a volcano.[112] A city of modern *engkantos*, Kanlaon has its own shops, residential areas, cars, trains, and *haciendas*, and is "in reality" an estate with its own *sentral*. The smoke from Kanlaon coincident with the milling season is the smoke emitted by the giant chimney of the concealed centrifugal mill. The enchanted factory produces sugar delivered "to America" by three golden vessels. Labor is drawn from the socially differentiated world of the spirits. Sometimes more workers are needed to finish certain tasks on time, so the *engkantos* draw workers from among ordinary human beings.

Some *hacienda* workers testify that they had at one point been drawn to work for the magical sugar mill after joining a job queue without knowing that the person who was processing the applications was a preternatural being, a *tamawo*. Once inside Kanlaon, it felt like a prison, for they could not leave without the express permission of the spirits. One laborer said he was "inside" Kanlaon for twenty-five years. Workers say that the Kanlaon spirit-city possesses a monetized economy, with a currency in the form of golden triangular objects. The wage rates paid by the magical sugar estate are said to be two to three times that paid by Negros employers. With their wages, workers could buy everything they needed inside Kanlaon.

As a vivid refraction of social realities, this mythical tale of sugar production depicts the facticity and the workers' acceptance of the modern centrifugal sugar complex. That spirit-beings are also engaged in a production system explicitly geared toward the U.S. export market (with its own magical dimension) suggests that Negros' absorption into the international ("foreign") capitalist system is acknowledged by sugar workers. Moreover, the existence of social inequality is accepted as normative, given that the same pattern of differentiation prevails even in the enchanted world of Kanlaon. Thus, social life built around sugar production has a solid legitimacy emanating from what workers would consider the highest source of authority, the spirit-world. The hierarchized reality is dominated by Sota, who personifies wealth, success, and the blessings of the gambling ethos, a spirit of legitimation of the structures of Mestizo Power.

At the same time, the Kanlaon myth offers an incisive critique of Negros society, giving vent to the palpable contradictions in the relations between commoditized labor and sugar capital. Ideal-typical of capitalism, job queues constitute an eloquent testimony to commodified labor, severed from the land and thrown into the market for survival, a testimony to the men, women, and children dependent on work in the sugar industry for sustenance and the social reproduction of labor power. The prisonlike hold of the Kanlaon spiritual economy signifies both the local sugar capitalists' control over the conditions of deproletarianized wage labor and the inescapable reality of sugar production as virtually the only viable source of livelihood in an export-oriented island dominated by a single crop. The workers' tropological narrative, however, is also a criticism, and a severe one in that, while the sale of labor power is accepted as inevitable and necessary, the remuneration of labor is deplored as extremely inadequate. Because of the marked wage differentials between the spiritual and the material worlds, some sugar workers would in fact have preferred to work for the Kanlaon spirits than for the dominant sugar capitalists of Negros.

Having little choice but to live in the material world of the Negros *haciendas,* sugar workers have accepted their lot with submission as well as quiet resistance. To fight Mestizo Power in the spiritual terrain is useless because of the ruling class's powerful ties with the spirit-world, whence they have acquired mystified wealth and the prowess to make themselves invulnerable to competitors and recalcitrant workers who might resort to sorcery and black magic. To fight through the courts would be even more daunting since, due to mestizo power's peremptory control of the state apparatus, the complainant could lose his job, land in jail, or be brutalized by hired thugs.

Somehow, the only hope would seem to lie in the Maka-ako or the Gino-o of the seventh layer, who might someday exact divine retribution.

Expressive of this hope are many legends told by elderly *hacienda* workers, one of which speaks of a caterpillar-like insect called Baringkot.[113]

One day, Baringkot was crawling in the field and could no longer bear his creaturely condition, for to move from one place to another was a terrible drudgery. So Baringkot pleaded with Gino-o to make him human and in return he promised to share his riches with everyone, especially the poor workers. Taking pity on Baringkot, Gino-o granted his wish and gave him wealth and a mansion of gold. But once he became human, Baringkot forgot about his promise.

Baringkot called the people to prepare a feast he would offer his friends. The people all worked very hard, fetching water, making firewood, and cooking the meal. Baringkot's vice, however, was to invite only the wealthy. The guests arrived and enjoyed themselves at the dinner table, while the workers in their hunger gaped at the food.

At high noon, an elderly person arrived. He told the guard he was thirsty and entered Baringkot's mansion through the front stairwell made of gold. Baringkot was furious that the elderly man had the gall to ascend the golden stairs. A boy was told to fetch a glass of water, but when he returned the old man was gone. Before the august guests could finish their meal, the elderly man came back on a golden horse and he himself was gold. When Baringkot saw him approaching, he ordered his workers to serve all the food there was to this new visitor, whom he said was more affluent than he was.

Baringkot invited the man to come up the golden stairwell, but the latter replied that he had had a surfeit of golden stairs and so decided to enter through the wooden stairs at the rear of the house. The elderly person then confronted Baringkot about his promise when he had pled to be given human form. When Baringkot could not answer, the old man ordered the workers to leave immediately. They had not sinned against the Gino-o for, although they had desired the sumptuous meal, they did not steal to eat. The workers sinned only against their own will because they wanted to eat but could not. The Gino-o told them to go home and prepare a simple meal of vegetables, and in that way live.

The old man went back to the house and, in anger that Baringkot reneged on his word, turned him back into a wormlike insect, this time with an even shorter length to make crawling difficult. Like Baringkot, all the guests were transformed into insects with tiny horns and cutaneous golden stripes, the only reminder of the fineries they had once worn. All of Baringkot's wealth vanished like smoke.

In this legend, the farm workers confess that they too have been tempted by the luxurious lifestyle of the landed elite. But the admission of some guilt begs the indulgence of the listener, for the poor workers' failing has been nothing more than to desire something natural, food, to satisfy their hunger. The story's focus is on the infidelity and baseness of the wealthy, which will not escape the day of reckoning. An indictment of sugar capitalist profligacy, the tale finds solace in the future but immanent justice of the Supreme Being, who identified himself with the lowly workers and their social distance from the affluent by entering through the rear of the rich man's house.

Although oppressed by the materiality of the social structure, *hacienda* workers have created in this legend a stoical rebuttal to Mestizo Power's manipulation of indigenous cosmology. But it is simply that, a rebuttal. The seventh layer, it seems, is rather distant and the Maka-ako is not engaged in everyday life. The sugar workers do see themselves as the pitiful underdogs relative to the affluent sugar capitalists. But, in the game of life, they just happen to be unlucky players. If a poor person can utilize magical connections with the *engkanto* for a little measure of social mobility, that person will seize the opportunity to obtain *suwerte* in whatever way. Social existence is powerfully perceived and lived as a gamble; stratification is the outcome of *dungan* rivalries that confound kin relations and intermingle social categories.

Despite the crisis of the mid-1970s and early 1980s, the *hacienda* workers of today congregate whenever they can to gamble. Using American playing cards easily obtainable from petty street traders in Bacolod, men and women, young and old intermingle: squatting around mats or bamboo-slatted sheds, *hacienda* workers play cards until sundown, the loyal bearers of tradition molded during the colonial epoch. Masonic Capitalism, after all, is a fun game. In the clash of spirits, one can bet and possibly win.

Abbreviations

AHN Archivo Historico Nacional, Madrid
AM Archivo de Marcilla, Navarre
APAF Archivo de la Provincia Agustiniana de Filipinas, Valladolid
BIA Bureau of Insular Affairs, U.S. War Department
B&R Emma Blair and James Robertson, eds., *The Philippine Islands 1493–1898*, 50 vols. (Cleveland, OH: The Arthur H. Park Company, 1903–1907).
BPI Bank of the Philippine Islands
FO Foreign Office, held at the Public Record Office, London
GPO Government Printing Office, Washington, DC
NL Newberry Library, Chicago
NOHC Negros Occidental Historical Commission, Bacolod
PIR Philippine Insurgent Records, PNL, Manila
PNA Philippine National Archives, Manila
PNB Philippine National Bank
PNL Philippine National Library, Manila
PR Provincial Records, in the PIR
RG Record Group, USNA
SD Selected Documents, in the PIR
s/n *sin numero* (unnumbered document at the AHN)
USNA United States National Archives, Washington, DC

Notes

Introduction

1. Nagano (1982), however, emphasizes the period's push toward "rationalization" centered around improved technology and the wage-labor system.

2. This statement reappears in McCoy 1984, 78. McCoy quotes approvingly the "counterhistory" written by Columban priests, which asserts that "Within thirteen years of his arrival at the Port City of Iloilo on 31 July 1856, Nicholas Loney had killed a city, raped a province, destroyed all local industry and initiative, and had set up an economic system which insured a life of increasing poverty for the vast majority of the people, and super profits for the rich" (75).

3. A more general statement on such duality is found in Ileto 1988.

4. For a fuller discussion of the theoretical concerns of this study, particularly the transcendence of the base-superstructure model, see Aguilar 1994a.

5. A specific instance in Philippine historiography of gross distortions arising from ignoring the structure–agency dynamic is the 1855 opening to world trade of provincial ports. See Aguilar 1994c.

CHAPTER 1 A Clash of Spirits

1. The captain-general, Mariano Fernandez de Folgueras, reported that 5,000 fell victim to the cholera epidemic. El Capitan General da cuenta de las novedades ocurridos en aquella Ysla con motivo de la Epidemia, coleramorbo, Manila 10 November 1820, AHN Legajo 5152, Expediente No. 3.

The first of a series of global cholera pandemics in the nineteenth century hit Asia from 1817 to 1824. About five months before the outbreak in Manila, an epidemic struck Siam (Thailand) in May 1820, claiming the lives of about 30,000 people (Terwiel 1987). In Java, an estimated 125,000 died during the cholera epidemic of 1821 (Boomgaard 1987). For a discussion of cholera epidemics in the nineteenth-century Philippines, see Huetz de Lemps (1990), Smith (1978), and De Bevoise (1995).

2. Cf. El Capitan General da cuenta de las novedades . . . , AHN Legajo 5152, Expediente No. 3. For a fantastic account of the massacre, see de la Gironiere (1962, chap. 1).

3. Along with the letter of Peter Dobell, the Russian consul, the decree of Folgueras is cited at length in B&R, vol. 51 (1907, 43–45).

4. The galleon trade reduced Manila to a mere transshipment point for the exchange of Chinese silk and Mexican silver. It was terminated in 1815.

5. El Capitan General de Filipinas manifiesta ser conveniente la remesa de Religiosos Españoles á aquellos dominios, AHN Legajo 5153, Expediente No. 57. Aguilar's letter of 25 November 1804 is quoted in the incumbent Captain-General Salazar's letter dated Manila 26 January 1837.

6. Describing them as "usually from the dregs of other nations," a civilian official accused foreigners in the 1820s of having "clandestinely introduced impious, revolutionary, and obscene books printed in the Spanish language, but pirated in France. . . ." (Manuel Bernaldez 1907, 207–208). In his secret report of the 1840s, the noted diplomat and "liberal" Sinibaldo de Mas looked unfavorably upon "the introduction of inopportune books," saying that he knew "of one who left in a house in a town in the province the history of the American Revolution." He recommended enforcing "the laws prohibiting foreigners from going to the provinces and not to open wide the doors to their admission to the capital. This policy is suspicious and unenlightened but still useful for *preserving the colony*" (de Mas 1963, 169; italics in original).

7. El Capitan General da cuenta de las novedades . . . , AHN Legajo 5152, Expediente No. 3.

8. Cf. Dobell (1907, 41); and Medidas para la admission y permanencia de estrangeros, AHN Legajo 2140, Expediente No. 10. See especially the Fiscal's *dictamen* of 9 November 1826 and the Real Tribunal de Comercio's memorandum dated 15 December 1826.

9. Reservado: El Capitan General de Filipinas Folgueras en carta de 10 de Noviembre 1820 da cuenta de las desgraciadas ocurrencias de aquel pais en Octubre anterior . . . , AHN Legajo 5152, Expediente No. 2. See especially the note from the Palace, 26 July 1821.

10. El Capitan General de Filipinas manifiesta ser conveniente la remesa de Religiosos Españoles . . . , AHN Legajo 5153, Expediente No. 57. The letters from two captains-general, dated Manila 12 April 1823 and 26 Janu-

ary 1837, expressed this opinion forcefully. On the history of the native clergy, cf. de la Costa (1969), and de la Costa and Schumacher (1980).

11. AHN Legajo 3150 contains several unnumbered *expedientes* (files) from 1820 to 1845 dealing with the topic "Misiones de Asia." Cf. especially the exhaustive report dated 19 February 1845.

12. There were limits to this liberal trend. In 1841 an English firm wanted to purchase the Dominican estate of Calamba, which had an annual production of 6,000 *cavans* of rice, 100 *quintals* of indigo, and 20,000 *pilones* of sugar. An indignant Dominican exclaimed: "But put the hacienda in English hands, and who will benefit? Only the foreigner, and the native will receive a daily pittance and he will return it in order to purchase cloths, food and foreign trinkets." Addressing the spectre of Protestantism, he concluded, "And within a few years Calamba would be a colony of pagans." The sale was not consummated (Cushner 1971, 208).

13. Contestaciones á ordenes expedidas por el Ministerio del Fomento, AHN Legajo 5153, Expediente No. 9. See especially the letter from Yntendencia General de Exercito, Manila 31 January 1833.

14. Jonathan M. Peele, comerciante Anglo-Americano en solicitud de permiso para residir, AHN Legajo 5153, Expediente No. 32. See the report from the Secretaria de la Seccion del Yndias del Consejo Real, Madrid 20 September 1834.

15. To avoid paying an additional 8 percent duty above the 6 percent charged the Spanish ships, merchants devised a system of transshipping imported goods from Europe to Spanish vessels in Hongkong or Singapore, which then brought these goods to Manila. Cf. Jagor 1965, 11; Mallat 1983, 465, 488–506; and William Farren to the earl of Aberdeen, No. 6, Manila 8 February 1845, FO 72/684.

16. In that year, the *Real Compañia de Filipinas,* organized in 1785 to stimulate agricultural production in the colony and trade with Spain, went bankrupt, thus forfeiting its exclusive trade rights.

17. Farren to the earl of Aberdeen, Manila 7 January 1845, FO 72/684.

18. For a succinct overview of the foreign merchant houses in nineteenth-century Manila from the perspective of the abaca trade, see Owen 1989, chap. 2. Note, however, that Owen tends to underemphasize the role of natives as middlemen in the first half of the century.

19. Farren to the earl of Aberdeen, Manila 3 January 1845, FO 72/684. Cf. statistical enclosure.

20. Farren to the earl of Aberdeen, Manila 19 July 1846, FO 72/708.

21. The philosopher George Santayana, son of Josefina Borras by a second marriage after Sturgis' death, was probably responding to this, then widely known, story by stressing in his autobiography that his mother's marriage to Sturgis was "rationally chosen" (Santayana 1944, 42). His account rectifies the misconception that Borras was a Spanish mestiza as, according

to him, she was born in Glasgow around 1826 of Catalan ancestry. In 1845, she sailed to the Spanish Philippines with her father, who very briefly held a position in one of the provincial governments. Orphaned at twenty and the only Spanish civilian on a remote island, she "adopted the native dress," (38) which gave her the appearance of a mestiza. After about two years, she went to Manila to live with a creole family. She was about twenty-three when she married the thirty-two-year-old Sturgis. Santayana describes his mother as follows: "if she was not a Protestant, at least she was no bigoted Catholic, but a stern, philosophical, virtuous soul" (42). Santayana's autobiography was recently republished; see Holzberger and Saatkamp 1986.

22. Two other chroniclers equated Laon, meaning Antiquity, with the Tagalog's Bathala Meycapal, but did not mention Laon as inhabiting the Negros volcano (Chirino 1969, 279; Colin 1906, 70).

23. Combining themes of life and death, reproduction and destruction, "complementary dualism" is a prevalent feature of thought in island Southeast Asia (Fox 1987, 520–527).

24. Variants of this reputedly Visayan folktale can be found among the Tinguian of northern Luzon and the Mandaya of Mindanao (M. C. Cole 1916, 65, 145, 201).

25. The *ng* in *dungan* is pronounced as in English "singer."

26. This custom was derided by a friar who wrote, "It is laughable to see them waken another who is sleeping like a stone, when they come up without making any noise and touching him very lightly with the point of the finger, will call him for two hours, until the sleeper finishes his sleep and awakens" (de San Agustin 1906, 211). The sleeper referred to could conceivably be a Spaniard, particularly a friar.

27. There is some parallelism between the concept of the *dungan* and its height among the hierarchical Visayan with the notion of the height of a man's heart as determined by *bèya* (knowledge) and *liget* (anger, passion, energy, force) among the more egalitarian Ilongot. See the fascinating ethnography of Michelle Rosaldo (1980).

28. The *ng* in *dunganon* is pronounced as in English "finger."

29. For an interpretation of the fluid hierarchy of precolonial Tagalog society, see Rafael 1988.

30. Suggestive of the relationship nurtured with the spirit-world, a chief in Bohol "kept many cups and small jars full of charms, together with other instruments for casting lots" and divination (Chirino 1969, 384). The Tagalogs offered sacrifices in "certain private oratories" owned by chiefs (de San Agustin 1906, 334). Until the late nineteenth century, the principal Bukidnon *datus* were in exclusive possession of a highly respected "idol called Tigbas" (Clotet 1906, 295, 304–305).

In seventeenth-century Mindanao, Sultan Kudrat, who practiced "sorcery" according to a Jesuit report, was the exemplar of a *datu*'s magical prowess. Among his extraordinary abilities, he could cause "the fish to enter his boat" and could make "a piece of artillery float on . . . water." He had tools for "good or evil augury." Because he talked "very familiarly with the devil," he became "a greater king than any of his forebears; for their fear of him is incredible, as they recognize in him one who has superior power to avenge himself" (Combes 1906, 138).

31. For the sociological connection between chief and shaman, cf., among the early Spanish accounts, Chirino (1969, 302) and More (1906, 204). For parallels in modern-day Bukidnon, see Cullen (1973, 8–9, 27–28).

32. On *datus* receiving "pay" for acting as judges, see de San Antonio 1906, 356–358. On the offering as a means of advanced appeasement, see Anon. 1979, 347 and Biernatzki 1973, 33–34.

CHAPTER 2   Cockfights and *Engkantos*

1. Until 1768, except for parish priests, some soldiers, and a handful of civilian officials, colonial law prohibited Spaniards from living in provinces beyond Manila. The friar was the only permanent colonial fixture in most areas outside the capital (Larkin 1972, 29).

2. Cited in Phelan 1959, 34. The same phrase was used in the conquest of Mexico (cf. Behar 1987, 34).

3. Some terms differ across ethnolinguistic groups, which is indicative of the absence of a national means of communication for most of the Spanish colonial epoch; other terms probably became generalized as a result of modern mass media. Nonetheless, the advent of colonial rule and increased mobility within given regions, I think, allowed for the common usage of these terms beyond the narrow confines of one village, town, or even province.

4. It was noted that the natives' "horror for Cafres and negroes . . . is so great that [they] would sooner suffer themselves to be killed than to receive them" (de San Agustin 1906, 254). Although *kaffir* is Islamic for "infidel," which might suggest that the entry to the islands of the word *kapre* might have predated Spanish colonialism, it is also probable that the dread of the *kapre* came with the cafres and other black slaves bought by Spaniards from Portuguese traders for use in the Spanish monastic estates in the Tagalog region, a practice that lasted until the 1690s (Cushner 1976, 47–48).

5. This list is certainly not exhaustive. There are also numerous terms to denote rites and implements, such as *santiguar, bentusa,* etc., which are of Spanish derivation.

6. For similar but less detailed descriptions arising from fieldwork in the Visayas, particularly Negros, see Hart 1966, 67, and Lieban 1962, 307.

7. Chapter 1, note 26 suggests that the natives perceived the Spaniards as possessing *dungan*.

8. See Colin 1906, 78, cited in Chapter 1. Cf. Rosaldo 1980, chap. 3.

9. On the golden age of Spanish Catholicism in the sixteenth century, cf. Bennassar 1979, 70–80. On the golden age of the Christianization of the Philippines, cf. Phelan 1959, 53–71. On the also literally golden century in Spain from the mid-sixteenth to the mid-seventeenth century, cf. Defourneaux 1970.

10. *Agnus Dei* is Latin for "Lamb of God."

11. An incident in early-seventeenth-century Binalbagan, Negros, demonstrated the process of substitution: a native chief who had a very sick two-year-old son offered the usual sacrifices, but "As he did not get what he was after, he begged father Fray Jacinto de San Fulgencio for a little water passed through the chalice. The father gave it to the sick child, and the latter was instantly cured" (L. de Jesus 1904, 244).

12. On the Santo Niño as a localized entity, and the icon's symbolism in contemporary Cebuano society, see Ness 1992, chap. 5. On the Santo Niño as a spirit in shamanic trances, cf. J. Bulatao 1984.

13. For a succinct discussion of egalitarian sex roles prior to Spanish colonialism, see Blanc-Szanton 1990, 354–358.

14. In present-day folklore, the fair-skinned child of a brown maiden is believed to have been sired by an *engkanto* (Demetrio 1970, 359).

15. The epidemiological havoc wrought by the Iberian conquest of the Americas is discussed by McNeill 1976, chap. 5.

16. The traditional forms of spirit appeasement were no longer possible in this context. Evidently, the generalized concept of pity and mercy *(awa, luoy)* denotes local constructs rooted in colonial Catholic imagery. In the preconquest age, the notion of pity and mercy had to give way to *dungan* contests, and even more so in war and plunder. Today, to show pity and mercy by preventing a child from experiencing *usug*, an adult recites the phrase *puwera usug*, clearly borrowing the Spanish command, ¡*fuera!* ("go away!").

17. For a focused discussion on gambling and its relation to Philippine state formation and electoral politics, see Aguilar 1994d.

18. On liminality in the ritual process, see V. Turner 1967, chap. 4.

19. For the association of cocks and other animals with magically endowed rulers in precolonial Southeast Asia, cf. Reid 1988, 189–195.

20. In 1858 an Austrian noted that cockfighting's "cruel, murderous issue is strangely in contrast with the mild, soft, timid character of the natives" (von Scherzer 1974, 238).

21. Primarily as fervent card players, women were just as engrossed in gambling. However, unlike the men whose loss of *datu*ship was, I imagine, analogous to castration, women were not driven by a compelling urge to identify with game fowls. Women were also prohibited by custom and the colonial state to enter the cockpit; cf. Capitulo 3 (chap. 3) of the 1861 cockfighting regulations in Artigas (1894, 154).

22. Natives later innovated with unlicensed cockfights, known as *tupada* (from the Spanish *topar*, to encounter) in inaccessible places.

23. But, in a profound continuity with the preconquest female priestly role, the Filipino Catholic priest has the image of being effeminate and unmanly (cf. Doherty 1964).

24. In the Visayas, three major uprisings occurred from the late sixteenth to the late eighteenth century: those led by Tamblot and Bankaw in Bohol and Samar in 1621, by Sumuroy in Samar in 1649, and by Dagohoy in Bohol in 1744. Angered at a curate's refusal to bury his brother, the legendary Dagohoy mobilized an estimated three thousand followers who attacked the friar and then fled to the mountains. When the rebellion ended some eighty-five years later, the following had swelled to twenty thousand people (Phelan 1959, 147–149).

25. On male circumcision, see de Loarca 1903, 119; Candish 1904, 296; de Morga 1904, 134; P. de Jesus 1906, 318–319; Dampier 1906, 35–42; Colin 1906, 88; and Ortiz 1906, 110. Of these sources, only Ortiz is positive about female circumcision, which, according to his report, used to be called *sonad*.

26. See n. 21 above.

27. In the early nineteenth century, de la Gironiere (1962, 43) described the *mabuting tao* as "most respected amongst his countrymen" and as "a real piratical chief; a fellow that would not hesitate to commit five or six murders in one expedition; but he was brave." Note how de la Gironiere is misunderstood by the editor. At the turn of the century, the unvanquished rebels in the final resistance against Spanish suzerainty, which spilled over to the war against American colonialism, were similarly known as *maayo nga lalaki* (interview on 8 September 1990 with Julian C., about seventy-eight years old, in Sagay, Negros Occidental, who talked about his grandfather as a magical anticolonial fighter who lived in a cave in Bantayan Island at the century's turn). In the 1990s, peasant men steeped in contemporary politics as well as magic are known as *magaling na lalaki* in Central Luzon (Fegan 1994). Whereas the colonial "good men" acted outside of the state framework, in the postcolonial period the exploits of the *magaling na lalaki* occur within it.

28. El Capitan General de Filipinas manifiesta ser conveniente la remesa de Religiosos Españoles . . . , AHN Legajo 5153, Expediente No. 57. Cf. Aguilar's 1804 letter quoted in Salazar's letter, Manila 26 January 1837.

29. The processes of landownership and transformation of production relations are discussed in the next chapter.

30. For the invention of the village in other parts of Asia, see Breman 1988.

31. The personal power of the *datu* was congruent with the ideal type of the Big Man in historical Melanesia (cf. Sahlins 1963).

32. Informed by formal anthropology, modern-day missionary analysis of

the changing *datu*ship in Mindanao suggests a remarkable continuity in that local leadership continues to be seen within a cosmological framework. The history of Bukidnon *datu*ship is inscribed, along with traditional law and ceremonial prescriptions, in the *giling*, "a black stick the length of one's forearm and hand." The *giling*'s sanctity emanates from its having been personally handed by a spirit called *dumalungdung* to a foremost *datu* of old (Biernatzki 1973, 19–20).

33. De Loarca (1903, 147, 149) observed that, among the Visayan *Pintados*, "If a *timagua* desires to live in a certain village, he joins himself to one of the chiefs . . . to whom he offers himself as his *timagua*. . . . For this service the chief is under obligation to defend the *timagua*, in his own person and those of his relatives. . . ."

34. The *gobernadorcillo* was an "elective" post during the first seventy-seven years of Spanish rule, but from 1642 it became appointive, although the officeholder was selected from a list of nominees annually drawn up by the *cabezas* and the parish priest (Phelan 1959, 124–125).

35. In a handful of cases, *indios*, predominantly Pampango, were gifted by the Spanish monarch with *encomiendas*, or tax farms, from the early seventeenth to the early eighteenth centuries (Santiago 1990).

36. Larkin (1972, 35–36) provides the highly revealing illustration of the town of Macabebe in Pampanga Province where from 1615 to 1765 the position of *gobernadorcillo* was rotated around thirteen different families only.

37. On the bilateral kinship systems in the Philippines, see the articles in Kikuchi 1989.

38. Until the end of the Spanish period, as is the practice in the Peninsula, a married woman kept her father's surname. Only under the Americans, with their Anglo-Saxon tradition, did a married woman assume the surname of her husband.

39. In contrast to the Catholic ban on cousin marriage to the third and fourth degrees, in the preconquest era first cousins could marry in some of the local societies.

40. In the Cagayan Valley, for example, despite an edict in 1642 proscribing trade and social intercourse with "the heathen, apostate, and fugitive [*indios*], negroes and Zambals, who inhabit the mountains and hills," the local *cabezas* and *gobernadorcillos* who were "largely drawn from the preconquest ruling elite" continued to trade salt, cotton, cloth, metal tools, and tobacco (the area's most valuable product) for mountain beeswax and gold, a pattern disrupted only by the temporary ban on local tobacco production from 1787 to 1800 (E. de Jesus 1978, 11, 88; E. de Jesus 1982, 27).

41. In the 1580s the lone bishop of the colony vehemently denounced this *encomendero* practice (D. de Salazar 1903, 222–224). In 1721, *encomiendas* were ordered not to be reassigned, paving the way for their dissolution (Phelan 1959, 97).

CHAPTER 3 Elusive Peasant, Weak State

1. Consider, for instance, the alliance built upon intermarriage that linked Cebu and Mactan at the time of Magellan's arrival in 1521 and its subsequent breakdown under the pressure of Legazpi's conquest in 1565 (W. H. Scott 1992a).

2. A recent restatement of this canon is found in Corpuz (1989, 38–39). Romance and an erroneous understanding of social relations in the *barangay* are evident in the following portrait, attributed to Jaime Bulatao, of the precolonial settlement: "everything was cozy within and everything dangerous without and, accordingly, the *kanayon* (relatives) [*sic*] had to stick together for their own safety." The *barangay* and the family allegedly constituted the Philippines' "lost Paradise" (Ramirez 1993, 25, 27).

3. See Chap. 2, n. 33.

4. Focusing exclusively on labor exchanges and failing to consider the elements of fear, gift exchange, magic, and warfare, Filipino historiography has portrayed the *barangay* as the romantic origin of an idealized *bayanihan* or "cooperative labor" (cf. Constantino 1975, 33; Corpuz 1989, 13–14).

5. Not grown extensively, wet rice was raised in swampland or the floodplain of rivers and lakes.

6. For example, on the number of children from one set of parents serving in the *datu*'s house, "when more [than one] are taken for service . . . they hold this as a grievance and a tyranny." Likewise, a servant in the *datu*'s house who had married was entitled to form a nuclear household, "unless the chief forces them to [return], which they consider as a great tyranny and offense" (Anon. 1979, 354).

7. Scott, for instance, says that "Most members of this [third] class live at such a low subsistence level that debt is a normal condition of their lives: it arises from outright loans for sustenance or from inability to pay fines, and its degree determines individual *oripun* rank" (W. H. Scott 1982, 123). Interestingly, he did not reconcile this need for "outright loans for sustenance" with his later description (W. H. Scott 1992b) of the exceedingly high carrying capacity of the environment that adequately fed permanent settlements.

8. I am here following the ideal-typical model set out by Collier (1988, chap. 2). As W. H. Scott (1994, 168) has also observed, "it was not uncommon for a suitor to offer himself to a prospective father-in-law if he could not meet the brideprice demanded."

9. Perhaps this would explain the substantial debt of gold owed by a Tagalog chief, together with his relatives and dependents, in the tenth century (Postma 1992).

10. As the most valuable object, gold was used to "pay" the *datu* for adju-

dicating cases and the shaman for priestly services (cf. Chirino 1969, 301; de San Antonio 1906, 356).

11. Debt peonage was reported as slavery in the Spanish accounts, a confusion that resulted in contradictory accounts in which what the Iberian colonizers thought were chattel slaves could own personal property as well as attain manumission through repayment of debt. There were chattel slaves prior to the Spanish conquest, not out of debt, but due to capture in raids. For the continuing debate on Philippine slavery, see W. H. Scott 1991.

12. Slavery, which in Spanish terminology included debt peonage, was officially abolished in 1591, but the practice was tolerated, especially to ensure Manila's food supply, until it was "finally outlawed" in 1692; both Spaniards and native elites had slaves, usually Muslims captured in war or bought from Portuguese Asian colonies (see Cushner 1976, 34, 38, 47–48, 120 n. 39, 123 n. 5; Larkin 1972, 37–38; Scott 1991).

13. Pace Bourdieu (1977, 159–197), whose argument on "social recognition" has privileged "objective, institutionalized mechanisms," I am arguing that individuals can "see through" certain aspects of their own societies as these undergo radical historical changes, in this case, the historical eclipse of the *datu*'s charisma.

14. Akin to the marriage banns, the announcement that a parcel of land was to be sold by a certain individual was made for four consecutive Sundays. This requirement apparently had some effect, especially in the Tagalog area, as land transactions were sometimes declared null and void if the proper announcements were not made (Cushner 1976, 26, 118 n. 12). The announcements were probably in Spanish, but the mention of a few key words could easily be imagined as spawning gossip and discussion.

15. In the Tagalog district, smallholding peasants disputed the claims to land by the ecclesiastical estates; many *indios,* however, lost their lands and protested that the tribute was hard to pay, given the tiny plots left to them (Cushner 1976, 21–22).

16. Although often a source of corruption and abuse by Spaniards and the native elite, tribute collection, until much of the eighteenth century, cannot sufficiently be proved to have created a new system of debt peonage as some have argued (e.g., Larkin 1972, 38–39). Flight was not a closed option at all.

17. Chinese gambling influences, particularly with the ascendancy of the Chinese mestizos, are definitely evident in local gambling concepts. It still holds, however, that the framework for their adoption is the Spanish colonial context and the native relationship to colonial power.

18. For a general discussion of earlier practices of infanticide, see Pedrosa 1983.

19. "Every Spaniard knows that there are certain [lottery ticket sellers]

who 'carry luck'—generally at the expense of some physical deformity . . ." (Bensusan 1910, 237).

20. In the 1760s a Spanish official observed that, "When [natives] gamble on a cockfight, a multitude of people assemble, and all of them carry ready money for this purpose" (de Viana 1907, 242).

21. For an excellent illustration, see the 1696 legal document that combined precolonial ideas of corporate kin control with the Spanish legal framework and thus empowered four members of the native elite in Maybunga, in the town of Pasig, "to take care and safeguard the property of the whole kin," "to lease *(magpabouis)* land," "to collect debts," "to remove land from any member of the kin who had not been faithful in holding land," and "to sell any property of the town [?] to a member of the kin, but not to anyone outside it." The document empowered the local elite to look after the legal defense of the land, but it was silent concerning what would be done with the lease payments and the debts collected. See Cushner (1976, 82–83).

22. It must be noted that most tenants on friar estates enjoyed exemption from corvee, which the colonial state liberally granted from the 1730s. The labor thus freed from mandatory state work was used gratuitously on the demesne maintained by some estates. In the early 1800s, however, grants of exemptions had either lapsed or been revoked (Roth 1982, 138–140), but other compelling reasons, such as the commercial movement after the opening of Manila in the 1830s and the changing population density and pressure on land, saw the retention of *indio* labor on the estates.

23. I owe this point to Zheng Liren, whose understanding of the term was informed by his grandmother, who lived part of her life in Negros and part in the Chinese mainland. Zheng also explains that, in the game of ping-pong, the word *acsa* is used, but as a derivative of the English word *outside*, which has added another layer of meaning to *acsa*.

24. Chinese merchants and artisans were themselves extremely anxious about their status in mainland society. For an analysis of the sociology of consumption of this stratum, see Clunas 1991.

25. De Loarca (1903, 183) reported the practice in which "two men formed a business partnership in which each placed the same amount of money" but only "one of them went to traffic with the money belonging to both."

26. Pearce's (1983) theorization of sharecropping, although in the context of generalized commodity production, sheds light on the possibility that "tenants meet landowners from a position of comparable economic power," impelled by the goal of the latter to gain access to labor.

27. Relativo al fomento del cultivo de la caña de azucar, AHN Legajo 429, Expediente No. 3. See Document No. 2: Al Sr. Ministro de Marina, Primera Secretaria del Despacho de Estado, Palacio 30 November 1842.

The Spanish ambassador even furnished Madrid with a copy of a sugar-growing contract used by the Dutch in Batavia.

28. Ibid. See note at the end of Document No. 2, dated 20 January 1843.

29. Ibid., see Document No. 4, Al Gobernador Capitan General de las Yslas Filipinas, Madrid 20 January 1843.

30. El Capitan General de Filipinas en carta de 12 de Julio de 1834 manifiesta las razones que en su sentir han contribuido al estado decadente que en el dia tiene el cultivo de la azucar y del añil, AHN Legajo 5153, Expediente No. 13.

31. Relativo al fomento . . . azucar, AHN Legajo 429, Expediente No. 3. See Documents No. 7, Ayuntamiento de Manila, 28 September 1843; No. 8, Real Tribunal de Comercio de Manila 23 December 1843; and No. 9, Real Sociedad Economica de Amigos de Filipinas, 22 June 1844.

32. Ibid., Document No. 7. Sugar production was estimated, based on one *balita* of land measuring 1,000 *brazas cuadradas* in Tondo Province and planted with 5,000 cane points, to have yielded 10 *pilones* valued at a conservative 2 pesos 6 *reales* each *pilon*. The production costs included lease of land (9 *reales*), cane points (2 pesos 4 *reales*), *"beneficio de la tierra y el completo del azucar"* (5 pesos 4 *reales*), *maestro Chino* or chief sugar clayer (2 *reales*), cost of *pilon* or *casco* (14 *cuartos*). The net profit per *balita* from sugar was placed at 15 pesos 4 *reales*, which was higher than the equivalent profit for rice of 10 pesos 4 *reales* 9 *granos* but less than that for indigo, 20 pesos 7 *reales*.

33. Ibid., Documents Nos. 8 and 9.

34. Data for 1845 showed the following sugar production in Filipinas, in units of (variable) *pilones* (averaging 110 pounds), which was also interchangeably used with *picos*:

		Pilones	Percent
Luzon	Pampanga	94,587	35.7
	Bulacan	51,930	19.6
	Bataan	13,571	5.1
	Pangasinan	10,811	4.1
	Laguna	7,029	2.6
	Tondo	6,019	2.3
	Cavite	2,206	0.8
Visayas	Cebu	63,582	24.0
	Iloilo	15,310	5.8
Totals		265,045	100.0

Source: Circular a las provincias que espresa para que los Gefes de ellas remitan noticia del numero de pilones de azucar que ha cosechado en cada pueblo el año procsimo anterior con la clarificacion que se espresa, 7 May 1846, PNA *Spanish Manila*, Bundle 7.

35. Relativo al fomento . . . azucar, AHN Legajo 429, Expediente No. 3; see Document No. 8. For an indication of the rise in sugar exports, with 1836 as base year, see Larkin 1993, 249.

36. The financial crisis in Madrid deepened in 1854: it required over a million dollars to be drawn from the Philippine treasury to assist Spain; another 180,000 dollars were obtained from the *Obras Pias* for the acquisition of three steamers for the mail and coast service to the Peninsula. Farren to the Earl of Clarendon, Manila 11 July 1854, FO 72/853.

37. Relativo al fomento . . . azucar, AHN Legajo 429, Expediente No. 3, Document No. 8.

38. Sugar prices in the Spanish Philippines were reputedly the lowest in the East. Farren to the earl of Aberdeen, Manila 3 January 1845, FO 72/684.

39. Born of the indelible images of that massacre, Farren, the British consul writing in 1848, expressed apprehensions about "the elements of barbarism in the native character," which might be incited by a new round of ecclesiastical bigotry appearing in *La Esperanza,* one of only two papers in Manila. It portrayed England as a country "insatiable in its selfishness, reckless of its starving poor, incendiary in its politics abroad, and oppressors of Catholicism at home." Farren to Palmerston, Manila 5 October 1848, FO 72/749.

40. See, for example, Sobre el robo y asesinato de dos subditos Anglo-Americanos, 1854, AHN Legajo 5165, Expediente No. 59. The two were involved in an abaca factory in Santa Mesa in the Manila area.

41. Relativo al fomento . . . azucar, AHN Legajo 429, Expediente No. 3, Document No. 8.

42. Ibid.

43. The decree issued on 23 September 1843, which applied to both *indio* and mestizo, required that the *cabeza de barangay* certify to the payment of tribute, and that the town magistrate, provincial governor, and parish priest be informed before a transfer of residence could be approved. Proponiendo reforma de la organizacion actual de las Cabecerias de Barangay, 1888, AHN Legajo 1488, Expediente No. 37. See Loney to Lord Stanley, Manila 31 January 1867 FO 72/1155.

44. In populous Java, however, the pass system introduced in 1816 appeared to have worked so well that, despite its revocation in 1863, "local administration . . . clung to what had become habit" (Breman 1990, 5).

45. Relativo al fomento . . . azucar, AHN Legajo 429, Expediente No. 3. In Document No. 8, the Tribunal bewailed: "¡Estraña y violenta situacion, en que juzgando la legislacion en la naturaleza de las cosas, los protegidos neofitos se han convertido en estafadores impunes, y en victimas los filantropicos protectores!"

46. For an analysis of contemporary social relations in rice farming, see Aguilar 1989.

47. Relativo al fomento . . . azucar, AHN Legajo 429, Expediente No. 3. Cf. El Secretario general del Consejo Real, 26 May 1846.

48. Comunicacion del Ministerio de Marina sobre mejorar la agricultura en aquellas Yslas, AHN Legajo 2159, Expediente No. 6. Cf. Señor Ministro de la Gobernacion de Ultramar al Gobernador Capitan General de las islas Filipinas, Madrid 15 September 1846.

49. Sobre creacion de una Escuela de Agricultura en Manila, 1887, AHN Legajo 609, Expediente No. 281.

50. Expediente relativo a la instalacion de la Granja-modelo de Visayas (La Carlota), 1887, AHN Legajo 473, Expediente No. 31; Sobre la ejecucion del mapa agronomico y formacion de las estadisticas agricola y pecuaria de la Isla de Negros, 1886; Legajo 473, Expediente No. 20; and Planos del proyecto de Granja Modela de Visayas, 1884, Legajo 476, Expediente No. 13.

51. From 1875 to 1878 close to 18,872 hectars were claimed throughout the archipelago in 121 *expedientes,* the average being 156 hectares per claim. Of the total area claimed, less than 8 percent was located in Negros, predominantly in the Pontevedra area. See Disposiciones adoptadas para la venta de terrenos baldios realengos, 1878, AHN Legajo 528, Expediente No. 35; and Concession de terrenos baldios del Estado—solicitada por varios particulares, 1876, AHN Legajo 529, Expediente No. 38.

52. Sobre venta de terrenos baldios, AHN Legajo 565, Expediente No. 8. See the letter of "sociedades extrangeras" to the Ultramar, Manila 19 November 1887; and the Governor-General's transmittal letter, Manila 24 December 1887. A copy of the regulations approved on 19 January 1883 may be found in: Venta de terrenos baldios del Estado, AHN Legajo 1488, Expediente No. 65.

53. Sobre creacion de colonias agricolas en Filipinas—Aplicacion al Archipelago de la Ley de colonias agricolas de la Peninsula de 3 de Junio de 1868, con las modificaciones oportunas, 1884, AHN Legajo 473, Expediente No. 17.

54. Sobre la conveniencia de adquirir en el Ymperio Chino niños de tierna edad con el objeto de aumentar la poblacion de Filipinas, AHN Legajo 5162, Expediente No. 24. At mid-century, civilian and religious authorities seriously considered a plan to purchase Chinese children "of tender age" to be brought to the colony as well-conditioned subjects and laborers. For the cost-benefit analysis of each child, see Sinibaldo de Mas, Legacion de España en China, to the Captain-General of Filipinas, Macao 20 November 1849. See also Primera Secretaria del Despacho de Estado to the Ministro de la Gobernacion del Reino, Madrid 31 January 1850.

55. See, for example, AHN Legajo 476, Expediente No. 27, containing the petition opposing Chinese immigration, with signatures from all over the archipelago, which the Governor-General forwarded to the Ultramar in

1886. Interestingly, there were no signatories from the major exporting districts of Iloilo, Negros, and Cebu.

56. See, for instance, the 1888 case of the deported Cuban Agustin Bao y Rol in AHN Legajo 5257, Expediente No. 43.

57. Ynformes . . . en el proyecto formado para la inmigracion de colonos españoles y braceros asiaticos, AHN Legajo 476, Expediente No. 29. The *Real Sociedad Economica de Amigos del Pais* bluntly stated in its letter from Manila, 31 October 1884: "la razon politica se opone á la inmigracion y se añade a modo de corolario que <u>España domina este Archipelago por reconocida superioridad de raza: y que los españoles son respetados y acatados como hombres superiores a los indigenas, pues lo son de hecho, y el dia en que perdieramos ese respeto, ese prestigio que nadie nos disputa, nuestra suerte estaria decidida</u>" (underscoring in original). Another reason adduced by the *Consejo de Administracion de Filipinas* on 8 April 1891, and by the *Real Sociedad Economica* on 7 January 1891, was the fact that the *indio* was cheaper at two *reales* per day for as much as ten hours of work, in contrast to the *jornalero Europeo*, who could work only five hours but demand one peso as daily wage.

58. Loney to Lord Stanley, Manila 31 January 1867, FO 72/1155.

59. Creacion del 3er. tercio de la Guardia Civil, con destino a Visayas, AHN Legajo 5328, Expediente No. 21; Sobre establecimiento de la Guardia civil en algunas provincias del Archipelago, 1867, AHN Legajo 5329, Expediente No. 32; and Filipinas: Guardia Civil, 1855, AHN Legajo 5360 (expediente s/n).

60. Sobre creacion de un Gobierno, una Alcaldia y una Administracion depositaria en la Ysla de Negros, 1861–1867, AHN Legajo 2243 (expediente s/n); and Visita girada por el G. G. á la Ysla de Negros y disposiciones tomadas por dicha autoridad modificando la division Politico-Militar en dicho territorio, AHN Legajo 5339, Expediente No. 84.

61. Colonial policy disorientation affected all of Spain's remaining colonies (Marias 1990, 366–370).

62. On the initial hesitation of foreign capital to move to the Iloilo-Negros area, see Aguilar 1994c, 86–90.

CHAPTER 4   The Formation of a Landed *Hacendero* Class in Negros

1. Loney to Farren, Iloilo 12 April 1857, FO 72/927.
2. Ibid.
3. Loney to Farren, Iloilo 10 July 1861, FO 72/1017.
4. Loney to Earl Russell, Iloilo 10 September 1864, FO 72/1087.
5. F. Ricketts to Lord Stanley, Manila 19 September 1868, FO 72/1193.
6. Ricketts to the earl of Granville, London 5 May 1873, FO 72/1355.
7. Loney to Farren, Iloilo 12 April 1857, FO 72/927.
8. Loney to Farren, Iloilo 10 July 1861, FO 72/1017.

9. Ibid.

10. Loney to Farren, Iloilo 12 April 1857, FO 72/927.

11. Loney to Farren, Iloilo 10 July 1861, FO 72/1017. McCoy (1977, 67–70) has shown a remarkable connection between Iloilo province's pre-1855 elite and the emerging sugar-planter class in Negros. Among the "family names" of former Molo *gobernadorcillos* found in Negros were those of Locsing, de la Rama, Maravilla, Araneta, Yusay, and Tiongco. In the case of former *cabezas de barangay* in Molo, the *naturales* or *indio* list included the "family names" of Villanueva, Arroyo, Maravilla, Hechanova, and Magalona, while the mestizo list included Lacson, Montelibano, Severino Yuson, and Araneta. However, because a mestiza who married an *indio* fell into the husband's legal category, it is not certain that the *naturales* had no mestizo admixtures (cf. Wickberg 1964, 65).

12. Loney 1964, 82; Loney to Farren, Iloilo 10 July 1861, FO 72/1017.

13. Ibid., 97; Loney to Farren, Iloilo 10 July 1861, FO 72/1017.

14. Loney to Farren, Iloilo 10 July 1861, FO 72/1017.

15. Ibid.

16. In 1844, Montilla was described as "having in his disposition ninety-five individuals" organized as two *cabecerias* fitted with a fort and cannons to fend off Moro raids. Rice, corn, mung beans, cotton, abaca, coconut, and other crops were grown in an area that stretched to a total of 500 hectares drawn from lands "abandoned" by former inhabitants. Montilla supplied the peasants with carabaos. Sugar was not mentioned as a crop. Espediente promovido por Don Agustin Montilla solicitando aprobacion de una estancia agricola—Pulupandan, PNA *Varias Provincias, Negros*.

17. Luzuriaga was a former *cobrador de tributos,* or tribute collector, against whom a complaint was lodged in 1844 by the *cabezas de barangay* of Bacolod; the case was later withdrawn. En la Casa Real del Pueblo de Himamaylan, Cavecera de la Provincia de Ysla de Negros . . . ante el Alcalde Mayor D. Jose Saenz de Vizmanos, 7 September 1844, PNA *Cabezas de Barangay, Negros Occidental*.

18. Loney to Farren, Iloilo 10 July 1861, FO 72/1017; and Cuesta 1980, 366.

19. Memoria de la Ysla de Negros, 31 May 1888, PNA *Memoria de Negros, Oriental y Occidental*. See the "Datos estadisticos conocidos."

20. The largest concentrations of *blancos* were found on the east coast of Negros: Bais (38), Dumaguete (31), and Tanjay (43); followed by towns in the western side of the island: Pontevedra (22), La Carlota (20), Minuluan (20), Bacolod (12), and Jimaymaylan (11). Cf. Figure 2.

21. Estado que demuestra el numero de individuos que se considera existen en cada pueblo de este distrito, comprendidos en las categorias que se espresan á continuacion, referentes á la riqueza rustica é industrial pecuaria,

segun los comprobantes que se acompañan, Bacolod 28 December 1881, PNA *Varias Provincias, Negros.*

22. The nine towns with at least 40 big taxpayers were: Ginigaran (193), Minuluan (135), Zamboanguita (98), Nueva Valencia (52), Calatrava (48), Bago (47), Bais (43), Cabancalan (43), and Valladolid (41). The seven towns with smaller concentrations of 20 to less than 40 such taxpayers were: Silay (38), Cadiz Nuevo (32), Jimamaylan (32), La Carlota (32), Cabancalan (29), Pontevedra (28), and Bacolod (22).

23. Some 64.8 percent of the 1,009 names fell in the 200 to 600 pesos tax category, while about 16.6 percent were in the over 600 to 1,000 pesos tax category.

24. The 187 residents assessed to pay over 1,000 pesos in taxes were from Ginigaran (48), Saravia (25), Valladolid (20), Bais (13), La Carlota (10), Minuluan (10), Victorias (9), Cadiz Nuevo (7), Jimaymaylan (6), Silay (6), Bago (4), Nueva Valencia (4), San Enrique (4), Suay (2), and one each from Argüelles, Bacolod, Calatrava, Granada, and Murcia.

25. Acta of Bacolod *principales*, 28 November 1881, PNA *Varias Provincias, Negros*.

26. Loney was keenly aware of the consequences of doing otherwise. In a letter to his father dated 23 October 1865, he noted the problems encountered by a member of the British consulate in Manila: "Though an estimable, gentlemanly person, his dislike of Spaniards is so great that they have discerned and resented it, never calling on him or exchanging the usual social amenities of that kind" (Loney 1964, 113).

27. Cronica semihistoria de Filipinas y en especial de las Yslas Visayas desde 1877 á 1887, folio 1, NL Ayer Collection MS 1390.

28. Mallat (1983, 301) further observed that the mixing of social categories was more prevalent in the opium dens where card games were played: "*Monte* is a game of chance prohibited by law; but it is played very fast and is easy to conceal . . . it is played especially in places where opium is smoked, and there all ranks, all sexes are mingled. . . ." In 1893 the three-year franchise of the opium monopoly in Negros was farmed out to Yap-Sangco of Silay for 91,800 pesos, surpassing the bid tendered by Nicasio Veloso Chiong-Tuico of Manila for 62,310 pesos and that by Ong-Songty of Bacolod for 26,700 pesos. Among the *comisionados* (agents) of Yap-Sangco were Gregorio Echauz Tan-Chipco, Tan Guanco, Yap-Chiengjo, Uy-Suatco, and Du-Coman. Espediente interior referente al arriendo de los fumadores de anfion de las provincias de Ysla de Negros Oriental y Occidental, 1893, PNA *Anfion, Negros Oriental and Occidental.*

29. Loney to Farren, Iloilo 10 July 1861, FO 72/1017; Loney 1964, 66.

30. Loney to Farren, Iloilo 12 April 1857, FO 72/927.

31. The *hacenderos* were described as "carece en absoluto de instruccion agricola, y podemos asegurar el noventa y cinco por ciento de ellos hu-

bieran sido muy utiles para cualquiera empresa, pero son nulidades en esta clase de industrias." Memoria descriptiva del estado de Agricultura, de la ganaderia é industria que de ellas se derivan, en Ysla de Negros con indicacion de los medios y procedimientos que pueden emplearse para mejorarlas; escrita para la Esposicion general de la Yslas Filipinas que ha de tener en Madrid el año de 1887, PNA *Memoria de Negros, Oriental y Occidental.*

32. Devoting "most of their time to either politics or other business," Pampanga *hacenderos* "entrust the whole responsibility of field operation to their '*aparceros*' [share tenants] who know no more than they do" (Ocampo 1923).

33. Carlos Locsin was at the forefront of this effort to systematize cost accounting. See Locsin 1924.

34. The illegalization of gambling in the United States in the 1890s made all other forms of gain appear normal and moral (Fabian 1990).

35. Varona 1965, 28th installment.

36. For a similar gamble with the state through manipulation of land laws in Nueva Ecija during the Spanish and American colonial periods, cf. McLennan 1984, 70–72.

37. Incidente sobre suspension del interdicto de despojo promovido por varios españoles residentes en la Carlota Ysla de Negros, 12 August 1876, PNA *Terrenos (Negros)*, Bundle 2. The signatories were Miguel Perez, Teodoro Gurrea, Lucas Rubin, Alejandro de la Viña, Domingo Tejido, Manuel Pacheco, Juan Rubin, Jose Maria Guaco, Pedro Camon, Manuel Lansirvia, and Jose Amonategui. Only de la Viña was positively identified as Creole, "*un español Filipino.*"

38. The petitioners added that if "valiant" *indios* were on the land, "matters did not terminate so peacefully" because the victims exacted vengeance according to the "natural instincts" of the "primitive race"; otherwise, the displaced *indios* simply retreated to the mountains (ibid.).

39. Broken down as follows: 300 hectares in La Carlota, 2,600 hectares in Bungayin, Isabela, an undetermined area in Antipolo, Pontevedra, and 1,600 hectares in La Castellana, then part of Isabela (ibid.).

40. Ibid.

41. Ramon Pastor, Gobernador Politico-Militar of Negros, to the Governor-General, Bacolod 31 January 1877, PNA *Terrenos (Negros)*, Bundle 2. See Enclosure No. 2.

42. Ibid.

43. Ibid. Cf. Enclosure No. 1, "Copia de escritura de composicion de terrenos," 3 July 1866; and "Copia de escritura de compra de tierras en el barrio de Borja," formalized on 7 August 1876.

44. Protocolo de escrituras publicas otorgadas y protocolizadas en este Juzgado de la Ysla de Negros, 1875, PNA *Protocolos*, Legajo 1744, folio 779ff.

45. Ramon Pastor, Gobernador Politico-Militar of Negros, to the Governor-

General, Bacolod 31 January 1877, PNA *Terrenos (Negros)*, Bundle 2. Cf. Enclosure No. 2.

46. Ibid. Cf. Enclosure No. 3, letter of Fr. Eustaquio Cascarro, San Enrique 31 October 1876.

47. Letter of Manuel Tomonong of La Carlota, dated Bacolod 27 October 1872, PNA *Terrenos (Negros)*, Bundle 2.

48. Ramon Pastor, Gobernador Politico-Militar of Negros, to the Governor-General, Bacolod 31 January 1877, PNA *Terrenos (Negros)*, Bundle 2.

49. Real Audiencia Territorial de Filipinas, Tribunal Pleno, espediente sobre deslinde de los Montes del Estado confinantes con la hacienda de San Bernardino de Don Teodoro Benedicto en la provincia de Ysla de Negros, 1883, PNA *Terrenos (Negros)*, Bundle 2. According to the final result of the investigation, the documents of purchase presented by Benedicto covered about 215 hectares only.

50. Ibid. Cf. testimony of Teodoro Benedicto on 14 March 1877.

51. Ibid. Cf. testimony of Vicente Agatis on 8 March 1877.

52. Escritura de reconocimiento de credito con hipoteca otorgada por Don Teodoro Gurrea á favor de Don Teodoro Benedicto, Bacolod 18 August 1875, PNA *Protocolos*, Legajo 1744, folios 646ff.

53. Real Audiencia Territorial de Filipinas, Tribunal Pleno, espediente sobre deslinde de los Montes del Estado confinantes con la hacienda de San Bernardino . . . , PNA *Terrenos (Negros)*, Bundle 2. Cf. Dictamen del Señor Fiscal dated Manila 15 June 1883, concurred in by the Tribunal Pleno de la Real Audiencia de Manila on 25 June 1883.

54. Instancia elevada á la Direccion General de Hacienda por Don Carlos Gemora, vecino de Ylog, provincia de Ysla de Negros, 1880, PNA *Terrenos (Negros)*, Bundle 2. Cf. *minuta* of letter from the Señor Yntendente general de Hacienda, Manila 4 June 1880.

55. Memorandum of Fernandez Cañete, Manila 13 October 1876, PNA *Terrenos (Negros)*, Bundle 2.

56. Order of the Governor-General to the Gobernador P. M. de Ysla de Negros, Manila 3 January 1877; and Enclosure No. 4, Ramon Pastor to the Governor-General, Bacolod 31 January 1877, PNA *Terrenos (Negros)*, Bundle 2.

57. Dictamen de la Comision nombrada por la Real Sociedad . . . pidiendo á S.M. por conducto del Excmo. Gobernador General, se reforme la legislacion vigente en estas Islas sobre interdictos, Manila 1 March 1880, PNA *Real Sociedad de Amigos del Pais*, Bundle 2.

58. Complaint by Don Francisco Jiz de Ortega representing Don Antonio Buenafe, to the Señor Juez de primera instancia del distrito de Ysla de Negros, Iloilo 26 February 1889, PNA *Terrenos (Negros)*, Bundle 2.

59. Alegato, Al Juzgado de primera instancia por Don Antonio Jayme y Ledesma, Abogado de Don Julian Giguiento, Bacolod 17 August 1891, PNA *Terrenos (Negros)*, Bundle 1.

60. Protocolo de escrituras publicas otorgadas y protocolizadas en este Juzgado de la Ysla de Negros, PNA *Protocolos,* Legajo 1722 (1860 to 8 November 1861) and Legajo 1723 (26 November 1861 to 1862). Records for most of September, October, and November 1861 have all been virtually erased. The fungus-infested records were formerly kept in Manila's Old Bilibid Prison, which was frequently under water.

61. PNA *Protocolos,* Legajo 1743 (January to June 1875) and Legajo 1744 (July to December 1875). Spaniards figured in only slightly over one-fifth of twenty-three land sales.

62. Calculable data for nineteen of twenty-one mortgage cases indicated an average transaction involving 45.5 hectares valued at 1,306.5 pesos, or about 26.5 pesos per hectare. The *protocolos,* however, yielded only a handful of notarized cases of *pacto de retroventa:* three in 1875 and six in 1890. Cf. Ibid. for 1875 and Legajo 1772 for 1890 data.

63. PNA *Protocolos,* Legajo 1772 (1890).

64. Ibid. Through this auction, Alejandro Montelibano acquired 507 hectares in Saravia, and the Spaniards Francisco and Lucas Rubin a composite 258 hectares in Cabancalan.

65. Proyecto de Ley de Credito agricola, AHN Legajo 2322, Expediente No. 9. Cf. Estado de los registros de la propiedad segun la Ley Hipotecaria, 1894 to 1896.

66. A similar pattern was observed in Pampanga, where one data set showed "estates" ranging from 10 to 1,033 hectares, with the average computed at 155 hectares (Larkin 1972, 76).

67. In 1950 about 1.5 percent of the individual farm holdings in Latin America exceeded 15,000 hectares and accounted for about half of all agricultural land (Farley 1972, 176). Another estimate reckoned the average *latifundio* as measuring about 41,000 hectares (Garcia 1966, 135). In Argentina in 1929, three families each owned more than 200,000 hectares and twelve each owned more than 100,000 hectares in the most fertile portions of Buenos Aires (Cole 1965, 99).

68. Lineas Telegraficas aereas de las Visayas, Panay, Negros y Cebu y cables para unirlas con Luzon, AHN Legajo 607, Expediente No. 259. Cf. "Comercio nacional y estrangero de Yloilo y agricultores de las Yslas de Panay y Negros" to the Ultramar, Iloilo 28 December 1892.

69. Loney to Farren, Iloilo 10 July 1861, FO 72/1017.

70. Ibid.

71. The steam engines had capacities ranging from 4 to 18 horsepower and the water-powered mills from 4 to at most 12 horsepower. "Memoria descriptiva del estado de Agricultura . . . en Ysla de Negros . . . escrita para la Esposicion general . . . en Madrid el año de 1887," PNA *Memoria de Negros, Oriental y Occidental.* See also Echauz 1894, 35, 37.

72. "Sobre los medios de colonizar la Ysla de Negros," AHN Legajo 447,

Expediente No. 15. Cf. Remigio Molto to Gobernador Superior Civil, Cebu 13 August 1864 and 17 August 1865.

73. Remigio Molto to Gobernador Superior Civil, Cebu 13 August 1864. Molto referred specifically to *Españoles Europeos,* or peninsulars, whom he described as "pobres pues que en su mayor parte proceden de licenciados del Ejercito, empleados de corto sueldo &ra."

74. Ibid. Molto also complained that in Iloilo Spaniards had been "relegated" to the coasting trade while, in international shipping, no Spanish vessels were being utilized and benefited by Iloilo's foreign trade.

75. Ibid., 1865 letter.

76. Eugenio Lopez, for instance, acquired 840 hectars in Cadiz Nuevo in a public auction following the bankruptcy of Russell and Sturgis; in 1889, he also bought 80 hectars in Cadiz from John MacNab, a British commercial agent, and in 1891 he acquired a 535-hectare *hacienda* in Pontevedra from James Fleming Macleod, a British trader (O. Lopez 1982, xlv).

77. Despite the terrible condition of Negros' roads, the *hacenderos* simply continued to rely upon corvee labor, even though it was already plainly deficient for maintenance of public works. Some found an argument. In 1889 the Spanish *hacienda* owners of La Carlota, Pontevedra, and San Enrique complained that the *"vias publicas"* to these interior towns were not properly maintained because the *gobernadorcillo* of San Enrique, required to serve as corvee supervisor, had paid more attention to "his pernicious habit of smoking opium." But just as the *hacenderos* were gambling, so did the *gobernadorcillo* enjoy his opium. Tovar to the Governor-General, Bacolod 7 March 1889, PNA *Varias Provincias, Negros.*

78. PNA *Protocolos,* Legajo 1722 and 1723 for 1861; Legajo 1743 and 1744 for 1875; and Legajo 1772 for 1890.

79. Memoria descriptiva del estado de Agricultura . . . en Ysla de Negros . . . escrita para la Esposicion general . . . en Madrid el año de 1887, PNA *Memoria de Negros, Oriental y Occidental.*

80. Ibid.

81. "Cronica semihistoria de Filipinas y en especial de las Yslas Visayas desde 1877 á 1887," NL Ayer Collection MS 1390.

82. Expediente exterior contra el vecion del pueblo de la Ysabela . . . Gavino Gasataya, 1891; Espediente principal referente á la alzada suscrita por D. Baldomero de la Rama impugnando de la providencia de la subalterna en la Ysla de Negros Occidental, 1892; Recurso de alzada interpuesta por D. Ramon Puentevella del pueblo de Minuluan; all in PNA *Varias Provincias, Negros.*

83. Asunto: Azucares de Filipinas, AHN Legajo 1492, Expediente No. 8. Cf. Real Sociedad Economica de Amigos del Pais to the Governor-General, Manila 7 February 1885.

84. Ibid. Cf. "los comerciantes y agricultores de Yloilo y Negros en las

Yslas Filipinas y 'El Porvenir de Bisayas'" to the Ministro de Ultramar, Iloilo 20 January 1885.

85. Ibid. Cf. "las casas mas importantes del comercio de esta plaza" to the Ministro de Ultramar, Manila 10 December 1884.

86. Ibid. Cf. Ministro de Ultramar to the Governor-General of Filipinas, Madrid 10 July 1885.

87. Mocion pidiendo franquicias para la produccion azucarera de Filipinas, AHN Legajo 5314, Expediente No. 480 (cf. Ortiga y Rey, Madrid 16 November 1884); Asunto: Azucares de Filipinas, AHN Legajo 1492, Expediente No. 8 (cf. Governor-General to the Ultramar, Manila 29 December 1884 and letter of Intendente General de Hacienda remitted by the Governor-General to the Ultramar, Manila 14 March 1885).

88. Asunto: Azucares de Filipinas, AHN Legajo 1492, Expediente No. 8. Cf. "los comerciantes y agricultores de Yloilo y Negros" to the Ultramar, Iloilo 20 January 1885.

89. Ibid.

90. Ibid. Cf. Ministro de Ultramar to the Governor-General of Filipinas, Madrid 28 March 1887.

91. See communications between U.S. Secretary of State T. F. Bayard and Spanish Envoy E. de Muruaga (United States 1888, 1030–5).

92. See communications between Minister of State Alejandro Groizard and Head of the U.S. Legation Hannis Taylor (United States 1895, 626–635; United States 1896, 1185–1186).

93. Lineas Telegraficas aereas de las Visayas, Panay, Negros y Cebu y cables para unirlas con Luzon, AHN Legajo 607, Expediente No. 259. Cf. Governor-General to the Ultramar, Manila 22 December 1879.

94. Ibid. Cf. Camara de Comercio de Manila to the Ultramar, Manila 6 October 1890.

95. Ibid. Cf. "Comercio nacional y estrangero de Yloilo y agricultores de las Yslas de Panay y Negros" to the Ultramar, Iloilo 28 December 1892.

96. "Expediente para que los telegramas que se expidan á las Yslas Filipinas sean justificadas—igual que los internacionales y abono de $35'55 importe de la media taza de los telegramas expedidos en la Ysla de Negros Occidental, á razon de 25 centimos de franco una," PNA *Administracion Central de Impuestos, Negros Occidental.* Cf. Direccion General de Administracion Civil to the Yntendente General de Hacienda, Manila 29 November 1897.

CHAPTER 5 "Capitalists Begging for Laborers"

1. Loney to Farren, Iloilo 10 July 1861, FO 72/1017.

2. "Memoria descriptiva del estado de Agricultura . . . en Ysla de Negros . . . para la Esposicion general . . . en Madrid el año de 1887," PNA *Memoria de Negros, Oriental y Occidental.*

3. Loney to Farren, Iloilo 10 July 1861, FO 72/1017.
4. Loney to Lord Stanley, Manila 31 January 1867, FO 72/1155.
5. Ibid.
6. In contrast to the long-distance migration of coolie labor across colonial Asia (cf. Breman 1990).
7. Loney to Lord Stanley, Manila 31 January 1867, FO 72/1155.
8. Expediente principal relativo á las disposiciones que deben adaptarse con respecto á la reforma organica de la Administracion de Hacienda de Ysla de Negros y al empadronamiento de tributantes inmigrantes de provincias vecinas, 1883, PNA *Administracion Central de Impuestos, Negros Occidental.* Cf. Estevanez to the Señor Ynspector de Hacienda, Don Manuel Lahora, Bacolod 4 June 1883.
9. Ibid. See note by the parish priest of Ginigaran at the back of the table "Noticia que manifiesta el estado del Distrito de Ysla de Negros, con respecto a sus producciones agricolas y poblacion en 1880."
10. Loney to Lord Stanley, Manila 31 January 1867, FO 72/1155.
11. Expediente principal . . . con respecto a la reforma organica de la Administracion de Hacienda de Ysla de Negros . . . , PNA *Administracion Central de Impuestos, Negros Occidental.* Cf. Estevanez to Lahora, Bacolod 4 June 1883.
12. In 1857 Loney (1964, 70–71) observed that "In fact, it is at all times difficult to get a sufficient permanent number of workmen here for any object. The [natives] are so woefully apathetic and so fond of their ease, that as soon as they earn a *real* or two, nothing will induce them to work while a cent remains."
13. Memoria descriptiva del estado de Agricultura . . . en Ysla de Negros . . . para la Esposicion general . . . en Madrid el año de 1887, PNA *Memoria de Negros, Oriental y Occidental.*
14. In rare instances, *indios* who had incurred mounting debts in the course of working for a *hacienda* were made to sign a contract to work as *jornalero, cabo,* or *maestro* for a specified sum and to promise that they would not leave the farm until the debt was paid. Four such cases were found among the notarized records of 1875, but none in 1861 and 1890. PNA *Protocolos,* Legajo 1744 (21 August 1875, folio 663ff.; and 1 September, 21 September, and 28 September 1875, folio 686ff, 731ff, and 774ff.).
15. Estadistica general de los negocios terminados y pendientes de despacho en la Real Audiencia de Filipinas, 1865, AHN Legajo 2243 (expediente s/n).
16. In 1886, however, the number of incendiary cases dropped to six, which further dropped to three plus one attempted arson in 1887. Memoria de la Ysla de Negros, 1888, PNA *Memorias, Negros.*
17. Ibid. The number of mestizos, Spaniards, and Chinese convicted of crimes in Negros paled in comparison to the number of *indios*. In terms of occupational categories, farmers and day laborers together accounted for

about 70 to slightly over 80 percent of those convicted in the 1870s and 1880s. Mechanics and domestic workers accounted for small proportions. A handful of civil servants *(empleados publicos)* and traders *(comerciantes)* were also involved. Only one farm property owner was implicated in the judicial statistics.

18. Remigio Molto to the Governor-General, Cebu 24 November 1863, PNA *Varias Provincias, Negros.*

19. Asked to comment on the proposal to regulate *hacienda* labor, the governor of Antique stated that, in his province, "no existen haciendas de grande estension, como en Ysla de Negros sucede, donde los braceros habitan en Barrios o agrupaciones de Casas pertenecientes al dueño de la finca." Manuel Castellon to the Gobernador Politico-Militar de Visayas, San Jose de Buenavista 8 August 1883, PNA *Cargadores.*

20. Maltreatment was also a major factor in conflicts between employers and domestic workers in Spanish Manila (cf. Bankoff 1992). On contemporary class relations and shaming, see Pinches 1991.

21. Memoria descriptiva del estado de Agricultura . . . en Ysla de Negros . . . para la Esposicion general . . . en Madrid el año de 1887, PNA *Memoria de Negros, Oriental y Occidental.* This practice also became entrenched in Pampanga (cf. Ross 1920; Latham 1923).

22. British India passed the Workmen's Breach of Contract Act in 1859 and the Dutch East Indies a coolie ordinance in 1880 (Breman 1990, 20–22).

23. Estevanez to the Political-Military Governor of the Visayas, Bacolod 12 May 1883, PNA *Cargadores.*

24. Ibid. Cf. "Proyecto de 'Reglamento de jornaleros y policia rural' para isla de Negros" dated 12 May 1883.

25. Ibid. See articles 16 and 17 of the "Proyecto de 'Reglamento de jornaleros. . . .'"

26. The documents specifying the response of the Manila government to the proposed regulations were not found in the archives. Evidently, no resolution was arrived at, for by early 1889 a different colonial governor was inquiring into the status of Estevanez' proposal. See Gobernador P.M. de Negros to the Gobernador Politico-Militar de Visayas, 30 January 1889, PNA *Memorias, Negros.*

27. Asunto: Azucares de Filipinas, AHN Legajo 1492, Expediente No. 8. Cf. "los comerciantes y agricultores de Yloilo y Negros . . ." to the Ministerio de Ultramar, Iloilo 20 January 1885.

28. Circular dictada por el Gobernador P.M. de Ysla de Negros en 23 de Abril del corriente año [1889], PNA *Varias Provincias, Negros.* See articles 2, 3, and 4.

29. Governor of the Visayas to the Governor-General, No. 710, Cebu 27 June 1889, PNA *Varias Provincias, Negros.*

30. Ibid.

31. Note from the Governor-General's office, Manila 10 June 1889, and from the Secretaria, Manila 23 September 1889, PNA *Varias Provincias, Negros*. The earlier note admitted that retention of *cedulas* would be advantageous to *hacienda* owners, but was contrary to law: "pero tambien es autorizar un hecho que, si necesario en la practica, no es legal."

32. Commentary from the Secretaria, Manila 23 September 1889, PNA *Varias Provincias, Negros*. As regards armed guards "as in the Peninsula," the Secretaria opined it was not a necessity because the *Reglamento de Cuadrilleros* had already authorized that *hacienda* guards were enlistable as volunteers in the roster of *Cuadrilleros* of each town.

33. Governor-General Valeriano Weyler to the Director General of Civil Administration, Manila 28 September 1889, PNA *Varias Provincias, Negros*.

34. Bandos sobre armas prohibidas, PNA *Spanish Manila*, Bundle 7. Cf. Governor-General Manuel Crespo's circular of 15 December 1854.

35. Estevanez to the Political-Military Governor of the Visayas, Bacolod 12 May 1883, PNA *Cargadores*. Cf. Article 9 of the "Proyecto de 'Reglamento de jornaleros y policia rural' para isla de Negros."

36. Ynstancia presentada por Don Leon Montinola ante el Gobierno P.M. de la Region Occidental de Ysla de Negros, contra al Cabeza de barangay Don Basilio Majinay, Jaro 5 September 1898, PNA *Varias Provincias, Negros*.

37. "Formerly . . . the general wage was 25 cents per week, with one meal per diem. . . . At Negros, field hands now get 62-1/2 cents per week, with food in addition." Loney to Farren, Iloilo 10 July 1861, FO 72/1017.

38. Molto to the Governor-General, Cebu 24 November 1863, PNA *Varias Provincias, Negros*.

39. Estevanez to the Political-Military Governor of the Visayas, Bacolod 12 May 1883, PNA *Cargadores;* Molto to the Governor-General, Cebu 24 November 1863, PNA *Varias Provincias, Negros*.

40. Giralt to the Señor Yntendente General de Hacienda, Bacolod 30 June 1889, PNA *Cedulas Personales, Negros Oriental*.

41. To balance the picture, some *hacenderos* actually reported criminals they found in their work force, but only when such individuals were seen as serious threats to the planters' interests. For example, Juan Jamora and Donato Severino, both of Silay, informed the *Guardia Civil* about Mariano Loceño (described as guilty of theft, robbery, arson, homicide, injuries, illegal detention, and escape from prison), who was then pursued and killed in Jamora's Hacienda Cotcot in Minuluan. Giralt to the Governor-General, Bacolod 17 April 1889, PNA *Varias Provincias, Negros*.

42. Expediente sobre la libertad concedida á varios individuos presos por la Guardia Civil en la Hacienda de San Juan pueblo de Cabangcalan (Isla de Negros), 1888, PNA *Varias Provincias, Negros*. Although referred to once as

dueño (owner), Alvarez was probably what most other documents in this *expediente* called him to be, an *encargado* (overseer).

43. Ibid. Cf. Gobernador P.M. de las Islas Visayas to the Governor-General, No. 46, Cebu 30 January 1889.

44. Ibid. See Captain of the Tercera Compañia del Tercer Tercio de la Guardia Civil dated 27 December 1888, cited in Capitania General de Filipinas to the Governor-General, Manila 1 February 1889.

45. Ibid. Political-Military Governor of the Visayas to the Governor-General, No. 48, Cebu 30 January 1889; and Francisco Costa (El Jefe de la Linea) to the Governor of Negros, Bacolod 2 January 1889.

46. Expediente sobre . . . varios individuos presos . . . en la Hacienda de San Juan . . . 1888, PNA *Varias Provincias, Negros*. Cf. Tovar to the Gobernador P.M. de las Islas Visayas, Bacolod 10 March 1889.

47. Ibid.

48. Ibid. Cf. Letter of Leon Bravo, El Alferez Comandante, Ilog 8 January 1889.

49. Ibid. See Governor-General to the Negros Governor, Manila 15 January 1890.

50. Ibid. See Governor-General to the Recollect Provincial, Manila 15 January 1890.

51. Echauz 1894, 44; Jagor 1965, 220; and Memoria descriptiva del estado de Agricultura . . . en Ysla de Negros . . . escrita para la Esposicion general . . . en Madrid el año de 1887, PNA *Memoria de Negros, Oriental y Occidental.*

52. Memoria descriptiva del estado de Agricultura . . . en Ysla de Negros . . . escrita para la Esposicion general . . . en Madrid el año de 1887, PNA *Memoria de Negros, Oriental y Occidental.*

53. Some reported that *agsas* obtained one-third of the product, a net share that probably included milling costs and interest on capital (Foreman 1899, 314; Jagor 1965, 220).

54. Loney to Farren, Iloilo 10 July 1861, FO 72/1017.

55. Cf. Memoria descriptiva del estado de Agricultura . . . en Ysla de Negros . . . escrita para la Esposicion general . . . en Madrid el año de 1887, PNA *Memoria de Negros, Oriental y Occidental.*

56. Although considered objectionable today, even the fact that children in their tender years were employable in the tasks of pasturing cattle and directing work animals in the circular motion of the crude mill cane crusher was probably an added attraction to settle in the *haciendas*. Molto to the Governor General, Cebu 24 November 1863, PNA *Varias Provincias, Negros;* "Ynstruccion publica," Memoria de la Ysla de Negros, 1888, PNA *Memorias, Negros.*

57. Governor of the Visayas to the Governor-General, No. 710, Cebu 27 June 1889, PNA *Varias Provincias, Negros.*

58. See the substantial difference in the proposed worker registration fee for day laborers and sharecroppers cited in the section, "Planter Competition and Further Antistate Strategies," of this chapter.

59. Tovar perceptively noted: "Es verdad que estos aparceros son gente de la clase de jornaleros la generalidad, que prefieren el sistema de aparceria . . . a trabajar a jornal por la mayor independencia que disfrutan." Memoria descriptiva del estado de Agricultura . . . en Ysla de Negros . . . para la Esposicion general . . . en Madrid el año de 1887, PNA *Memoria de Negros, Oriental y Occidental*.

60. Interview with Clemente L., eighty-eight years old, Bacolod 2 August 1990.

61. Last will and testament of Pedro Pullicar, with his legitimate wife Juana Gustilo, Saravia 4 September 1891, PNA *Terrenos (Negros)*, Bundle 2. Cf. the tenth article.

62. Escritura de venta otorgada por Don Raymundo Alunan á favor de Don Efigenio Lizares, Bacolod 2 September 1890, PNA *Protocolos*, Legajo 1772, folio 755ff.

63. Interviews with: Constancio G., sixty-six; Tomas J., seventy-nine; Aurora J., eighty, on 4 and 13 September 1990 in Bago; Montano M., seventy-six, on 6 September 1990 in Bago; Julieta M., seventy-eight, and Florencio S., seventy-four, on 6 and 7 September 1990 in Sagay; and Teofisto S., seventy-one, on 11 October 1990 in Isabela.

64. The gambling habit of Negros migrants was not unique. For example, during the prosperous times of the abaca industry in the nineteenth century, Owen (1984, 91) observed that "the workers of Kabikolan had sufficient funds to gamble regularly at cockfighting and cards."

65. This hiring policy of Negros planters contrasted with the offering of fun money (for drink, opium, prostitutes, and gambling) in other parts of colonial Asia "just before the contract expired" to bind coolies to stay (Breman 1990, 22).

66. The two phrases are cited in, respectively, Wickberg 1965, 7 n. 10 and Guggenheim 1982, 5.

67. Compared to Spanish America. A solicitud de Don Pasqual Baylon, Asentista de Gallos de la Provincia de la Pampanga, quejandose contra Don Pantaleon Angelo de Miranda, Don Juan Agustin del Aguila é Ycaza, Don Tomas Gallegos, y otros por el perjuicio que le sigue, con motivo de haver puesto Plazas de Gallos en sus Haciendas, arreglandose a la Real orden de 6 de Abril de 1828: en que concede el privilegio de una Gallera gratis a los propietarios de grandes Haciendas, 1829, PNA *Galleras de Manila*.

68. Bando sobre juegos, issued by Captain-General Andres Garcia Camba, Manila 7 March 1838, PNA *Spanish Manila*, Bundle 7.

69. Decreto sobre juegos prohibidos, issued by Captain-General Narciso Claveria y Zaldua, Manila 11 October 1847, PNA *Spanish Manila*, Bundle 7.

70. Reglamento para la represion de juegos prohibidos en Filipinas, Aranjuez 3 May 1863, PNA *Control of Gambling,* Bundle 2.

71. Wickberg (1965, 13 n. 84) estimated that annual revenue from the cockpit franchise payments ranged from 100,000 to 200,000 pesos, a low estimate considering that, as Norman (1895, 178) reported, Manila alone had six cockpit owners one of whom claimed to have paid "68,600 [Mexican] dollars a year." See also the rather low revenue data from cockpits compiled by Bankoff (1991, 280) for selected years from 1780 to 1894.

72. Presidencia of the Real Audiencia de Manila to the Governor-General, Manila 21 October 1881, PNA *Varias Provincias, Negros.*

73. See, for example, Rush 1990.

74. Bernaldez (1907, 237) reported in 1827: "In almost all the provinces of the islands very little money circulates, and in some of them there is not even what is necessary in order that the natives can pay the government taxes; and from this has proceeded the necessity of commuting the tribute from money to kind. The Spanish pesos go from and return to the provinces rapidly...."

75. Memoria de la Ysla de Negros, 31 May 1888, PNA *Memoria de Negros, Oriental y Occidental;* Copias de las Cuentas de Arbitrios Correpondientes al año de 1853, PNA *Varias Provincias, Negros.*

76. If contemporary *haciendas* can provide a clue, tenants' huts in the late nineteenth century were sometimes constructed around a small *hacienda* core, but some were also dispersed in tiny clusters where tenants could stay closer to their fields. The recollections of an American officer stationed in Isabela portrayed the Negros *hacienda* as "a community in itself—a feudal community of which the *hacendero* was the overlord. The *hacendero*'s house, like a baron's fortress of the Middle Ages, stood in the center of the buildings and dependents' huts" (White 1928, 116–117). This pattern was atypical, as Tables 6 and 7 suggest. Moreover, a number of the big *hacenderos* did not actually live in Negros. Eugenio Lopez, for example, lived in Silay from the early 1860s until about 1875, after which he moved back to Jaro in Iloilo Province (O. Lopez 1982, xliv–xlvi).

77. In the early 1900s, it was reported that "There is one house near Bacolod that cost $90,000, and another a few miles farther in the country that cost $120,000" (Hughes 1902, 650).

78. Agapito V., sixty years old, a farmer and *hacienda* worker from Isabela, Negros Occidental, interviewed on 12 October 1990.

79. The *imol*'s gambling had become a detestable vice that the famed Lopez de Kabayao family of pianists and violinists sought to mend in the 1940s by offering free concerts featuring the music of Beethoven, Mozart, Liszt, Chopin, and Haydn, which they hoped would transform the tenants and laborers of their Hacienda Kalubkub into "diligent, economical and honourable" citizens (Regalado 1948).

CHAPTER 6   Toward Mestizo Power

1. Ereccion de un Monumento para Memoria de los Señores Legazpi y Urdaneta, PNA *Rare Document.*

2. Carta circular de Fr. Gabriada á los curas parrocos de la Isla de Negros, Costa Oriental, Dauin 3 April 1891, AM Legajo 59, No. 1.

3. Reduccion de infieles en el Norte de la Ysla de Luzon, AHN Legajo 5318, Expediente No. 193. See de Rivera 1881, 14, contained in this *expediente.*

4. Ibid. Cf. de Rivera 1881, 12–13.

5. Ibid. Cf. Primo de Rivera to the Ultramar, Manila 26 March 1881.

6. Ibid. Cf. various reports on the "reduccion de infieles" in several Luzon provinces: Isabela, Cagayan, and Nueva Vizcaya in the north, and Camarines Sur in the southeast.

7. Ibid. Cf. de Rivera 1881, 48–55.

8. Ibid. Cf. Primo de Rivera to the Ultramar, Manila 26 March 1881.

9. Creacion de Misiones en la Ysla de Negros, solicitada por el Rdo. Provincial de Recoletos, AHN Legajo 2303, Expediente No. 179 (cf. Fr. Mariano Bernad to the Governor-General, Manila 6 December 1893); Fr. Mauricio Ferrero to the Bishop of Jaro, Bacolod 25 April 1889, AM Legajo 57, No. 1; Ferrero to the Recollect Provincial, Bacolod 14 March 1892, AM Legajo 59, No. 2.

10. Creacion de Misiones en la Ysla de Negros . . . , AHN Legajo 2303, Expediente No. 179. Cf. Ferrero to the Governor-General, Bacolod 27 January 1894.

11. Ibid. Cf. Fr. Mariano Bernad del Pilar to the Governor-General, Manila 6 December 1893.

12. Interview with two grandchildren of Fernando Cuenca, Ymelda M., eighty-four, and Francisco D., seventy, in Bacolod on 8 July 1990.

13. This account is based on the 25th to the 28th installments of Varona 1965.

14. Ibid., 25th installment.

15. Ibid., 27th installment.

16. Ibid., 28th installment.

17. Ibid., 27th installment.

18. Ibid.

19. Ibid.

20. Ibid., 25th installment.

21. Although workers served as the source for this reconstructed legend, beliefs in the spirit-world are not restricted to the working class, as many *hacenderos* and wealthy people share in the same fundamental beliefs. This cultural legacy knows no class boundaries.

22. Interviews with: Julian C., eighty-two, on 4 September 1990, Bago City;

Constancio G., sixty-six, on 4 September 1990, Bago; Restituta P., seventy-three, on 23 October 1990, Bago; Rudy V., forty-one, and Celsa V., forty-three, on 18 September 1990, Talisay.

23. Interviews with workers in one de la Rama *hacienda* in Bago suggest that the *agsa* arrangement was followed in earlier times.

24. See the section "*Hacendero* Gambling in the Negros Sugar Cockpit" in Chapter 4.

25. Found in Daet, Camarines Norte, the monument contains neither Rizal's image nor his statue (Anderson 1993). On Masonry and the revolution, cf. Schumacher 1991, chaps. 11 and 12, and Kalaw 1956, chaps. 5 and 6.

26. Masonry in Iloilo has had a long history, but in Negros the first lodge was organized only in 1920. Before that date, Negros *hacenderos* who were Masons were formally affiliated with the lodge at Iloilo. The Bacolod Lodge has since counted among its members the following prominent members: Rafael Alunan, Eusebio Luzuriaga, Nicolas Lizares, Vicente T. Ramos, Ildefonso Coscolluela, Rosendo Locsin, Ricardo Nolan, Olimpio de la Rama, Carlos Dreyfus, and Arturo Villanueva. The list is long, but the point is that many of the wealthiest families in Negros Occidental have been represented in the roster of the local Masonry chapter (Estacion n.d.). Countless notables of the Philippine ruling class have also been associated with Masonry. To name only a few, the list includes Jose Rizal, Manuel Quezon, Emilio Aguinaldo, Aglipay, Felipe Buencamino, Rafael Palma, Teodoro M. Kalaw, Jose Abad Santos, Conrado Benitez, Emilio Virata, and Wenceslao Trinidad (cf. Kalaw 1956).

27. Masonry's avowed goal is simply the separation of church and state. They do have secret rituals, practices, and modes of recognizing a fellow Mason; nonetheless, it is claimed that the organization has no secret purposes. There is a feeling that a resurgence in the persecution of Masons is on the rise in the Philippines, but apparently the Catholic Church has been lenient in Negros Occidental. Interview with Gil O., Jr., fifty-three years old, on 3 October 1990, Bacolod.

28. Pueblo de San Joaquin, Provincia de Iloilo, Seccion Geografica Historica y Estadistica formada en el año 1882, por Fr. Tomas Santaren, APAF 356/1-c. I am grateful to Alice Magos for providing me with a copy of this document.

29. Ibid., especially the second half of the answer to Pregunta 45. See also Echauz 1894, 143.

30. Creacion del Tercer tercio de la Guardia Civil, con destino á Visayas, AHN Legajo 5328, Expediente No. 21. Cf. El Conde de Clonard to Governor-General, Cebu 16 March 1882; and the Governor-General to the Ultramar, Manila 31 May 1882.

31. Pueblo de San Joaquin . . . por Fr. Tomas Santaren, APAF 356/1-c.

32. Literally meaning "the refuge of the people in time of crisis" (Magos 1992, 35), Bangotbanwa could well be the "Bangos banna" in the Augustinian record, which the Spanish chronicler assumed to be the *babaylans'* "Pope." The Augustinian account also mentioned that the *babaylans* were "descendants of the family of the Estellas," a "family name" which might well refer to Bangotbanwa's given name, "Estrella."

33. Expediente sobre la muerte dada á cuatro Cabecillas de Babaylanes . . . , PNA *Guardia Civil,* Bundle No. 32 (1888–1889A). Cf. Comandante encargado del Despacho, Tercer Tercio de Guardia Civil, to the Governor-General, No. 138, Cebu 13 April 1889.

34. Expediente contra los vecinos de Yloilo Pedro Padernal [et al.] por afiliados á la Secta de babailanes, PNA *Sediciones y Rebeliones,* Vol. 19, Book 1 (1874–1898). Cf. Gobernador Politico-Militar of Iloilo to the Governor-General, Iloilo 1 April 1898.

35. Expediente sobre la muerte dada á cuatro . . . Babaylanes . . . , PNA *Guardia Civil,* Bundle No. 32 (1888–1889A). Cf. Comandante, Tercer Tercio de Guardia Civil, to the Governor-General, No. 158, Cebu 23 April 1889.

36. Superior decreto sobre el bandolerismo, Manila 15 January 1897, PNA *Guardia Civil,* Bundle No. 39 (1897–A); and Sucesos contra el orden publico independientes de la rebelion, AHN Legajo 5356, Expediente No. 5. Cf. "Reservado," Governor-General to the Ministro de Ultramar, Manila 13 March 1898.

37. Tovar to the Governor-General, Bacolod 31 October 1888, PNA *Varias Provincias, Negros.* Cf. the appended news item from *El Eco de Panay* dated 25 October 1888.

38. Ibid. Cf. the news report in *El Porvenir de Bisayas* dated 28 October 1888, similarly appended. Reportedly found on his person was a list of one hundred names targeted for liquidation, including the local chief of the *Guardia Civil.*

39. Negros Governor to the Governor-General, Bacolod 10 November 1888, PNA *Varias Provincias, Negros.*

40. Comandante, Tercer Tercio de Guardia Civil, to the Governor-General, No. 142, Cebu 13 April 1889, PNA *Guardia Civil,* Bundle No. 32 (1888–1889A).

41. Ricardo Monet, Coronel de Guardia Civil, to the Governor-General, No. 48 and No. 50, Jaro 12 May 1897, PNA *Guardia Civil,* Bundle No. 39 (1987–A).

42. De Mas (1963, 160) recommended that "Spaniards ought to be distinguished by the use of clothes which neither natives nor mestizos may wear. . . . Natives should not use any clothing other than that which they themselves have selected: open shirt and a straw head gear."

43. Libro de cosas notables por Fr. Angel Martinez, Jimamaylan 31 December 1897, AM Legajo 57, No. 4.

44. Governor of Negros Occidental to the Comandante General de Panay y Negros, 15 December 1896, in Anon. (n.d., 12–13).

45. Ibid., 11–12.

46. Negros Governor Luis Martinez to the Governor-General, Bacolod 25 January 1897, PNA *Guardia Civil,* Bundle No. 39 (1897–A). Cf. appended letter of Juan Adarves, Primer Teniente Comandante de Guardia Civil, Ilog 22 January 1897.

47. Ibid. Cf. letter of Adarves, Ilog 23 January 1897, appended in the same.

48. In late-nineteenth-century Iloilo, a *babaylan* called Juan Perfecto formed his own insignia by incorporating his initials, J.P., with the words *Agnus Dei* and the seal of the seminary in Jaro. PNL *Noble Collection,* 10, 1676. Cf. McCoy 1982b, 169.

49. Cosas notables de este Ministerio de Manjuyod . . . por Fr. Juan Perez, Majuyod 18 March 1898, AM Legajo 56, No. 3.

50. Negros Governor Eduardo Subinza to the Governor-General, Bacolod 1 August 1884, with two enclosures, PNA *Sediciones y Rebeliones,* Vol. 3, Book 1 (1884–1896).

51. Expediente instruido contra el gobernadorcillo de esta Cabecera D. Prudencio Mariño por injurias contra los Señores D. Ygnacio Gonzales y otros, PNA *Varias Provincias, Negros.* Cf. Benito Perez de Tagle et al. to the Gobernador Politico-Militar of Negros Oriental, Dumaguete 30 April 1891.

52. This account of the Spanish defeat in Negros is based on Fuentes 1919, chaps. 5 and 6, and Araneta 1898.

53. Aniceto Lacson to the Señor Presidente del Gobierno Revolucionario de Filipinas, Bacolod 7 November 1898, PIR, SD 13, Folder 315.1; Stewart 1908, 185 and 199.

54. The appointment of Araneta as general and governor was issued by the Presidente del Gobierno Revolucionario de Filipinas on 12 November 1898, with Araneta assuming the post on 8 December. Circular of Diego de la Viña, Delegado de la Guerra, to the local military chiefs of Negros Oriental, Dumaguete 17 December 1898, PNL *Noble Collection,* 29, 4679.

55. The myths spun around Juan Araneta were pieced together from innumerable interviews, but the following informants, all from Bago, provided the most interesting details: Julian C., eighty-two, 4 September 1990; Constancio G., sixty-six, 4 and 13 September 1990; Tomas J., seventy-nine, 4 September 1990; Aurora J., eighty, 13 September 1990; Carmen M., seventy-nine, 21 August 1990; Montano M., seventy-six, 6 September 1990; Pedro M., sixty-eight, 6 September 1990; and Jose P., seventy-two, 5 September 1990.

56. The *kambang* idea seems to parallel the Javanese notion of magical men's ability to concentrate opposites (Anderson 1990, 28–31).

57. Interview with Juan Araneta's granddaughter Emma A., eighty-five, in Bacolod on 25 October 1990.

58. By 1908, Araneta reportedly owned "in the neighborhood of 5,000 hectares of land, 1,000 hectares of which was mountain land," with his two pet estates each comprising 1,000 hectares (Stewart 1908, 187).

59. Interview with Emma A., eighty-five, in Bacolod on 25 October 1990.

60. Certification by Melecio Severino y Yorac, Secretario del Gobierno Republicano Federal del Canton de Ysla de Negros, Bacolod 28 November 1898, PIR, SD 9, Folder 77.3.

61. For the full text of this historic message, see Fuentes 1919, 127–129.

62. After seeking U.S. protection, Negros leaders then went on to establish the Cantonal Government, which was affiliated with Luzon and which subsequently followed the order from Malolos issued on 30 January 1899 to release the friar-prisoners (Fuentes 1919, 171).

63. Ynstrucciones o deberes del Comandante del sitio de Cambalay en medio del monte de Ysabela de Jimamalaylan [sic], Juan Araneta and Dionisio Segovela y Papa, Bacolod 19 December 1898, PNL *Noble Collection*, 25, 4141–4143.

64. Appointment of Generals and formation of *plantilla* by Dionisio Papa, Alabhid o Paraiso 2 March 1899, PIR, SD 58, Folder 970.8 (also PNL *Noble Collection*, 29, 4655–4657).

65. Acta of the Republica Filipina de Negros, Alabhid or Paraizo 2 May 1899, and Filomeno Auit, Comandante del Ejercito, to Arcadio Maxilom y Molero, Jefe de Operaciones en Cebu, Kabalanan 26 April 1901, PIR, PR 91, "Daily Account of Operations in Negros" Folder; Undated and unsigned circular requesting contributions to the revolution (original Ilonggo in the PIR file), PIR, SD 58, Folder 970.11 (also PNL *Noble Collection*, 14, 2387–2388); and Dionicio Papa to Rufo Oyos, Campamento General 13 June 1901, PIR, SD 58, Folder 970.4–4 (also PNL *Noble Collection*, 29, 4646–4648).

66. Leandro Fullon to the Ejercito Expedicionario para Negros, San Juan de Buenavista 25 September 1899, PNL *Noble Collection*, 26, 4299–4300.

67. Dionisio Papa to Jose de la Viña, Paraiso 3 August 1900, PNL *Noble Collection*, 29, 4669–4670.

68. Acta of the Republica Filipina de Negros, Alabhid or Paraizo 2 May 1899, PIR, PR 91, "Daily Account of Operations in Negros" Folder.

69. Sample pledges in Ilonggo by followers of Dionisio Papa, with Spanish translations, Paraizo 6 April 1901, PNL *Noble Collection*, 25, 4132–4135.

70. Undated and unsigned circular requesting contributions to the revolution (original Ilonggo in the PIR file), PIR, SD 58, Folder 970.11 (also PNL *Noble Collection*, 14, 2387–2388).

71. Dionisio Papa to Jose de la Viña, Paraiso 3 August 1900, PNL *Noble Collection*, 29, 4679; Calistro Segubila to Rufo Oyos, Cartagena 9 November 1900, PIR, SD 58, Folder 970.4–1 (also PNL *Noble Collection*, 29, 4645–4646).

72. Catipunan: Ynstrucciones para los Presidentes locales y Delegados de

los pueblos, por Dionisio Papa, Sipalay 20 July 1900, PIR, SD 58, Folder 970.6 (also PNL *Noble Collection,* 29, 4637–4639).

73. Catipunan: Ynstrucciones para los Presidentes locales . . . , PIR, SD 58, Folder 970.6 (also PNL *Noble Collection,* 29, 4637–4639).

74. Dionisio Papa to Rufo Uyos, Calibon 19 May 1900, PIR, SD 58 Folder 970.4 (also PNL *Noble Collection,* 29, 4644–4645).

75. Dionisio Papa to Jose de la Viña, Paraiso 3 August 1900, PNL *Noble Collection,* 29, 4669–4670.

76. A concise biography of Diego de la Viña is found in Rodriguez 1983, 69–80.

77. Jose de la Viña to the Senior Juez de la Corte Suprema de Justicia, La Castellana n.d., PNL *Noble Collection,* 29, 4671–4672.

78. PNL *Noble Collection,* 29, 4669–4670: Dionisio Papa to Jose de la Viña, Paraiso 3 August 1990.

79. Nota del interprete oficial F. R. Fabie, PNL *Noble Collection,* 29, 4672.

80. Interview with Clemente L., eighty-eight, in Bacolod, on 2 and 28 August 1990.

81. Ibid. A contemporary of Jose de la Viña's son Emiliano, Clemente avowed to have been a close friend of the family.

82. Ibid.

83. Ibid.

84. On Visayan sorcery, see Lieban 1967.

CHAPTER 7   The American Colonial State

1. Lacson to the "Honorable Mr. President of the Great Republic of the United States," Bacolod 27 May 1899, USNA RG 350, Entry 5, Box 151.

2. Otis to the U.S. Army's Adjutant General, Manila 23 July 1899, USNA RG 350, Entry 5, Box 151.

3. Charles Denby, Acting President, to John Hay, Secretary of State, Manila 15 June 1899, USNA RG 59, Entry 739.

4. Otis to the Adjutant General, Manila 23 July 1899, USNA RG 350, Entry 5, Box 151.

5. Charles Denby, Acting President, to John Hay, Secretary of War, Manila 25 July 1899, USNA RG 59, Entry 739.

6. Melecio Severino, the Civil Governor, objected to the reduction of his office to a mere sinecure and, together with the American-appointed attorney-general Dionisio Mapa, reportedly devised a plot against the government. But a cabinet official betrayed the plan to the military governor and a round of arrests took place. Severino and Mapa continued in office, however, as no substantial evidence was found to implicate them (Romero 1974, 161).

7. Acta Numero 23, Sesion Ordinaria, 5 June 1908, NOHC *Minutes of the Provincial Board of Negros Occidental for 1908,* pp. 541–542.

8. Interview with Carlos Hilado, former editor of *The Commoner*, in Bacolod, 9 August 1990. Cf. McCoy 1982a, 338.

9. For a fuller discussion of gambling and electoral politics, see Aguilar 1994d, 176–183.

10. Cablegram to Secretary of War at Washington from Manila, 23 August 1906, USNA RG 350, Entry 5, Box 469.

11. Act No. 1537, dated 8 September 1906, limiting the number of horse-racing days, USNA RG 350, Entry 5, Box 469.

12. Cable from Governor-General Smith to the Secretary of War, Manila 18 February 1908, and Memorandum for the Secretary of War from the Bureau of Insular Affairs, Washington, D.C. 24 February 1908, USNA RG 350, Entry 5, Box 454.

13. Dr. H. P. Clarke to President Theodore Roosevelt, Indianapolis 19 February 1908, USNA RG 350, Entry 5, Box 454 (emphasis in original). Clarke noted: "While it does not seem possible that the Chief Executive would take pains to interfere with the carnival festivities of the Filipinos, this might be a good occasion to observe that many Americans are ardent devotees of the ancient sport of cocking, one of the very few amusements of man in which our animal pets are willing participants."

14. Act. No. 1909 of the First Philippine Legislature, Manila 19 May 1909, USNA RG 350, Entry 5, Box 454.

15. Chief of the Bureau of Insular Affairs to Rev. Ansel E. Johnson, Washington, D.C. 29 November 1910, USNA RG 350, Entry 5, Box 454.

16. Ibid.

17. Philippine Commission Act No. 1757, 9 October 1907, USNA RG 350, Entry 5, Box 469.

18. Orden Ejecutiva No. 13, 18 February 1910, and Proclama No. 1, 1 March 1910, under "Asuntos Varios," NOHC *Minutes of the Provincial Board of Negros Occidental for 1910*, p. 11.

19. Circular of the Office of the Governor-General to all Provincial Treasurers, Manila 11 June 1912, USNA RG 350, Entry 5, Box 469.

20. Order of Acting Governor-General Newton Gilbert to the Secretary of Commerce and Police, Manila 4 June 1912, USNA RG 350, Entry 5, Box 469.

21. Ordenes Ejecutivas Nos. 8, 9, and 15, respectively dated 20 March, 4 April, and 17 July 1923, NOHC *Minutes of the Provincial Board of Negros Occidental, 1923*.

22. Ibid. Orden Ejecutiva No. 28, 14 December 1923. See also Minutes of Regular Session, 21 December 1923, folio 695.

23. Orden Ejecutiva No. 8 and No. 25, dated 20 March and 7 December 1923, respectively, NOHC *Minutes of the Provincial Board of Negros Occidental, 1923*.

24. A gambling enthusiast commented: "try as the government might, it

can not successfully suppress gambling as it can never do away with wine, women and song" (*The Commoner* 1936b).

25. Feeding into a lucrative sellers' market, the value of carabaos escalated from 50 to 150 Mexican dollars (Philippine Commission 1903, 18; 1906b, 87).

26. The locust menace continued to occur periodically, and in 1932 the Philippine Legislature allocated 200,000 pesos for their extermination (*Sugar News* 1932a).

27. NOHC *Minutes of the Provincial Board of Negros Occidental, 1913* (Session of 14 August), p. 621; *Sugar News* (1923b).

28. In the 1920s, direct offshore loading along the Negros coast had become possible, but the technology's full potential was not exploited until the 1931 strike by stevedores at the Iloilo waterfront (McCoy 1982a, 328–343).

29. "Sugar Cane" (unsigned memorandum on the Philippine sugar industry, c. 1904), pp. 12–13, USNA RG 350, Entry 5, Box 373.

30. The exchange rate was pegged at two Philippine pesos to one U.S. dollar.

31. NOHC *Minutes of the Provincial Board of Negros Occidental for 1910*. Cf. sessions of 11 March 1910 (pp. 126–127) and 10 June 1910 (pp. 468–469).

32. The ceiling of 35,000 pesos was granted to twelve individual applicants. In one batch of loan applications, the Montillas of Isabela obtained a total of 155,600 pesos via nine separate applications. NOHC *Minutes of the Provincial Board of Negros Occidental for 1913*, pp. 42, 49, 109, 120, 181–182, 204, 257–259, 285–286, 293–297, 312–317, 346–351, 356–358, 371, 399–404, 439–440, 469–470, 495–499, 579, 595–597, 614–615, 638–639, 687, 699–701, 711–712, 764–766, 820, 970–971, and 1029.

33. From 1911 to 1913, land values rose dramatically and lands accessible to town centers could not be obtained for less than 300 pesos per hectare (NOHC *Minutes of the Provincial Board of Negros Occidental for 1913*, p. 616). This was in marked contrast to 1903 when, it was said, "There is always saleable land in western Negros, the price of which varies from 25 to 150 pesos per hectare . . ." (L. Locsin 1904, 694).

34. Forbes to the Insular Treasurer, Manila 13 June 1913, USNA RG 350, Entry 5, Box 373.

35. The BPI was directed, among others, to *(a)* provide not more than 2 pesos per picul of the estimated crop, *(b)* charge not more than 8 percent interest on these loans, and *(c)* deposit the harvested crop on which a loan had been made in warehouses, which would issue receipts called *quedanes* endorsable to the government. Later at the PNB, crop advances were granted at the rate of 3 pesos per picul of estimated crop without fertilizer and 4 pesos per picul with fertilizer application.

36. Memorandum of Ben F. Wright to the Governor-General, "Re Offer

to Buy the Interests of the National Bank in Five Negros Sugar Centrals," Manila 9 June 1927, p. 16, USNA RG 350, Entry 5, Box 522.

37. Forbes to Lindley M. Garrison, Secretary of War, Manila 27 August 1913, USNA RG 350, Entry 5, Box 373.

38. The Secretary of War "remained strongly of the opinion that it would be much better to have these loans handled in a purely commercial way than to have the government enter into the matter by loans to the bank." Frank McIntyre to Newton W. Gilbert, Acting Governor-General, Washington, D.C. 9 August 1912, USNA RG 350, Entry 5, Box 373.

39. In 1917, with the European war and the consequent shipping problems, it became a matter of policy to sell Philippine muscovado sugar to the China market. In 1918, however, shipments were also sent to the United States to relieve the sugar shortage there.

40. Forbes wrote in his journal that the land limit "was a device of the [beet] sugar people to prevent the proper development of the Islands, fearing they would become formidable as a sugar producer in competition with existing lines of trade . . ." (Horn 1941, 240).

41. The compatibility of American protectionism and Filipino elite nationalism was also evident in the Chinese exclusion law, which was applied by the U.S. Congress to the Philippines through the 1902 Organic Law. Philippine elites were fearful of the competition that immigrant Chinese merchants might pose, while American labor feared the prospects of cheap Chinese coolie labor migration.

42. Truman G. Palmer to Gen. Clarence G. Edwards, Washington, D.C. 16 February 1911, USNA RG 350, Entry 5, Box 373.

43. Forbes had planned "to send the cable ship *RIZAL* to Negros to invite all of the most prominent hacienderos [sic] to the number of perhaps a hundred or more and bring them to Mindoro in order that they may see a practical modern sugar mill." Forbes to Gen. Clarence Edwards, Manila 21 September 1911, USNA RG 350, Entry 5, Box 373.

44. The de la Rama Centrals in Talisay (1912, 100 tons), Bago (1913, 300 tons), and Escalante (1913, 300 tons), the Kanlaon Central in La Castellana (1914, 200 tons) (later bought by a Filipino company and moved to Cabiao, Nueva Ecija), and the Spanish-owned San Isidro Central in Kabankalan (1914, 400 tons). These and subsequent data on Philippine centrifugal mills were culled from the listings of Fairchild and Palmer, and reconciled with archival data. Note that the rated capacities shown refer to original capacities before upgrading. See Fairchild 1925, 1929; Palmer 1925; Venancio Concepcion to the PNB Board of Directors, Manila 21 June 1920, and the table "Comparative Statement of the Sugar Centrals," USNA RG 350, Box 521.

45. Under consideration since 1910, the plan for the San Carlos mill took time to materialize due to the land-size limitation. George R. Carter to

W. C. Forbes, Honolulu, Hawaii 8 August 1911, USNA RG 350, Entry 5, Box 373.

46. Statement of Gen. McIntyre before the U.S. Congressional Committee on Ways and Means, 21 January 1929, p. 16, USNA RG 350, Entry 5, Box 523. Notice how the Americans transformed the local word *sentral* into an English word, "central." The appropriation of *sentral* into English is most evident in the popular use of the phrase "Bank Centrals."

47. Circular of Ellis Cromwell to all Internal Revenue Agents, Manila 23 June 1912; and Ignacio Villamor, Attorney-General, to the Governor-General, Manila 21 June 1912, USNA RG 350, Entry 5, Box 469.

48. Translation of news item "For the Improvement of Philippine Sugar" that appeared in *La Democracia* (Manila, 24 January 1914), USNA RG 350, Entry 5, Box 373.

49. Opinions of the Attorney-General on Act. No. 2479, Manila 24 April 1915 and 16 June 1915, USNA RG 350, Entry 5, Box 520.

50. Translation of "The Fiasco of the Sugar Centrals," which appeared in *Consolidacion Nacional* (Manila, 12 March 1916), USNA RG 350, Entry 5, Box 530.

51. Chas. Walcutt, Jr., to George Fairchild, Washington, D.C. 1 October 1918, citing the governor-general's cable, USNA RG 350, Entry 5, Box 521. The PNB later absorbed the old Agricultural Bank (Sa-onoy 1982, 48). The year 1916 also witnessed the formation of the Philippine Senate.

52. Chief of Bureau of Insular Affairs to Governor-General Francis Burton Harrison, Washington, D.C. 10 December 1915, USNA RG 350, Entry 5, Box 520; and A. Kopp to McIntyre, New York 29 September 1922, USNA RG 350, Entry 5, Box 521.

53. The Guanco mill in Hinigaran (1915, 250 tons), later bought by an American and moved to Calumpit, Bulacan; the Nueva Apolonia Sugar Factory in Vallehermoso (1918, 90 tons); and the Ossorios' North Negros mill in Manapla (1918, 400 tons).

54. The 800-ton Victorias mill of the Ossorios, which opened in 1921, and the 400-ton central built by the Lopezes at Sagay in 1927.

55. The Pampanga Sugar Mills (1918, 2,500 tons) in Del Carmen, Pampanga and the Hawaiian-Philippine Co. (1920, 1,500 tons) in Silay, Negros Occidental.

56. Bearin, Palma, Guanco, and Kanlaon. Venancio Concepcion to the PNB Board of Directors, Manila 21 June 1920, USNA RG 350, Entry 5, Box 521.

57. The Isabela Sugar Co. (1919), the Bacolod-Murcia Milling Co. (1920), the Ma-ao Sugar Central Co. in Bago (1920), the Talisay-Silay Milling Co. (1920), the Binalbagan Estate, Inc. (1921), and the Pampanga Sugar Development Co. (1921). Memoranda from Frank McIntyre to the Secre-

tary of War, Washington, D.C. 4 and 23 October 1922, USNA RG 350, Entry 5, Box 521.

58. Memorandum to Governor-General Wood from Ben F. Wright, Special Bank Examiner, on the PNB Sugar Centrals, Manila 28 December 1922, p. 3, USNA RG 350, Entry 5, Box 521.

59. Memorandum from Frank McIntyre to the Secretary of War, Washington, D.C. 23 October 1922, USNA RG 350, Entry 5, Box 521.

60. Ibid.

61. Ibid.

62. Varona 1965, 67th installment.

63. Ibid., 67th and 72d installments; *Philippines Herald* 1924.

64. Memorandum of Ben F. Wright to the Governor-General "Re Offer to Buy the Interests of the National Bank in Five Negros Sugar Centrals," Manila 9 June 1927, p. 11; and Hallgarten and Co. to the Governor-General, New York 29 July 1927, USNA RG 350, Entry 5, Box 522.

65. Memorandum on "Bank Sugar Centrals, 1925–26 Crop," c. April 1927, USNA RG 350, Entry 5, Box 521; Rafael Corpus, PNB President, to Governor-General Leonard Wood, Manila 27 May 1927, USNA RG 350, Entry 5, Box 522.

66. For the sugar year ending October 1927, the six Bank Centrals reduced their debt to 39 million pesos and paid 3.5 million pesos in interest. Memorandum of Frank McIntyre to the Secretary of War "Report for the sugar year ending October 31, 1927, of the six Bank sugar centrals," Washington, D.C. 16 February 1928, USNA RG 350, Entry 5, Box 522.

67. Confidential letter of Ben F. Wright to Mr. E. W. Wilson, Iloilo 8 March 1921, USNA RG 350, Entry 5, Box 521.

68. Ibid.

69. Ibid.

70. First annual report of the Philippine Sugar Centrals Agency, July 1, 1921 to June 30, 1922, Manila, USNA RG 350, Entry 5, Box 521.

71. What can be considered the planter's version of this controversy suggested that the American manager "had ordered a slowdown in milling operations, forcing the planters to revive the old muscovado mills to save their canes, otherwise they would lose much of the crop due to the deterioration of the canes. Others declared a strike against the central and allowed 600,000 tons of cane to rot in the fields. The board of directors of the mill had no choice but to terminate the services of the manager or face the next crop year without any cane to grind" (Sa-onoy 1982, 51).

72. Memorandum of Ben F. Wright to the Governor-General, "Re Offer to Buy the Interests of the National Bank in Five Negros Sugar Centrals," Manila 9 June 1927, p. 4, USNA RG 350, Entry 5, Box 522.

73. Ibid., 5.

74. Wood to John W. Weeks, Secretary of War, Manila 25 August 1923, USNA RG 350, Entry 5, Box 521.
75. McIntyre to Wood, Washington, D.C. 25 October 1922, USNA RG 350, Entry 5, Box 521.
76. Undated memorandum for the record by McIntyre, "To be filed with letter of Wright to Governor-General Wood, 28 December 1922," USNA RG 350, Entry 5, Box 521.
77. Wood to McIntyre, Manila 1 June 1923, USNA RG 350, Entry 5, Box 521.
78. McIntyre to Wright, Washington, D.C. 8 July 1927, USNA RG 350, Entry 5, Box 522.
79. Memorandum of McIntyre to the Secretary of War, "Sale of the Philippine National Bank's sugar centrals," Washington, D.C. 25 January 1926, USNA RG 350, Entry 5, Box 521.
80. A. W. Mellon, Secretary of the Treasury, to the Secretary of War, Washington, D.C. 22 June 1927, USNA RG 350, Entry 5, Box 522.
81. Wood to McIntyre, Manila 1 June 1923, USNA RG 350, Entry 5, Box 521.
82. Ibid.
83. Memorandum of Wright to Wood, Manila 28 December 1922, p. 5, USNA RG 350, Entry 5, Box 521.
84. E. W. Wilson to Judge Raphael Corpus, PNB President, San Francisco 18 October 1924, USNA RG 350, Entry 5, Box 521; emphasis in original. Wilson resigned from the PNB in 1924 "because there was so much difference between Governor Wood's and my opinion. . . ." Wilson to McIntyre, San Francisco 6 October 1924, USNA RG 350, Entry 5, Box 521.
85. J. H. Foley, Agent of the Philippine National Bank New York Agency, to McIntyre, New York 26 January 1926, USNA RG 350, Entry 5, Box 521.
86. Confidential cable of James Ross to McIntyre, Manila 12 May 1927, USNA RG 350, Entry 5, Box 522.
87. Memorandum of Wright to the Governor-General, "Re Offer to Buy the Interests of the National Bank in Five Negros Sugar Centrals," Manila 9 June 1927; Wright to McIntyre, Manila 11 June 1927, USNA RG 350, Entry 5, Box 522.
88. McIntyre to A. J. Miller, Washington, D.C. 23 August 1927, USNA RG 350, Entry 5, Box 522.
89. Charles D. Orth to Governor-General Henry Stimson, 20 December - 1927, USNA RG 350, Entry 5, Box 522.
90. During the Second World War, a Basque *hacendero* with the Negros guerilla movement was surprised "to find in this region, at every inlet or bay, plantations owned by some American families . . . who had arrived in the Philippines during the Spanish War—former soldiers of the U.S. Army who had retired to these hidden places to establish their homes and families.

In Maricalum and Asia there were the Pfleider family; in Mambulao, the Watkins; at Hinobaan, the Wantingtons . . . and at Basay, there were the descendants of the Fords and Browns" (de Uriarte 1962, 70).

91. American capital includes that of Miguel Ossorio, a Spanish-Filipino who acquired U.S. citizenship in the late 1920s, resulting in a change of classification of the capital represented by the two Ossorio mills, North Negros and Victorias Milling (Nagano 1988, 177).

92. On 7 October 1933 the Philippine Legislature "declined to accept" the Hare-Hawes-Cutting Act, but on 1 May 1934 it accepted the Tydings-McDuffie Act, which fundamentally was the same as the earlier legislation. For the relationship of Philippine independence to the sugar industry, cf. Friend 1965, 1963; Castro 1965; Constantino 1975, 322–341; and Larkin 1993, chap. 5.

93. Operating at below capacity and unable to continue unless at a loss, sugar industry people predicted that twenty of the forty-five centralized sugar mills would close down due to the sugar limitation provision of the 1933 Independence Act (*Sugar News* 1933b).

94. Warned Vicente Lopez, president of the International Chamber of Commerce of Iloilo, "within a few hours after independence is granted, unless a free-trade extension arrangement accompanies the grant of political freedom," the Red Cross would have to be called in to bring relief to the people of Negros and Iloilo (*Sugar News* 1931).

95. But such rivalry was intrinsic to the gambling worldview the *Sugar News* (1919) concomitantly encouraged.

96. The *sentral* was also asked to share proportionately the cost of fertilizers "in as much as the central benefits from the application of fertilizers," a demand that did not consider improvements and maintenance costs by the mill (*Sugar News* 1930e).

97. Statement made by Miguel Ossorio in July 1930 (*Sugar News* 1930c).

98. The U.S. federal law that reimbursed planters subjected to mandatory reduction of production was echoed by a law passed by the Philippine Legislature, which provided, for a total of six years, an annual allotment of sugar quotas according to destination: the U.S. export market, domestic consumption, emergency reserve, and (the lowest priority) export to the world market (*Sugar News* 1934a).

99. Following the Isabela case, Binalbagan gave 5 percent in the form of shares of stocks, Talisay and La Carlota returned the equivalent of 25 centavos per picul, and Ma-ao and San Carlos gave in to the 60 percent partition (Larkin 1993, 186–187).

100. For the whole country, lessee-planters accounted for only one-third of the total 15,848 sugar planters enumerated; Negros accounted for 43 percent of all lessee-planters (Alunan 1939, 13).

101. Varona (1965), 76th installment.

102. Ibid.; *Sugar News* (1930d).

103. On the hollowness of most planters' concern for workers once their share had been raised, see J. C. Locsin 1969 and G. Bulatao 1971.

104. Interview with Juan G., ninety, in Bacolod on 21 August 1990; and with Clemente L., eighty-eight, in Bacolod on 2 August 1990. The latter recounted that Spanish planters were subjected to "discrimination," such as in the tardy assignment of vehicles to haul cane to the mill. They also had lower priority in selling cane by-products, such as molasses and alcohol, to the sugar mills.

105. In the Filipino's characteristic way of courting danger, such activities invited successful raids by the Japanese. The guerilla leaders decided to restrict those gatherings to Sundays and "legal holidays" (Hofileña 1990, 116).

106. The quota restrictions contained in the 1946 Bell Trade Act were loosened by the 1955 Laurel-Langley Agreement, which pegged the quota flexibly to a percentage of the U.S. market and, after the 1961 Cuban revolution, were reversed as Cuba's lost quota had to be filled in by other exporters, the new Philippine quota jumping to 1,040,000 short tons.

107. As owner of the carabao, the tenant was said to be "free to work as the laborer sees fit," except when he was "required on the hacienda"; his labor was then compensated with three free meals a day (Pitcairn 1925, 32). The tenant could also use the carabao for rice farming on sharecropped land, from which the tenant obtained six to eleven cavans of rice (Runes 1939, 16).

108. Bureau of Labor data showed that the average daily wage rate of unskilled agricultural workers in 1925 was 78 and 57 centavos, respectively, in Negros Occidental and Oriental, compared to 1.02 pesos in Pampanga, 1.19 pesos in Laguna, and 92 centavos in Batangas (Cruz 1935, 66).

109. At one point, planters instituted a labor registration scheme to prevent flight, akin to the foiled plan contemplated during the Spanish period. First put into effect in Silay-Saravia in February 1926 using the fingerprint method, labor registration was speedily adopted in other parts of Negros Occidental (*Sugar News* 1926b, 1926d).

110. The workers who supplied the details of the tales concerning Esteban de la Rama and his daughter discussed below were interviewed in different villages of Bago and Talisay: Carmen M., seventy-nine, 21 August 1990; Julian C., eighty-two, 4 September 1990; Constancio G., sixty-six, 4 September 1990; Jose P., seventy-two, 5 September 1990; Montano M., seventy-six, 6 September 1990; Rudy V., forty-one, 18 September 1990; Celsa V., forty-three, 18 September 1990; and Restituta P., seventy-three, 23 October 1990.

111. This section on Intrencherado is based on Sturtevant 1976 (158–174, 267–268); interviews with Clemente L., eighty-eight, in Bacolod on 7 August 1990, and with Maria L. B., seventy-eight, in Victorias on 22 Septem-

ber 1990; and the private papers of Jose C. Locsin, available at the University of St. La Salle, Bacolod.

112. Details concerning the Kanlaon story were supplied by Florencio S., seventy-one, interviewed in Sagay on 6 and 7 September 1990; and Marieta N., fifty-seven, in La Castellana on 7 October 1990.

113. Interview with Julieta M., about seventy-eight years old, in Sagay on 8 September 1990. Told in Cebuano, the story of Baringkot is genderless. To facilitate its retelling, I use the male pronoun.

References

This bibliography includes manuscripts, dissertations/theses, newspaper articles, and other published works. Citations for unpublished archival material are to be found in the Notes.

Abrams, Philip
    1982 *Historical Sociology.* Ithaca, NY: Cornell University Press.

Ackerman, Susan, and Raymond Lee
    1981 "Communication and Cognitive Pluralism in a Spirit Possession Event in Malaysia." *American Ethnologist* 8(4): 789–799.

Adas, Michael
    1980 "'Moral Economy' or 'Contest State'?: Elite Demands and the Origins of Peasant Protest in Southeast Asia." *Journal of Social History* 13(4): 521–546.

Aguilar, Filomeno, Jr.
    1994a Nouns That Sail through History: Reified Categories and Their Transcendence in Marx's Social Theory." Working Papers No. 120, Department of Sociology, National University of Singapore.
    1994b "Sugar Planter–State Relations and Labour Processes in Colonial Philippine *Haciendas.*" *Journal of Peasant Studies* 22(1): 50–80.
    1994c "Beyond Inevitability: The Opening of Philippine Provincial Ports in 1855." *Journal of Southeast Asian Studies* 25(1): 70–90.
    1994d "Of Cocks and Bets: Gambling, Class Structuring, and State For

mation in the Philippines." In James Eder and Robert Youngblood, eds., *Patterns of Power and Politics in the Philippines: Implications for Development.* Tempe, AZ: Program for Southeast Asian Studies, Arizona State University.

   1989 "The Philippine Peasant as Capitalist: Beyond the Categories of Ideal-Typical Capitalism." *Journal of Peasant Studies* 17(1): 41–67.

Alcina, Francisco Ignacio
   1960 "The Muñoz Text of Alcina's History of the Bisayan Islands: Part 1, Book 3" [1668]. Preliminary translation by Paul S. Lietz. Chicago: Philippine Studies Program, Department of Anthropology, University of Chicago (typescript).

Alunan, Rafael, et al.
   1939 "Report of the National Sugar Board to His Excellency the President of the Philippines." Manila: National Sugar Board (mimeo).

Anderson, Benedict
   1993 "Replica, Aura, and Late Nationalist Imaginings." *Qui Parle* 7(1): 1–21.
   1991 *Imagined Communities: Reflections on the Origin and Spread of Nationalism.* Rev. ed. London and New York: Verso.
   1990 "The Idea of Power in Javanese Culture" [1972]. In Benedict Anderson, *Language and Power: Exploring Political Cultures in Indonesia.* Ithaca and London: Cornell University Press.

Anima, Nid
   1977 *In Defense of Cockfighting.* Quezon City: Omar Publications.
   1972 *Death in the Afternoon and Far into the Night.* Quezon City: Omar Publications.

Anonymous ["An Englishman"]
   1907 "Remarks on the Philippine Islands, 1819–22" [1828]. In B&R, vol. 51.

Anonymous ["The Boxer Codex"]
   1979 "The Manners, Customs and Beliefs of the Filipinos of Long Ago" [undated, late sixteenth century]. In Mauro Garcia, ed., *Readings in Philippine Prehistory.* Manila: Filipiniana Book Guild.

Anonymous
   1922 "An Agriculturist's Report." *Sugar News* (March): 118–119.
   1898 "Una Interview." *La Republica Filipina* (26 November): 1–2.
   n.d. *Sucesos de Negros desde Octubre de 1896 á Febrero de 1897.* Printed compilation of Guardia Civil correspondence available in the library of the University of Sto. Tomas, Manila.

Araneta, Juan
   1898 "Rendición de Bakolod." *La Independencia* (28 December): 1–2.

Artigas, Manuel
  1894 *El Municipio Filipino: Compilacion de cuanto se ha prescrito sobre este particular é historia municipal de Filipinas desde los primeros tiempos de la dominacion española.* Tomo I. Manila: Imprenta de D. J. Atayde y Compañia.

Asociacion de Hacenderos de Silay-Saravia
  n.d. *Recopilacion de Datos de la Cosecha de 1929–30 y Resumen General Conmemorativo de Diez Anos de Relaciones Con La Hawaiian Philipine Company.* Iloilo: Imprenta La Defensa.

Bankoff, Greg
  1992 "Servant–Master Conflicts in Manila in the Late Nineteenth Century." *Philippine Studies* 40(3): 281–301.
  1991 "Redefining Criminality: Gambling and Financial Expediency in the Colonial Philippines, 1764–1898." *Journal of Southeast Asian Studies* 22(2): 267–281.

Barnshaw, Tom
  1924 "Impressions of Negros." *Sugar News* (November): 661–664.

Basco y Vargas, Jose
  1907 "Agriculture in Filipinas: A Decree by Basco in 1784." In B&R, vol. 52.

Behar, Ruth
  1987 "Sex and Sin, Witchcraft and The Devil in Late-Colonial Mexico." *American Ethnologist* 14(1): 34–54.

Benitez, Conrado
  1954 *History of the Philippines.* Rev. ed. Boston: Ginn and Company.

Bennassar, Bartolome
  1979 *The Spanish Character: Attitudes and Mentalities from the Sixteenth to the Nineteenth Century.* Translated by Benjamin Keen. Berkeley, Los Angeles and London: University of California Press.

Bensusan, S. L.
  1910 *Home Life in Spain.* New York: Macmillan.

Bernaldez, Manuel
  1907 "Reforms Needed in Filipinas" [1827]. In B&R, vol. 51.

Biernatzki, William
  1973 "Bukidnon Datuship in the Upper Pulangi River Valley." In Alfonso de Guzman II and Esther Pacheco, eds., *Bukidnon Politics and Religion.* Quezon City: Institute of Philippine Culture, Ateneo de Manila University.

Blair, Emma, and James Robertson, eds. (B&R)
  1904–1907 *The Philippine Islands 1493–1898.* 50 vols. Cleveland, OH: The Arthur H. Clark Company.

Blanc-Szanton, Cristina
   1990 "Collision of Cultures: Historical Reformulations of Gender in the Lowland Visayas, Philippines." In Jane Monnig Atkinson and Shelly Errington, eds., *Power and Difference: Gender in Island Southeast Asia.* Stanford, CA: Stanford University Press.

Boomgaard, Peter
   1987 "Morbidity and Mortality in Java, 1820–1880: Changing Patterns of Disease and Death." In Norman Owen, ed., *Death and Disease in Southeast Asia: Explorations in Social, Medical and Demographic History.* Singapore: Oxford University Press.

Bourdieu, Pierre
   1977 *Outline of a Theory of Practice.* Translated by Richard Nice. Cambridge and New York: Cambridge University Press.

Brass, Tom, and Henry Bernstein
   1992 "Introduction: Proletarianisation and Deproletarianisation on the Colonial Plantation." *Journal of Peasant Studies* 19(3/4): 1–40.

Breen, Timothy
   1977 "Horses and Gentlemen: The Cultural Significance of Gambling among the Gentry of Virginia." *William and Mary Quarterly,* 3d series, 34(2): 239–257.

Breman, Jan
   1990 *Labour Migration and Rural Transformation in Colonial Asia.* Amsterdam: Free University Press.

   1988 *The Shattered Image: Construction and Deconstruction of the Village in Colonial Asia.* Dordrecht, The Netherlands: Foris Publications for the Centre for Asian Studies Amsterdam.

Brill, Gerow
   1901 "Statement of Gerow D. Brill, 25 June 1901." In *Report of the Philippine Commission.* Part 2, Appendix FF. Bureau of Insular Affairs, U.S. War Department. Washington, DC: GPO.

Bulatao, Gerry
   1971 "The 'Swindle of the Century' Or, How the Sugar Act of 1952 was Circumvented in the Victorias Milling District." In Antonio Ledesma, Gerry Bulatao, Nini Abarquez, Felix Pasquin, and Rufino Suplido, eds., *Liberation in Sugarland: Readings on Social Problems in the Sugar Industry.* Manila: KIBAPIL.

Bulatao, Jaime
   1984 "When Roman Theology Meets an Animistic Culture: Mysticism in Present-day Philippines." *Kinaadman* 6(1): 102–111.

Byres, T. J.
   1983 "Historical Perspectives on Sharecropping." *Journal of Peasant Studies* 10(2/3): 7–40.

References

Callahan, William
  1984 *Church, Politics, and Society in Spain, 1750–1874.* Cambridge, MA, and London: Harvard University Press.

Calleo, David, and Benjamin Rowland
  1973 *America and the World Political Economy: Atlantic Dreams and National Realities.* Bloomington: Indiana University Press.

Candish, Thomas
  1904 "Expedition of Thomas Candish" [c. 1625]. In B&R, vol. 15.

Castro, Amado
  1965 "Philippine–American Tariff and Trade Relations, 1898–1954." *Philippine Economic Journal* 4(1): 29–56.

Chirino, Pedro
  1969 *Relacion de las Islas Filipinas* [1604]. With English translation by Ramon Echevarria. Manila: Historical Conservation Society.

Christian, William, Jr.
  1981 *Local Religion in Sixteenth-Century Spain.* Princeton, NJ: Princeton University Press.

Clark, James Hyde
  1896 *Cuba and the Fight for Freedom.* Philadelphia: Globe Bible Publishing Co.

Clotet, Jose Maria
  1906 "Letter from Father Jose Maria Clotet to the Reverend Rector of the Ateneo Municipal" [Talisayan, 11 May 1889]. In B&R, vol. 43.

Clunas, Craig
  1991 *Superfluous Things: Material Culture and Social Status in Early Modern China.* Oxford: Polity Press.

Cole, John
  1965 *Latin America: An Economic and Social Geography.* London: Butterworths.

Cole, Mabel Cook
  1916 *Philippine Folk Tales.* Chicago: A. C. McClurg and Co.

Colin, Francisco
  1906 "Native Races and Their Customs" [Excerpts from *Labor Evangelica*, 1663]. In B&R, vol. 40.

Collier, Jane Fishburne
  1988 *Marriage and Inequality in Classless Societies.* Stanford, CA: Stanford University Press.

Combes, Francisco
  1906 "The Natives of the Southern Islands" [1667]. In B&R, vol. 40.

Commoner, The
  1940a "Mayor Mariño to Solve Unemployment Problem in Town." 22 February, 10.

1940b "Laborers Burn Sugar Cane Fields South." 4 January, 6.
1939a "Labor Strikes in Occidental Negros." 9 February, 5.
1939b "Alunan versus Gambling." 2 February, editorial page.
1937a "15,649 Hectares Already Reserved for Tabla Valley Homesteaders." 3 June, 7.
1937b "Pres. Makilan Will Solve Unemployment Problem of Bacolod." 20 May, 7.
1936a "A Commoner's Topics." 19 November, 10.
1936b "A Commoner's Topics." 7 May, 10.
1936c "Special Investigator Makes Preliminary Report to Department of Labor." 23 April, 8, 13.
1936d "Shall We Compromise with Evil?" 5 March, 14.
1936e "Tabla Valley Movement Vindicated." 13 February, 8–9.
1935a "Bacolod Jueteng." 10 October, 5.
1935b "For Bridge Addicts." 15 August, 26.
1935c "A Commoner's Topics." 15 August, 10.
1935d "Acting President Ciocon Answer Pres. Makilan." 15 August, 13.
1935e "A Challenge." 1 August, editorial page.
1935f "Suspended Municipal President of Bacolod Deplores Scandalous Existence of Jueteng." 1 August, 2.

Concepcion, Venancio
  1927 *"La Tragedia" del Banco Nacional Filipino*. Manila: The author.

Constantino, Renato
  1985 *Synthetic Culture and Development*. Quezon City: Foundation for Nationalist Studies.
  1975 *The Philippines: A Past Revisited*. Quezon City: The author.

Corcuera, Sebastian, and Fausto Cruzat
  1907 "Ordinances of Good Government" [1642 and 1696]. In B&R, vol. 50.

Corpuz, Onofre
  1989 *The Roots of the Filipino Nation*. Vol. 1. Quezon City: Aklahi Foundation.

Costenoble, H.
  1923 "Methods of Cultivation in the San Carlos District." *Sugar News* (March): 111–114.

Cruz, Hermenegildo
  1935 *Labor Conditions in the Philippine Islands*. Bureau of Labor, Department of Commerce and Communications. Manila: Bureau of Printing.

Cuesta, Angel Martinez
- 1980 *History of Negros.* Translated by Alfonso Felix, Jr., and Sor Caritas Sevilla. Manila: Historical Conservation Society.
- 1974 *Historia de la Isla de Negros, Filipinas 1565–1898.* Madrid: Raycar, S.A. Impresores.

Cullamar, Evelyn
- 1986 *Babaylanism in Negros: 1896–1907.* Quezon City: New Day Publishers.

Cullen, Vincent
- 1973 "Bukidnon Animism and Christianity." In Alfonso de Guzman II and Esther Pacheco, eds., *Bukidnon Politics and Religion.* IPC Papers No. 11. Quezon City: Institute of Philippine Culture, Ateneo de Manila University.

Cullinane, Michael
- 1982 "The Changing Nature of the Cebu Urban Elite in the Nineteenth Century." In Alfred McCoy and Ed. de Jesus, eds., *Philippine Social History: Global Trade and Local Transformations.* Quezon City: Ateneo de Manila University Press.

Cushner, Nicholas
- 1976 *Landed Estates in the Colonial Philippines.* Monograph Series No. 20. New Haven, CT: Yale University Southeast Asia Studies.
- 1971 *Spain in the Philippines: From Conquest to Revolution.* Quezon City: Institute of Philippine Culture, Ateneo de Manila University.

Dampier, William
- 1906 "Dampier in Philippines" [1697]. In B&R, vol. 39.

Daston, Lorraine
- 1991 "Marvelous Facts and Miraculous Evidence in Early Modern Europe." *Critical Inquiry* 18(1): 93–124.

De Bevoise, Ken
- 1995 *Agents of Apocalypse: Epidemic Disease in the Colonial Philippines.* Princeton, NJ: Princeton University Press.

de Comyn, Tomas
- 1969 *State of the Philippines in 1810* [1820]. Manila: Filipiniana Book Guild.

Defourneaux, Marcelin
- 1970 *Daily Life in Spain in the Golden Age.* Translated by Newton Branch. London: George Allen and Unwin.

de Jesus, Ed.
- 1982 "Control and Compromise in the Cagayan Valley." In Alfred McCoy and Ed. de Jesus, eds., *Philippine Social History: Global Trade and Local Transformations.* Quezon City: Ateneo de Manila University Press.
- 1978 "The Tobacco Monopoly in the Cagayan Valley, 1786–1881." In

Ruth McVey, ed., *Southeast Asian Transitions: Approaches through Social History*. New Haven and London: Yale University Press.

de Jesus, Luis
    1904 "General History of the Discalced Religious of St. Augustine" [1681]. In B&R, vol. 21.

de Jesus, Pablo
    1906 "Letter to Gregory XIII" [1580]. In B&R, vol. 34.

de la Concepcion, Juan
    1904 "General History of the Philipinas" [1788]. In B&R, vol. 21.

de la Costa, Horacio
    1969 "The Development of the Native Clergy in the Philippines." In Gerald Anderson, ed., *Studies in Philippine Church History*. Ithaca, NY: Cornell University Press.

de la Costa, Horacio, and John Schumacher
    1980 *The Filipino Clergy: Historical Studies and Future Perspectives*. Loyola Papers No. 12. Quezon City: Loyola School of Theology, Ateneo de Manila University.

de la Gironière, Paul
    1962 *Twenty Years in the Philippines* [1854]. Manila: Filipiniana Book Guild.

de Loarca, Miguel
    1903 "Relación de las Yslas Filipinas" [1582]. In B&R, vol. 5.

de Mas, Sinibaldo
    1963 *Report on the Condition of the Philippines in 1842* [1842]. Manila: Historical Conservation Society.

Demetrio, Francisco
    1968 "The *Engkanto* Belief: An Essay in Interpretation." *Philippine Sociological Review* 16(3/4): 136–143.

Demetrio, Francisco, ed.
    1970 *Dictionary of Philippine Folk Beliefs and Customs*. Cagayan de Oro City: Xavier University.

de Morga, Antonio
    1904 "Sucesos de las Islas Filipinas" [1609]. In B&R, vol. 16.

de Rivera, Fernando
    1881 *Documentos Referentes á la Reducción de Ynfieles é Inmigración en las Provincias de Cagayan y La Ysabela*. Manila: Imprenta del Colegio de Santo Tomas.

De Rooy, Piet
    1990 "Of Monkeys, Blacks and Proles: Ernst Haeckel's Theory of Recapitulation." In Jan Breman, ed., *Imperial Monkey Business: Racial Supremacy in Social Darwinist Theory and Colonial Practice*. Amsterdam: VU University Press.

de Salazar, Domingo
  1903 "Affairs in the Philipinas Islands" [1583]. In B&R, vol. 5.
de Salazar, Vicente
  1906 "Dominican Missions, 1670–1700" [1742]. In B&R, vol. 43.
de San Agustin, Gaspar
  1906 "Letter on the Filipinos" [1720]. In B&R, vol. 40.
de San Antonio, Juan Francisco
  1906 "The Native Peoples and Their Customs" [1738]. In B&R, vol. 40.
de San Nicolas, Andres
  1904 "General History of the Discalced Augustinian Fathers" [1664]. In B&R, vol. 21.
de Uriarte, Higinio
  1962 *A Basque among the Guerrillas of Negros.* Translated by Soledad Lacson Locsin. Bacolod: Civismo Weekly.
de Viana, Francisco Leandro
  1907 "Memorial of 1765." In B&R, vol. 48.
Diaz, Zollo
  1969 *"Memorias": The Autobiography of Zollo S. Diaz.* Translated by A. A. Ortega. Quezon City: Phoenix Press.
Diogenes
  1910 "The Present and Future of the Islands of Panay and Negros." *Philippine Resources* 1(7): 25–28.
Dobell, Peter
  1907 "Letter dated 28 November 1820." In B&R, vol. 51.
Doeppers, Daniel
  1986 "Destination, Selection and Turnover among Chinese Migrants to Philippine Cities in the Nineteenth Century." *Journal of Historical Geography* 12(4): 381–401.
  1984 *Manila 1900–1941: Social Change in a Late Colonial Metropolis.* Quezon City: Ateneo de Manila University Press.
Doherty, John
  1964 "The Image of the Priest: A Study in Stereotyping." *Philippine Sociological Review* 12(1/2): 70–76.
Echauz, Robustiano
  1894 *Apuntes de la isla de Negros.* Manila: Tipo-Litografia de Chofre y Compaóia.
Elson, R. E.
  1984 *Javanese Peasants and the Colonial Sugar Industry: Impact and Change in an East Java Residency, 1830–1940.* Singapore: Oxford University Press.

Estacion, Jose
   n.d. *History of Kanla-on Lodge No. 64, F. & A.M.* Bacolod City: The author.

Fabian, Ann
   1990 *Card Sharps, Dream Books, and Bucket Shops: Gambling in 19th-Century America.* Ithaca, NY, and London: Cornell University Press.

Fairchild, George
   1929 "A Brief History of the Philippine Sugar Industry." In *Handbook of the Philippine Sugar Industry*. 2d ed. Manila: Sugar News Press Company.
   1925 "A History of the Sugar Industry of the Philippine Islands." Part 2. *Sugar News* (September).

Fallows, James
   1987 "A Damaged Culture." *Atlantic Monthly* (November): 49–58.

Farley, Rawle
   1972 *The Economics of Latin America: Development Problems in Perspective.* New York: Harper and Sons.

Farriss, N. M.
   1968 *Crown and Clergy in Colonial Mexico, 1759–1821: The Crisis of Ecclesiastical Privilege.* London: The Athlone Press.

Fast, Jonathan, and Jim Richardson
   1979 *Roots of Dependency: Political and Economic Revolution in 19th Century Philippines.* Quezon City: Foundation for Nationalist Studies.

Fay, Bernard
   1935 *Revolution and Freemasonry 1860–1800.* Boston: Little, Brown, and Company.

Fegan, Brian
   1994 "Entrepreneurs in Votes and Violence: Three Generations of a Peasant Political Family." In Alfred McCoy, ed., *An Anarchy of Families: State and Family in the Philippines.* Quezon City: Ateneo de Manila University Press.

Fisher, Lillian Estelle
   1939 "Early Masonry in Mexico (1806–1828)." *Southwestern Historical Quarterly* 42(3): 198–214.

Forbes-Lindsay, C. H.
   1906 *The Philippines: Under Spanish and American Rules.* Philadelphia: John C. Winston Co.

Foreman, John
   1899 *The Philippine Islands.* 2d ed. New York: Charles Scribner's Sons.

Fox, James
   1987 "Southeast Asian Religions: Insular Cultures." In Mircea Eliade, ed., *The Encyclopedia of Religion.* Vol. 13. New York: Macmillan.

References

Friend, Theodore
  1965 *Between Two Empires: The Ordeal of the Philippines, 1929–1946.* New Haven, CT: Yale University Press.
  1963 "The Philippine Sugar Industry and the Politics of Independence, 1929–1935." *Journal of Asian Studies* 22(2): 179–192.

Fuentes, Cornelio
  1919 *Datos para la Historia: Apuntes Documentados de la Revolucion en toda la Isla de Negros.* Iloilo: El Centinela.

Garcia, Antonio
  1966 *El Problema Agrario en America Latina y los Medios de Informacion Colectiva.* Quito, Ecuador: Centro International de Estudios Superiores de Periodismo para America Latina (CIESPAL).

Godelier, Maurice
  1978a "Economy and Religion: An Evolutionary Optical Illusion." In J. Friedman and M. J. Rowlands, eds., *The Evolution of Social Systems.* Pittsburgh, PA: University of Pittsburgh Press.
  1978b "Politics as 'Infrastructure': An Anthropologist's Thoughts on the Example of Classical Greece and the Notions of Relations of Production and Economic Determination." In J. Friedman and M. J. Rowlands, eds., *The Evolution of Social Systems.* Pittsburgh, PA: University of Pittsburgh Press.

Goncharov, Ivan
  1974 "Voyage of the Frigate 'Pallada.'" In *Travel Accounts of the Islands (1832–1858).* Manila: Filipiniana Book Guild.

Gonzaga, Violeta
  1991 *The Negrense: A Social History of an Elite Class.* Bacolod: Institute for Social Research and Development, University of Saint La Salle.
  1987 "Capital Expansion, Frontier Development and the Rise of Monocrop Economy in Negros (1850–1898)." Occasional Paper No. 1. Bacolod: SRC-Negrense Studies Program.

Gordon, G. G.
  1933 "Present Status of the Philippine Sugar Industry." *Sugar News* (December): 605–610.

Graham, D. E.
  1939 "The Big Problem the Sugar Industry Faces." *Sugar News* (September): 371–372.

Greenleaf, Richard
  1991 "Recent Historiography of the Mexican Inquisition: Evolutions of Interpretations and Methodologies." In Mary Elizabeth Perry and Anne J. Cruz, eds., *Cultural Encounters: The Impact of the Inquisition in*

Spain and the New World. Berkeley, Los Angeles, Oxford: University of California Press.

1969 "The Mexican Inquisition and the Masonic Movement: 1751–1820." *New Mexico Historical Review* 44(2): 92–117.

1966 "North American Protestants and the Mexican Inquisition, 1765–1820." *Journal of Church and State* 8(2): 186–199.

Guggenheim, Scott
1982 "Cock or Bull: Cockfighting, Social Structure, and Political Commentary in the Philippines." *Pilipinas* 3(1): 1–35.

Hart, Donn
1967 "Buhawi of the Bisayas: The Revitalization Process and Legend Making in the Philippines." In Mario Zamora, ed., *Studies in Philippine Anthropology: Essays in Honor of H. Otley Beyer.* Quezon City: Alemar-Phoenix Press.

1966 "The Filipino Villager and His Spirits." *Solidarity* 1(4): 65–71.

1964 "Guerilla Warfare and the Filipino Resistance on Negros Island in the Bisayas, 1942–1945." *Journal of Southeast Asian History* 5(1): 101–125.

1954 "Barrio Caticugan: A Visayan Filipino Community." D.S.Sc. diss., Syracuse University.

Hartendorp, A. V. H.
1958 *History of Industry and Trade of the Philippines.* Manila: American Chamber of Commerce of the Philippines.

Hawes, Gary
1987 *The Philippine State and the Marcos Regime: The Politics of Export.* Ithaca, NY, and London: Cornell University Press.

Hayne, A. P.
1904 "Report of the Director, College of Agriculture and Experiment Station at La Granja Modelo, La Carlota 31 August 1903." In *Fourth Annual Report of the Philippine Commission, 1903.* Part 2, Appendix I. BIA. Washington, DC: GPO.

Henry, Yves
1929a "Technical and Financial Conditions of the Production of Sugar in the Philippines." Part 4. *Sugar News* (July): 483–492.

1929b "Technical and Financial Conditions of the Production of Sugar in the Philippines." Part 3. *Sugar News* (February): 124–130.

1929c "Technical and Financial Conditions of the Production of Sugar in the Philippines." Part 2. *Sugar News* (January): 18–20.

1928 "Technical and Financial Conditions of the Production of Sugar in the Philippines." Part 1. *Sugar News* (October): 727–734.

Hofileña, Josefina Dalupan
  1990 "Wartime in Negros Occidental, 1942–1945." Master's thesis, University of the Philippines at Diliman.
Holzberger, William, and Herman Saatkamp, Jr.
  1986 *Persons and Places: Fragments of Autobiography.* Cambridge, MA, and London: MIT Press.
Hord, John
  1910 "A Few Notes on the Visayas." *Philippine Resources* 1(7): 10–16.
Horn, Florence
  1941 *Orphans of the Pacific: The Philippines.* New York: Reynal and Hitchcock.
Huetz de Lemps, Xavier
  1990 "Les Philippines face au 'Fantôme du Gange': Le Choléra dans la seconde moitié du XIXe siècle." In *Annales de Démographie Historique.* Paris: Société de Démographie Historique.
Hughes, Robert
  1902 "Statement of Brig. Gen. Robert P. Hughes." In *Affairs in the Philippine Islands.* Hearings before the Committee on the Philippines of the United States Senate. 57th Cong., 1st sess., 1901–1902. Senate Document Vol. 23, Doc. No. 331, Part 1. Washington, DC: GPO.
Ikehata, Setsuho
  1990 "Popular Catholicism in the Nineteenth-Century Philippines: The Case of the Cofradia de San Jose." In *Reading Southeast Asia.* Ithaca, NY: Cornell Southeast Asia Program.
Ilag, Leodegario
  1964 "Farm Management Analysis of Some Sugarcane Farms in the Victorias Mill District. Philippines, 1961–1962." Master's thesis, University of the Philippines at Los Baños.
Ileto, Reynaldo
  1992 "Religion and Anti-Colonial Movements." In Nicholas Tarling, ed., *The Cambridge History of Southeast Asia.* Vol. 2. Cambridge: Cambridge University Press.
  1988 "Outlines of a Non-linear Emplotment of Philippine History." In Lim Tech Ghee, ed., *Reflections on Development in Southeast Asia.* Singapore: Institute of Southeast Asian Studies.
  1979 *Pasyon and Revolution: Popular Movements in the Philippines, 1840–1910.* Quezon City: Ateneo de Manila University Press.
Jagor, Fedor
  1965 *Travels in the Philippines* [1873]. Manila: Filipiniana Book Guild.
  1907 "Agricultural Conditions in 1866." In B&R, vol. 52.

Jayme, Antonio
  1906 "Report of the Governor of Occidental Negros, Bacolod 15 July 1905." In *Sixth Annual Report of the Philippine Commission, 1905*. Part 1, Appendix H. BIA. Washington, DC: GPO.
Kalaw, Teodoro
  1956 *Philippine Masonry: Its Origin, Development, and Vicissitudes up to the Present Time (1920)*. Translated by Frederic Stevens and Antonio Amechazurra. Manila: McCullough Printing Company.
Keller, A. F.
  1926 "Keller Makes a Gloomy Report on Centrals." *Manila Times*, 24 January 1926. Also in *Sugar News* (February): 101–105.
Kerkvliet, Benedict
  1977 *The Huk Rebellion: A Study of Peasant Revolt in the Philippines*. Berkeley, Los Angeles, London: University of California Press.
Kikuchi, Yasuchi, ed.
  1989 *Philippine Kinship and Society*. Quezon City: New Day Publishers.
Kitching, Gavin
  1988 *Karl Marx and the Philosophy of Praxis*. London and New York: Routledge.
Krenn, Michael
  1990 *U.S. Policy Toward Economic Nationalism in Latin America, 1917–1929*. Wilmington, DE: SR Books.
Lachica, Eduardo
  1971 *Huk: Philippine Agrarian Society in Revolt*. Manila: Solidaridad.
Lacsamana, Ramon, ed.
  1939 *Negros Occidental 1938–1939 Year Book and Special Supplement, Inauguration of the City of Bacolod*. N.p.: The editor.
Lansang, Angel
  n.d. *Cockfighting University: Past and Present Outlook of our National Pastime*. Manila: Unnamed publisher.
Larkin, John
  1993 *Sugar and the Origins of Modern Philippine Society*. Berkeley, Los Angeles, London: University of California Press.
  1972 *The Pampangans: Colonial Society in a Philippine Province*. Berkeley and Los Angeles: University of California Press.
Latham, A. B.
  1923 "Report of the Committee on Labor for Luzon." In *Compilation of Committee Reports for the First Annual Convention, 6 to 12 October 1923*. Manila: Philippine Sugar Association.
Lea, Henry Charles
  1908 *The Inquisition in the Spanish Dependencies*. New York: The Macmillan Company.

Legarda, Benito, Jr.
 1972 "American Entrepreneurs in the 19th Century Philippines." *Bulletin of the American Historical Collection* 1(1): 25–52.
 1962 "Introduction." In Paul de la Gironière's *Twenty Years in the Philippines* [1854]. Manila: Filipiniana Book Guild.
 1955 "Foreign Trade, Economic Change, and Entrepreneurship in the Nineteenth Century Philippines." Ph.D. diss., Harvard University.

Leonard, Thomas
 1910 "The Sugar Country." *Philippine Resources* 1(10): 17–21.

Le Roy, James
 1968 "Philippine Life in Town and Country." Book 1 in *The Philippines Circa 1900*. Manila: Filipiniana Book Guild.
 1914 *The Americans in the Philippines*. Vol. 1. Boston and New York: Houghton Mifflin Company.

Lieban, Richard
 1967 *Cebuano Sorcery: Malign Magic in the Philippines*. Berkeley and Los Angeles: University of California Press.
 1962 "The Dangerous *Ingkantos:* Illness and Social Control in a Philippine Community." *American Anthropologist* 62(2): 306–312.

Lim, Linda Y. C.
 1978 "Women Workers in Multinational Corporations: The Case of the Electronics Industry in Malaysia and Singapore." Michigan Occasional Paper No. 9, University of Michigan.

Locsin, Carlos
 1926 "Registration of Agricultural Laborers." *Sugar News* (March): 136–143.
 1925 "Cane Cultivation and Animal Husbandry on Occidental Negros." In *Compilation of the Committee Reports for the Third Annual Convention, 5 to 10 October 1925*. Manila: Philippine Sugar Association.
 1924 "Profit or Loss." *Sugar News* (July): 463–465.
 1923 "Report of the Committee on Agricultural Costs and Field Data." In *Compilation of Committee Reports for the First Annual Convention, 6 to 12 October 1923*. Manila: Philippine Sugar Association.
 1921 "The New Era in Negros." *Sugar News* (November): 466–469.

Locsin, Jose C.
 1969 "Christian Justice and the Sugar Industry." *Philippines Free Press*, 26 April, 10, 63–67.

Locsin, Leandro
 1904 "Memorandum of L. Locsin Rama, Provincial Governor, to F. Lamson-Scribner, Bureau of Agriculture, Bacolod September 1903." In *Fourth Annual Report of the Philippine Commission, 1903*. Part 2, Exhibit B. BIA. Washington, DC: GPO.

1902 "Annual Report of the Governor of Occidental Negros, P.I., Bacolod 19 December 1901." In *Affairs of the Philippine Islands*. Hearings before the Committee on the Philippines of the United States Senate. 57th Cong., 1st sess., 1901–1902. Senate Document Vol. 24, Doc. No. 331, Part 2. Washington, DC: GPO.

Loney, Nicholas
1964 *A Britisher in the Philippines, or The Letters of Nicholas Loney*. Manila: National Library.

Lopez, Manuel
1908 "Report of the Governor of Negros Occidental, Bacolod 24 July 1907." In *Eighth Annual Report of the Philippine Commission, 1907*. Part 1, Exhibit N. BIA. Washington, DC: GPO.

Lopez, Oscar
1982 "Historical Introduction to the Lopez Family." In *The Lopez Family: Its Origins and Genealogy*. Pasig, Metro Manila: Eugenio Lopez Foundation.

Luzuriaga, Jose Ruiz de
1900 "Testimony of Señor Luzuriaga, Manila 11 September 1899." In *Report of the Philippine Commission*. Vol. 2. Bureau of Insular Affairs, U.S. War Department. Washington, DC: GPO.

Lynch, Frank
1979 "Big and Little People: Social Class in the Rural Philippines." In Mary Hollnsteiner, ed., *Society, Culture and the Filipino*. Quezon City: Institute of Philippine Culture, Ateneo de Manila University.

Lynch, John
1973 *The Spanish American Revolutions, 1808–1826*. New York: W. W. Norton and Co.

Lyon, William
1903 "Report of William S. Lyon, In Charge of Seed and Plant Introduction." In *Third Annual Report of the Philippine Commission, 1902*. Appendix O. BIA. Washington, DC: GPO.

McCoy, Alfred
1992 "Sugar Barons: Formation of a Native Planter Class in the Colonial Philippines." *Journal of Peasant Studies* 19(3/4): 106–141.

1984 *Priests on Trial: Father Gore and Father O'Brien Caught in the Crossfire between Dictatorship and Revolution*. Ringwood, Victoria, Australia: Penguin Books.

1982a "A Queen Dies Slowly: The Rise and Decline of Iloilo City." In Alfred McCoy and Ed. de Jesus, eds., *Philippine Social History: Global Trade and Local Transformations*. Quezon City: Ateneo de Manila University Press.

1982b "*Baylan:* Animist Religion and Philippine Peasant Ideology." *Philippine Quarterly of Culture and Society* 10(3): 141–194.

1977 "Ylo-ilo: Factional Conflict in a Colonial Economy, Iloilo Province, Philippines, 1937–1955." Ph.D. diss., Yale University.

McLennan, Marshall

1984 "Changing Human Ecology on the Central Luzon Plain: Nueva Ecija, 1705–1939." In Alfred McCoy and Ed. de Jesus, eds., *Philippine Social History: Global Trade and Local Transformations.* Quezon City: Ateneo de Manila University Press.

1969 "Land and Tenancy in the Central Luzon Plain." *Philippine Studies* 17(4): 651–682.

McNeill, William

1976 *Plagues and Peoples.* New York: Anchor Press.

Magos, Alicia

1992 *The Enduring Ma-aram Tradition: An Ethnography of a Kinaray-a Village in Antique.* Quezon City: New Day Publishers.

Magsalin, Pedro

1948 "The Sugar Industry and Labor." *Planters' Review* (Bacolod), 7 May, 2, 4.

Mallat, Jean

1983 *The Philippines: History, Geography, Customs, Agriculture, Industry and Commerce of the Spanish Colonies in Oceania* [1846]. Translated by Pura Santillan-Castrence. Manila: National Historical Institute.

Manila Daily Bulletin

1925 "Land Bank for Islands Urged." 8 June.

1918 "Sugar Industry Still Thrives." 9 October.

Marcelo, Salvador

1953 "The Sugar Act of 1952." In Casiano Anunciacion, ed., *A Handbook of the Sugar and Other Industries in the Philippines.* Manila: Sugar News Press.

Marco, Jose

1912 *Reseña Historica de la Isla de Negros desde los tiempos mas remotos hasta nuestros dias.* Manila: Imprenta Tipografica de "La Vanguardia."

Marias, Julian

1990 *Understanding Spain.* Translated by Frances Lopez-Morillas. Ann Arbor: University of Michigan Press.

Marx, Karl

1963 *The Eighteenth Brumaire of Louis Bonaparte* [1852]. New York: International Publishers.

Milne, O. G. C.

1925 "Labor Conditions in Luzon." In *Compilation of the Committee Reports* for the Third Annual Convention, 5 to 10 October 1925. Manila: Philippine Sugar Association.

Mintz, Sidney
    1985 *Sweetness and Power: The Place of Sugar in Modern History*. New York: Elisabeth Sifton Books-Viking.

Mirasol, Jose
    1950 "The Philippine Sugar Industry." In *Philippine Sugar Year Book 1950*. Manila: National Federation of Sugarcane Planters.

More, Quirico
    1906 "Letter of Father Quirico More, to the Father Superior of the Mission" [Davao, 20 January 1885]. In B&R, vol. 43.

Murray, Frank
    1970 "Local Groups and Kin Groups in a Tagalog Tenant Rice-Farmers' Barrio." Ph.D. diss., University of Pittsburgh.

Myrick, Conrad
    1969 "Some Aspects of the British Occupation of Manila." In Gerald Anderson, ed., *Studies in Philippine Church History*. Ithaca, NY: Cornell University Press.

Nagano, Yoshiko
    1997 "The Agricultural Bank of the Philippine Government, 1908–1916." *Journal of Southeast Asian Studies* 28(2): 301–323.
    1988 "The Oligopolistic Structure of the Philippine Sugar Industry during the Great Depression." In Bill Albert and Adrian Graves, eds., *The World Sugar Economy in War and Depression 1914–40*. London and New York: Routledge.
    1982 "Formation of Sugarlandia in the Late 19th Century Negros: Origin of *Underdevelopment* in the Philippines." Research and Working Papers Series No. 12, Third World Studies Program, University of the Philippines.

Nash, June
    1979 *We Eat the Mines and the Mines Eat Us: Dependency and Exploitation in Bolivian Tin Mines*. New York: Columbia University Press.

Ner, Jose
    1898 "El Porque del Levantamiento de Negros Occidental." *La Republica Filipina*, 18 November, 2.

Ness, Sally Ann
    1992 *Body, Movement, and Culture: Kinesthetic and Visual Symbolism in a Philippine Community*. Philadelphia: University of Pennsylvania Press.

Norman, Henry
    1895 *The Peoples and Politics of the Far East*. New York: Charles Scribner's Sons.

Ocampo, Ricardo
- 1923 "Letter to Atherton Lee." In *Compilation of Committee Reports for the First Annual Convention, 6 to 12 October 1923.* Manila: Philippine Sugar Association.

Ong, Aihwa
- 1987 *Spirits of Resistance and Capitalist Discipline: Factory Women in Malaysia.* Albany: State University of New York Press.

Onrubia, L. R.
- 1925 "Labor on Luzon Island." *Sugar News* (January): 20–25.

Ortiz, Tomas
- 1906 "Superstitions and Beliefs of the Filipinos" [ca. 1731]. In B&R, vol. 43.

Owen, Norman
- 1984 *Prosperity without Progress: Manila Hemp and Material Life in the Colonial Philippines.* Quezon City: Ateneo de Manila University Press.

Palmer, Truman
- 1925 "Philippine Sugar Centrals, 1925." *Concerning Sugar* (September). Washington, DC: United States Sugar Manufacturers' Association.
- 1920 *Concerning Sugar.* Loose Leaf Service. Washington, DC: Bureau of Statistics, United States Sugar Manufacturers' Association.

Pardo de Tavera, T. H.
- 1905 "Discovery and Progress." In *Census of the Philippine Islands (Taken under the Direction of the Philippine Commission in the Year 1903).* Vol. 1. Washington, DC: U.S. Bureau of the Census.

Pearce, R.
- 1983 "Sharecropping: Towards a Marxist View." *Journal of Peasant Studies* 10(2/3): 42–70.

Pedrosa, Ramon
- 1983 "Abortion and Infanticide in the Philippines during the Spanish Contact." *Philippiniana Sacra* 18(52): 7–37.

Phelan, John Leddy
- 1959 *The Hispanization of the Philippines: Spanish Aims and Filipino Responses, 1565–1700.* Madison: University of Wisconsin Press.

Philippine Commission
- 1907 "Report of the District Director, Third District, Bureau of Constabulary." In *Seventh Annual Report of the Philippine Commission, 1906.* Part 2, Appendix 3 to Exhibit A. BIA. Washington, DC: GPO.
- 1906a "Report of the Secretary of Finance and Justice to the Philippine Commission." In *Sixth Annual Report of the Philippine Commission, 1905.* Part 4. BIA. Washington, DC: GPO.

1906b "Report of the Officer Commanding Third District, Philippines Constabulary, Iloilo 19 July 1905." In *Sixth Annual Report of the Philippine Commission, 1905*. Part 3, Exhibit 3, Appendix A. BIA. Washington, DC: GPO.

1904 "Report of the Civil Governor for the Period Ending December 23, 1903." In *Fourth Annual Report of the Philippine Commission, 1903*. Part 1. BIA. Washington, DC: GPO.

1903 "Report of the Secretary of the Interior 31 August 1902." In *Third Annual Report of the Philippine Commission, 1902*. Part 1. BIA. Washington, DC: GPO.

Philippine Statistical Review
1936 "Registration of Motor Vehicles by Provinces, 1935." 3(1): 63.

1935a "Capital Investment in Leading Philippine Farm Industries." 2(4): 310.

1935b "Registration of Motor Vehicles by Provinces, 1933–1934." 2(2): 40–41.

Philippines Herald, The
1924 "Filipino One of the Best Sugar Planters in the World says De la Rama." 14 June.

Philippines Monthly, The
1912 "Planter's Prospects Bright." (December): 11.

Pigafetta, Antonio
1969 *First Voyage Around the World* [1522]. Manila: Filipiniana Book Guild.

Pinches, Michael
1991 "The Working Class Experience of Shame, Inequality, and People Power in Tatalon, Manila." In Benedict Kerkvliet and Resil Mojares, eds., *From Marcos to Aquino: Local Perspectives on Political Transition in the Philippines*. Honolulu: University of Hawai'i Press; Quezon City: Ateneo de Manila University Press.

Pitcairn, R. C.
1925 "Labor Situation on Negros." In *Compilation of the Committee Reports for the Third Annual Convention, 5 to 10 October 1925*. Manila: Philippine Sugar Association.

1922 "Natural Methods of Increasing Production (Occidental Negros Conditions)." *Sugar News* (October): 471–477.

Pitt, Harold
1909 "Leading Agricultural Products of the Philippines." *Philippine Resources* 1(1): 9–14.

Postma, Antoon
    1992 "The Laguna Copper-Plate Inscription: Text and Commentary." *Philippine Studies* 40(2): 183–203.
Pritchett, G. H.
    1926 "Agricultural Conditions in Negros." *Sugar News* (July): 449–460.
Rafael, Vicente
    1988 *Contracting Colonialism: Translation and Christian Conversion in Tagalog Society under Early Spanish Rule.* Ithaca, NY, and London: Cornell University Press.
Ramirez, Mina
    1993 *Understanding Philippine Social Realities through the Filipino Family: A Phenomenological Approach.* 2d ed. Manila: Asian Social Institute.
Ramos, Maximo
    1971 *Creatures of Philippine Lower Mythology.* Quezon City: University of the Philippines Press.
Raon, Jose
    1907 "Ordinances of Good Government" [1768]. In B&R, vol. 50.
Reed, Robert
    1967 "Hispanic Urbanism in the Philippines: A Study of the Impact of Church and State." *Journal of East Asiatic Studies* 11:1–222.
Regalado, C.
    1948 "La Agricultura y La Musica." *Civismo,* 16 May, 2.
Regidor, Antonio Ma.
    1982 "Masonry in the Philippines" [1896]. *The Cabletow:* The Official Organ of the Grand Lodge of the Philippines (July–August): 2–25.
Regidor, Antonio, and J. Warren Mason
    1925 *Commercial Progress in the Philippine Islands* [1905]. Manila: The American Chamber of Commerce of the Philippine Islands.
Reid, Anthony
    1988 *Southeast Asia in the Age of Commerce 1450–1680.* Vol. 1, *The Land Below the Winds.* New Haven and London: Yale University Press.
Rivera, Temario
    1982 "Rethinking the Philippine Social Formation: Some Problematic Concepts and Issues." In *Feudalism and Capitalism in the Philippines: Trends and Implications.* Quezon City: Foundation for Nationalist Studies.
Rivers, W. C.
    1908 "Report of the District Director, Third Constabulary District." In *Annual Report of the Philippine Commission, 1907.* Part 2. Bureau of Insular Affairs, U.S. War Department. Washington, DC: GPO.

Rizal, Jose
   1958 *Noli Me Tangere.* Reprint of Berlin ed. [1886]. Quezon City: R. Martinez and Sons.
Rodriguez, Caridad
   1983 *Negros Oriental and the Philippine Revolution.* Dumaguete City: The Provincial Government of Negros Oriental.
Romero, Ma. Fe Hernaez
   1974 *Negros Occidental between Two Foreign Powers (1888–1909).* Bacolod: Negros Occidental Historical Commission.
Rosaldo, Michelle
   1980 *Knowledge and Passion: Ilongot Notions of Self and Social Life.* Cambridge: Cambridge University Press.
Rosell, Pedro
   1906 "Letter from Father Pedro Rosell to the Father Superior of the Mission" [Caraga, 17 April 1885]. In B&R, vol. 43.
Ross, H. B.
   1920 "The Philippine Labor Problem." *Sugar News* (March): 1–4.
Roth, Dennis
   1982 "Church Lands in the Agrarian History of the Tagalog Region." In Alfred McCoy and Ed. de Jesus, eds., *Philippine Social History: Global Trade and Local Transformations.* Quezon City: Ateneo de Manila University Press.
Runes, I. T.
   1939 *General Standards of Living and Wages of Workers in the Philippine Sugar Industry.* Manila: Philippine Council, Institute of Pacific Relations.
Rush, James
   1990 *Opium to Java: Revenue Farming and Chinese Enterprise in Colonial Indonesia, 1860–1910.* Ithaca, NY: Cornell University Press.
Sahlins, Marshall
   1972 *Stone Age Economics.* New York: Aldine de Gruyter.
   1963 "Poor Man, Rich Man, Big-Man, Chief: Political Types in Melanesia and Polynesia." *Comparative Studies in Society and History* 10(3): 285–303.
Santayana, George
   1944 *Persons and Places: The Background of My Life.* New York: Charles Scribner's Sons.
Santiago, Luciano
   1990 "The Filipino *Indios Encomenderos* (ca. 1620–1711)." *Philippine Quarterly of Culture and Society* 18(3): 162–184.
Sa-onoy, Modesto
   1982 "The Americans and the Negros Sugar Industry." *Journal of History* (Manila) 27(1/2): 40–66.

References

Sawyer, Frederic
  1900 *The Inhabitants of the Philippines.* New York: Charles Scribner's Sons.

Sayer, Derek
  1987 *The Violence of Abstraction: The Analytic Foundations of Historical Materialism.* Oxford and New York: Basil Blackwell.

Schul, Norman
  1967 "Hacienda Magnitude and Philippine Sugar Cane Production." *Asian Studies* 5(2): 258–273.

Schumacher, John
  1991 *The Making of a Nation: Essays on Nineteenth-Century Filipino Nationalism.* Quezon City: Ateneo de Manila University Press.

Schurman, Jacob Gould
  1902 *Philippine Affairs: A Retrospect and Outlook.* New York: Charles Scribner's Sons.

Schurman, J. G., et al.
  1900 "Preliminary Report of the Commission, 2 November 1899." In *Report of the Philippine Commission to the President, 31 January 1900.* Vol. I, Exhibit I. Bureau of Insular Affairs, U.S. War Department. Washington, DC: GPO.

Scott, James
  1976 *The Moral Economy of the Peasant: Rebellion and Subsistence in Southeast Asia.* New Haven, CT, and London: Yale University Press.

Scott, William Henry
  1994 "Prehispanic Filipino Concepts of Land Rights." *Philippine Quarterly of Culture and Society* 22(3): 166–173.
  1992a "Why Did Tupas Betray Dagami?" In William Henry Scott, *Looking for the Prehispanic Filipino and Other Essays in Philippine History.* Quezon City: New Day Publishers.
  1992b "Sixteenth-Century Visayan Food and Farming." In William Henry Scott, *Looking for the Prehispanic Filipino and Other Essays in Philippine History.* Quezon City: New Day Publishers.
  1991 *Slavery in the Spanish Philippines.* Manila: De La Salle University Press.
  1982 "Filipino Class Structure in the Sixteenth Century" [1978]. In William Henry Scott, *Cracks in the Parchment Curtain and Other Essays in Philippine History.* Quezon City: New Day Publishers.

Shiraishi, Takashi
  1990 *An Age in Motion: Popular Radicalism in Java, 1912–1926.* Ithaca, NY, and London: Cornell University Press.

Simonena, Marcelino
  1974 *Father Fernando Cuenca of St. Joseph, Augustinian Recollect* [c. 1930s]. Translated by Ma. Soledad L. Locsin. Bacolod City: Negros Occidental Historical Commission.

Smith, Peter [Peter Xenos]
  1978 "Crisis Mortality in the Nineteenth Century Philippines: Data from Parish Records." *Journal of Asian Studies* 38(1): 51–76.
Stewart, Alonzo
  1908 "Agricultural Conditions in the Philippine Islands." In *Senate and House Documents and Reports: Hearings*, 60th Cong., 1st sess., Document No. 535. Washington, DC: GPO.
Stoler, Ann Laura
  1985 *Capitalism and Confrontation in Sumatra's Plantation Belt, 1870–1979*. New Haven and London: Yale University Press.
Stower, H. Gifford
  1924 "Labor for Negros and Luzon." In *Proceedings of the Second Annual Convention of the Philippine Sugar Association as Reported by Arnold H. Warren, 1 to 7 October 1924*. Manila: Philippine Sugar Association.
Stower, H. Gifford, and Santiogo Urquijo
  1924 "Report of the Committee on Labor for Negros and Luzon." In *Compilation of Committee Reports for the Second Annual Convention, 1 to 7 October 1924*. Manila: Philippine Sugar Association.
Sturtevant, David
  1976 *Popular Uprisings in the Philippines, 1840–1940*. Ithaca, NY, and London: Cornell University Press.
Sugar News
  1939a "Larger Planters' Share Urged by President." October, 431.
  1939b "Planter Yulo Advocates 60–40 Share." July, 299.
  1939c "Survey of Muscovado and Other Low-Grade Sugar in the Philippines 1937–1938." May, 198.
  1935a "American Central Becomes Filipino." December, 626–627.
  1935b "Sugar Plantation Laborers Favored by Ruling." June, 296.
  1935c "Planters to Receive P2.40 Benefit Payments." January, 52.
  1934a "The Marketing Agreement" and "Theory and Practice." November, 621–622, 625–626.
  1934b "Eight-Hour Labor Law Extended." April, 229.
  1934c "The Eight-Hour Labor Law." March, 127–128.
  1934d "Sugar Cane Tenancy Bill Becomes Law." January, 52.
  1934e "Eight-Hour Working Law." January, 28–29.
  1934f "Governor General Murphy's Veto Message on Sugar Bill." January, 17–18.
  1933a "When the Manapla Factory Made the Millionth Picul Bag of Sugar." December, 633–640.
  1933b "Twenty Centrals Must Close If Hawes-Cutting Bill is Accepted." February, 72.

References

1932a "P200,000.00 Made Available For Locust Campaign." November, 776–777.
1932b "Petition of Isabela Planters Granted Conditionally." March, 171–173.
1932c "Increasing the Planters' Participation." February, 69.
1931 "Iloilo Chamber President Predicts Crash of Sugar Industry." October, 747.
1930a "Sugar Planters Appeal to Governor General." October, 621–622.
1930b "What is a Planter?" September, 485–486.
1930c "The Great Guessing Contest: Ossorio says Freedom Due in 10 Years." August, 467.
1930d "Central Danao Changes Hands." July, 409–410.
1930e "News from the Planters." June, 332–335.
1929 "Mindoro Sugar Estate to be Sold at Auction"; "Mindoro Sale May Start Big Legal Battle"; "Bank Explains Mindoro Sale"; "Mindoro Sale Postponed by Archbishop"; "New Firm to Buy Mindoro"; "Mindoro Sale is Postponed"; "Sale of Mindoro Estate has been Set for May 13, Wire to Manila Informs"; "Nolting says Postponement Indefinite." May, 372–377.
1926a "San Jose Plantation Now Sold." June, 426–427.
1926b "Items of Interest." June, 421–422.
1926c "Profits in Sugar." April, 261–262.
1926d "Registration of Agricultural Laborers." March, 136–143.
1926e "Port Contract Declared Legal." February, 118–119.
1926f "The Pampanga Central's Lesson." February, 59.
1925 "Farm Tenant Disaffection." April, 181–182.
1924a "Administration Cane and Profits." December, 697–698.
1924b "Sugar Prices." June, 273–274.
1924c "Agrarian Disputes." April, 165–166.
1924d "The Binalbagan Mess." March, 110–111.
1923a "Locust Extermination Campaign Planned." September, 463–465.
1923b "Items of Interest." July, 355.
1923c "Items of Interest." March, 143.
1923d "Protection of Centrals Urged." February, 84–85.
1923e "Items of Interest." January, 45–46.
1922a "Statistics Concerning Negros from the Census of the Philippine Islands for 1918." December, 599–601.
1922b "Proceedings, Second Annual Conference of Sugar Men, October 3, 1922." November, 551.

1922c "Locust Extermination in Negros." September, 443.
1922d "Cane Knives Versus Bolos." August, 365–368.
1922e "Locusts Infest 24 Provinces." July, 339.
1922f "Pulupandan Real Deep Water Port." May, 236–238.
1922g "Are Tractors Too Expensive?" March, 132.
1922h "Muscovado—A Relic of Antiquity." March, 101–102.
1922i "Take a Chance." February, 57–58.
1921a "Can Power Farming be Continued?" May, 162–163.
1921b "Wage Adjustment." March, 85–87.
1921c "Some Facts." January, 7–8.
1920a "Notes from Negros: Concepcion says Negros Will Roll in Wealth." July, 23.
1920b "Do You Know Your Hacienda?" April, 8–9.
1919 "Philippine Sugar Planters' Association." October, 6.

Takahashi, Akira
  1969 *Land and Peasants in Central Luzon: Socio-Economic Structure of a Bulacan Village*. Tokyo: Institute of Developing Economies.

Tambs, Lewis
  1965 "The Inquisition in Eighteenth-Century Mexico." *The Americas* 22(2): 167–181.

Tan, Michael
  1987 *Usug, Kulam, Pasma: Traditional Concepts of Health and Illness in the Philippines*. Quezon City: Alay Kapwa Kilusang Pangkalusugan (AKAP).

Tarling, Nicholas
  1963 "Some Aspects of British Trade in the Philippines in the Nineteenth Century." *Journal of History* 11(3/4): 287–327.

Taussig, Michael
  1980 *The Devil and Commodity Fetishism in Latin America*. Chapel Hill: University of North Carolina Press.

Taylor, W. C.
  1905 "Report of the Third District, Philippines Constabulary, Iloilo, 30 June 1904." In *Fifth Annual Report of the Philippine Commission, 1904*. Part 3. Bureau of Insular Affairs, U.S. War Department. Washington, DC: GPO.
  1903 "Report of the Third District, Philippines Constabulary, Cebu, 11 September 1902." In *Third Annual Report of the Philippine Commission, 1902*. Part 1, Exhibit G. Bureau of Insular Affairs, U.S. War Department. Washington, DC: GPO.

Terwiel, B. J.
   1987 "Asiatic Cholera in Siam: Its First Occurrence and the 1820 Epidemic." In Norman Owen, ed., *Death and Disease in Southeast Asia: Explorations in Social, Medical and Demographic History*. Singapore: Oxford University Press.

Thomas, S. P. R.
   1902 "Life in a Philippine Province." *Philippine Review* 2(6): 209–216.

Thompson, Edward
   1977 "Folklore, Anthropology, and Social History." *Indian Historical Review* 3(2): 247–266.

Timberman, David
   1991 *A Changeless Land: Continuity and Change in Philippine Politics*. Singapore: Institute of Southeast Asian Studies.

Turner, Terence
   1988 "Commentary: Ethno-ethnohistory: Myth and History in Native South American Representations of Contact with Western Society." In Jonathan Hill, ed., *Rethinking History and Myth: Indigenous South American Perspectives on the Past*. Urbana and Chicago: University of Illinois Press.

Turner, Victor
   1974 *Fields, Dramas and Metaphors: Symbolic Action in Human Society*. Ithaca, NY: Cornell University Press.
   1967 *The Forest of Symbols: Aspects of Ndembu Ritual*. Ithaca, NY: Cornell University Press.

Turton, Andrew, and Shigeharu Tanabe, eds.
   1984 *History and Peasant Consciousness in South East Asia*. Osaka: National Museum of Ethnology.

United States of America (United States)
   1941 *Census of the Philippines, 1939*. Vol. 2. Manila: Bureau of Printing.
   1940 *Census of the Philippines, 1939*. Vol. 1, Part 3. Manila: Bureau of Printing.
   1921 *Census of the Philippines, 1918*. Vol. 3. Manila: Bureau of Printing.
   1905 *Census of the Philippine Islands 1903*. Vol. 2. Washington, DC: U.S. Bureau of the Census.
   1896 *Papers Relating to the Foreign Relations of the United States, with the Annual Message of the President, Transmitted to Congress December 2, 1895*. Part 2. Washington, DC: GPO.
   1895 *Papers Relating to the Foreign Relations of the United States, with the Annual Message of the President transmitted to Congress December 3, 1894*. Washington, DC: GPO.

1888 *Papers Relating to the Foreign Relations of the United States, for the Year 1887, Transmitted to Congress, with a Message of the President, June 26, 1888.* Washington, DC: GPO.

Varona, Francisco
1965 *Negros: Historia Anecdotica de su Riqueza y de sus Hombres* [1938]. Translated by Raul L. Locsin. Serialized in the *Western Visayas Chronicle*, June–September.

Velez, Natalio
1925 "Labor Situation on Negros According to Hacenderos' Viewpoint." In *Compilation of the Committee Reports for the Third Annual Convention, 5 to 10 October 1925.* Manila: Philippine Sugar Association.

Vila, Jose, et al.
1906 "Conditions of the Islands, 1701." In B&R, vol. 44.

Villanueva, Hermenegildo
1909 "Report of the Governor of Oriental Negros, Dumaguete 11 July 1908." In *Report of the Philippine Commission, 1908.* Part 1, Exhibit C. BIA. Washington, DC: GPO.

von Scherzer, Karl
1974 "Narrative of the Circumnavigation of the Globe by the Austrian Frigate 'Novara.'" In *Travel Accounts of the Islands (1832–1858).* Manila: Filipiniana Book Guild.

Walker, Herbert
1926 "Negros Letter." *Sugar News* (May): 319.

Welborn, W. C.
1908 "Statement of W. C. Welborn, Bureau of Agriculture." Hearings before the Committee on Ways and Means, House of Representatives, 23–28 January and 3 February 1905. In *Eighth Annual Report of the Philippine Commission.* Part 2, Appendix. BIA. Washington, DC: GPO.

Wernstedt, Frederick
1953 "Agricultural Regionalism on Negros Island, Philippines." Ph.D. diss., University of California Los Angeles.

White, John
1962 "Manila in 1819." *Historical Bulletin* 6(1): 81–109.

White, John
1928 *Bullets and Bolos: Fifteen Years in the Philippine Islands.* New York and London: The Century Co.

Wickberg, Edgar
1965 *The Chinese in Philippine Life, 1850–1898.* New Haven, CT, and London: Yale University Press.
1964 "The Chinese Mestizo in Philippine History." *Journal of Southeast Asian History* 5(1): 62–100.

Williams, Raymond
    1973 *The Country and the City.* London: Chatto and Windus.

Wolf, Diane
    1992 *Factory Daughters: Gender, Household Dynamics, and Rural Industrialization in Java.* Berkeley, Los Angeles, and Oxford: University of California Press.

Wolters, Oliver
    1982 *History, Culture, and Region in Southeast Asian Perspectives.* Singapore: Institute of Southeast Asian Studies.

Worcester, Dean
    1914 *The Philippines Past and Present.* Vol. 1. New York: Macmillan Co.
    1898 *The Philippine Islands and Their People.* New York: Macmillan Co.

Yulo, M.
    1909 "Report of the Governor of Occidental Negros, Bacolod 28 July 1908." In *Report of the Philippine Commission, 1908.* Part 1, Exhibit C. BIA. Washington, DC: GPO.

Zafra, Urbano
    1926 "Digest of the Proceedings of the Fourth Annual Convention of the Philippine Sugar Association." *Sugar News* (October): 740–764.

Zaragoza, Dominador
    1982 *Defiance (The Human Side of the Negros Guerillas).* 2d ed. Quezon City: Businessday Corporation.

Index

AAA (Agricultural Adjustment Administration) benefit payments, 212, 218, 221, 271n. 98
Abrams, Philip, 8
agency, human, 5, 7–9, 10, 153, 231n. 5
Agnus Dei, 40, 170, 236n. 10, 262n. 48
Agricultural Bank: American period, 196–197, 266n. 32; Spanish period, 118
agsa tenancy. *See* sharecropping
Aguilar, Rafael, 16
Alcaldes Mayores (provincial governors), 54, 73, 85
Alcina, Francisco, 55, 69, 72, 73
American colonial state: centrifugal sugar mills, 197–203, 211–212; corporatism, 189–191, 197, 200; and gambling, 191–194, 200; Philippine independence, 209–214; revival of sugar industry, 194–198
anting-anting: in cockpit, 48; of foreigners, 186–187; of *hacenderos*, 161, 184, 186, 223; Latin phrases, 51–52; and resistance, 72, 168. *See also* shaman

Araneta, Juan, 106, 171–178, 180–181, 186, 196, 262n. 54, 263n. 58
Augustinians, 16, 165–166
Ayuntamiento de Manila, 84–86

babaylan. *See* shaman
Bank of the Philippine Islands (BPI), 196–197, 266n. 35
Basco y Vargas, Jose, 81–82
Bataan, 242n. 24
beet sugar, 120, 198, 267n. 40
Benedicto, Teodoro, 104, 110–113, 171, 248nn. 39, 49
bet: in gambling, 47–48, 50, 74, 109–110; in society, 60–62, 192, 214, 215
Bohol, 39, 40, 234n. 30, 237n. 24
Bourdieu, Pierre, 80, 240n. 13
Bukidnon, 38, 64, 237n. 32
Bulacan, 84, 242n. 34, 268n. 53
Bureau of Agriculture, Philippines, 194–195, 197–198
Bureau of Insular Affairs, U.S., 198–199, 201, 204
Butuan, 38, 44
buwis, 66–67, 75, 241n. 21

cabeza de barangay (village head), 54, 57, 59, 71, 243n. 43, 246nn. 11, 17
capital-as-evil, 11, 15, 21–22. *See also* devil
cash advance: crop loan system, 196–197, 266n. 35; on production, 25, 70, 76, 107–108, 141–143; to workers, 130–131, 134, 136, 138–139, 223
Cavite, 77, 242n. 34
Cebu: Chinese trade, 98; natives/friars, 16, 38, 39, 40, 80; revolution, 172; spirit-beings 34, 38; sugar production, 24, 84, 242n. 34; Visayas government, 118
cedulas personales, 129, 137, 143, 157, 255n. 31
Chinese, ethnic: anti-Chinese, 16, 21, 60, 224, 244n. 55, 267n. 41; child immigrants, 91, 244n. 54; domestic trade, 84, 97–99; and gambling, 147, 193, 240n. 17; mixed marriages, 59–60; Negros residents, 101; opium farmers, 247n. 28; and sharecropping, 76, 241nn. 23–24; sugar experts, 85, 242n. 32
Chinese mestizo: and *babaylan,* 168; as legal category, 57, 60, 71, 105, 160, 189; middlemen-traders, 21, 26, 62, 78, 84; migration to Negros, 98–100, 246n. 11; moneylending and sharecropping, 78–82, 87–90; nationalist imagination, 105, 161; non-traditional aristocracy, 16, 60–61, 80, 90; vs. Spanish planters, 102, 104–106, 138–139; vs. Spanish rule, 162–164, 169–170. See also *hacendero;* Mestizo Power
Chirino, Pedro, 38, 54
cholera: 1820 epidemic and massacre, 15–16, 20–22, 61, 86, 232n. 2, 243n. 39; other epidemics, 31, 166–167, 231n. 1
cockfighting: cosmic-colonial meanings of, 47–50, 236nn. 19–21; money to bet, 79, 241n. 20; social categories in, 106, 146–147, 149; Spanish regulations, 49, 52–53, 72, 236n. 21, 237n. 22, 258n. 71; under U.S. rule, 192–193, 265n. 13; and World War II, 214
colonial state. *See* American colonial state; Spanish colonial state
colonias agricolas, 91
Concepcion, Venancio, 202–203, 205, 213
Consejo Real (Royal Council), 2–4, 90–92
Constantino, Renato, ix, 6, 65–66, 182, 208
Corpuz, O.D., 239n. 2
criminality, 132, 133 table 4, 134 table 5, 253nn. 16–17, 255n. 41
Cuba: American sugar interests in, 204, 206; exiles from, 91, 245n. 56; Philippine sugar compared with, 83, 85, 120–124, 198
Cuenca, Fernando, 158–160
Cuesta, Angel Martinez, 102, 158
Cullamar, Evelyn, 6, 182
cultural ambivalence: amid power encounters, 46–47, 53, 58–59; imitative behavior, 46, 60–62, 105, 107–108; and manipulation of spirit-beliefs, 176, 187; transcendence of, 223

datu: and cockfighting, 48; *hacenderos* act like, 160, 163, 173–176, 185, 191, 223; impact of conquest on, 44–45, 54, 57–59, 69–70, 238n. 40, 240n. 13, 241n. 21; preconquest, 29–30, 54–57, 59, 64–67, 69–70, 234n. 30, 235n. 32, 237nn. 31–32; U.S. treated as, 179. See also man of prowess
debt: and abscondment, 85, 89–90, 129, 131, 180, 253n. 14; as appeasement, 130–135, 235n. 32; 5-pesos limit, 70, 88, 92, 131; on friar estates, 76, 79; of the inside, 89–90; peonage, 69, 240nn. 11–12, 16; precolonial, 64, 68–69, 239nn. 7, 9; and sharecropping, 5, 75, 79,

89–90, 142–145, 241n. 21; of sugar capitalists, 195–196, 202, 205, 269n. 66
de Comyn, Tomas, 71, 83, 87, 147
de la Gironière, Paul, 186–187, 237n. 27
de la Rama, Esteban, 223–224
de la Rama, Isidro, 110, 159–164, 169, 186, 187, 223
de la Viña, Diego, 183, 184, 264n. 76
de la Viña, Jose, 183–186, 223
de Mas, Sinibaldo, 83, 232n. 6, 261n. 42
Demetrio, Francisco, 33, 40, 41, 42
de Morga, Antonio, 54
de San Agustin, Gaspar, 39
devil: imagery, 11; and Latin, 51; science of the, 21; selling soul to, 73, 161–163; *yawa* as, 161–163, 223
dungan: bravery and, 53, 56; of Chinese mestizos/*hacenderos*, 61, 110, 117, 160–161, 173, 175, 186, 210, 215, 228; collective, 44, 45, 170; debt peonage and, 69–70; of friars, 34, 39, 41, 165, 234n. 26; Ilongot notion compared, 234n. 27; of natives, 28–29, 30, 55, 130, 234n. 27; and power encounter, 28–29, 39, 89–90, 236n. 16; and sharecropping, 89–90; and *usug,* 28–29, 236n. 16
Dutch East Indies, 83, 91, 232n. 1, 242n. 27, 243n. 44, 254n. 22

Echauz, Robustiano, 126, 129, 138, 141, 154
economic protectionism, 24, 85–86, 156, 198–199
electoral politics, 189–190, 192, 193, 215, 236n. 17
encomienda, 238nn. 35, 41, 240n. 16
engkanto: burglary by, 145; diseases caused by, 34, 40, 41, 44, 45; friars as, 33–34, 58; and Latin, 51; of Mount Kanlaon, 225–226; native attraction to, 42, 170; offspring of, 43, 236n. 14; as source of power, 170; and spiritual conquest, 33–38

Estevanez, Ramon, 130, 135–136, 254n. 26
exclaustración, 23
extrangeros (foreigners): and nationalism, 205, 223; Spaniards as, 87; Spanish images of, 15, 17, 118, 181, 232n. 6, 233n. 12. *See also* foreign capital; foreign merchants

Fabian, Ann, 11
family: of Americans, 270n. 90; and Bank Centrals, 205; and *barangay,* 65, 239n. 2; *engkanto* as, 33; invention of, 57–58, 61, 219, 238n. 39; and sugarcane cultivation, 127–128, 141, 144, 145, 153, 219
firearms: 138, 139–140, 185, 191, 255n. 32
Forbes, Cameron, 196–197, 199–200, 201, 267nn. 40, 43
foreign capital: American period, 197–202, 204–208, 214, 268nn. 53, 55, 271n. 91; Spanish period, 85–86, 93, 105–108, 117–125, 245n. 62. *See also* foreign merchants
foreign merchants: allies of *hacenderos,* 105, 117–125; contest with Friar Power, 25–26, 61–62, 162; inquisitorial labels on, 17, 21; presence in Spanish colony, 20, 25, 61, 156–157. See also *extrangeros;* foreign capital
Foreman, John, 107–108, 130
Freemasonry. *See* Masonry
Friar Power: encounter with indigenous spirit-world, 39–43, 70; foreign merchants vs., 25–26, 30, 61–62, 162, 233n. 12; as gluttony, 162–163; *hacenderos* vs., 159–165, 171–176, 223; holy water and, 39, 61; imitation-cum-subversion of, 50–51, 71–72, 165–170 (*see also* shaman); sacraments and, 38, 40. *See also* friars
friars: art of domination, 16–17, 22, 32, 157, 158; concubinage, 37, 42–43, 50, 159; in *datu*'s role, 76–77;

economic control, 58, 76–77, 162–163; as *engkanto*, 33–34, 42–43, 58; friar estates, 76–79, 199, 233n. 12, 235n. 4, 240n. 15, 241n. 22; and gambling, 73; and native labor, 128, 139–140, 157; personal rule of, 34, 53–54, 58, 235n. 1, 238n. 34; as prisoners, 173, 263n. 62; as shamans, 38–40, 42, 236n. 11. *See also* Augustinians; Friar Power; Recollects

gambling: capital accumulation impaired by, 107, 109–110, 212, 218; and debt, 79, 89–90, 107–108, 202–203; to entice labor, 146–149, 154–155, 257n. 65; historic events as, 176–180; and intermingling of categories, 105–107, 117, 149–154, 228, 247n. 28; manipulation of spirit-beliefs as, 176, 185–188; manipulation of state apparatus as, 59, 73, 200–204; and money, 74, 109, 139, 146, 148–149, 154, 241n. 20; as passion, 53, 257n. 64, 258n. 79, 265n. 24; price speculation, 108–110; regulations concerning, 109, 147–148, 191–194, 200, 272n. 105; as relationship to colonial power, 46–47, 72, 240n. 17; as worldview, 72–74, 106, 109, 200. *See also* bet; cockfighting
Gaston, Ives Leopold de Germain, 100, 138
geographic mobility (of natives): end of, 219–220; *indocumentados* (undocumented migrants), 128–129, 137, 139; internal passport system, 87, 128, 243nn. 43–44; proposed measures in Negros, 136–138, 272n. 109; weak state control of, 71, 81, 85, 146. *See also* debt
Giralt, Fernando, 137–138
gobernadorcillo (town magistrate): individuals, 112, 166, 171, 246n. 11, 251n. 77; position, 57, 58, 110, 238nn. 34, 36, 40

Godelier, Maurice, 67
Guardia Civil (Civil Guards): formation of, 92, 157; and fugitives, 139–140, 255n. 41; and gambling, 148; and revolution, 167–172
Gurrea, Teodoro, 112, 248n. 37

hacendero (sugar planter): akin to *datu*, 160, 163, 173, 175–176, 185; Americans as, 251n. 76, 270n. 90; antistate strategies of, 106–107, 120, 128–130, 138–141, 148 (*see also* gambling); bankruptcy of, 195–196; capital accumulation mystified, 161–164, 188, 223–226 (*see also* inquisitorial labels); class dominance, 4–5, 127–138, 185–188, 215, 218–219, 223–228; compared with *hacendado*, 116–117; leaseholders as, 153, 213, 271n. 100; new Negros elite, 99, 102, 104–106, 127–129, 246n. 11; no agricultural expertise, 108–109, 211, 247n. 31, 248n. 32; as revolutionary, 170–180, 190; rivalries, 108, 113, 198, 210–211; Spaniards as, 100–102, 104–106, 110–113, 115, 118, 139, 214, 272n. 104; struggle with sugar mills, 209–215. *See also* Chinese mestizo; Mestizo Power
hacienda: direct administration of, 126–127, 220; labor recruitment, 4, 126–135, 146–149, 215–223, 225–226, 254n. 19; settlement pattern, 149–153, 258n. 76. See also *hacendero;* sharecropping
Hawai'i, 198–199, 202, 268n. 55
Henry, Yves, 211
Hindu world, 35
Holy Week, 17, 45, 108, 165, 174, 184

Ileto, Reynaldo, 5, 107, 231n. 3
Iloilo: Chinese mestizos, 78, 97–99, 127, 246n. 11; criminality, 132; merchant houses, 119–121, 124, 143; port, 4, 9, 12, 93, 98, 100,

231n. 5, 266n. 28; sugar industry, 24, 120–121, 196, 242n. 34; textile industry, 78, 97–100; workers from, 127–130
indio: conflict with Chinese mestizo, 60–61, 80, 90 (*see also* Chinese mestizo; Mestizo Power); domination by friar, 16–17, 22, 32, 157, 158 (*see also* Friar Power; friars); as individuated peasant, 69–72, 75–77, 81–82, 129, 145–146 (*see also* cash advance; debt; geographic mobility); indolence, 87, 99 (*see also* labor recruitment; race); marriage with Chinese, 59–60; submission-cum-resistance, 12, 38–43, 45, 48, 50–53, 60, 226, 237n. 24 (*see also* cockfighting; shaman)
infanticide, 73, 240n. 18
Inquisition, 17–18, 36–37
inquisitorial labels, 17–18, 20–22, 25, 163, 181
Inspeccion General de Montes, 112
Insular Auditor, 197
interest rates (on loans), 85, 107, 118–119, 195–197, 202

Jagor, Fedor, 101

Kanlaon, Mount, 27, 41, 165, 174, 184, 224–226
Kitching, Gavin, 8
Kusug Sang Imol, 191, 204, 225

labor recruitment: and fair treatment, 89, 135, 144, 254n. 20; illegal, 128–129, 139–40; problems of, 126–135; proposed measures on, 136–138; worker registration, 136. See also geographic mobility; sharecropping; wage labor
Lacson, Aniceto, 172–173, 189–190
Laguna, 77, 199, 242n. 34
land, agricultural: colonial elite's acquisition of, 75, 110–116, 241n. 21, 244n. 51, 248nn. 36, 38, 250n. 62; as commodity, 71, 74–75, 240n. 14; foreign ownership, 91, 118, 198, 251n. 76, 267nn. 40, 45; friar estates, 76–80, 82, 199, 233n. 12, 235n. 4, 240n. 15, 241n. 22; *interdicto de despojo,* 110, 113; Latin America compared, 116, 147, 250n. 67, 257n. 67; leaseholding, 76, 213, 271n. 100; *pacto de retroventa,* 250n. 62; preconquest, 66; valuation, 115–116, 196, 202, 266n. 33. See also sharecropping
Larkin, John, 4, 182, 191, 238n. 36
Latin, 41, 51–52
law, 73, 75, 88, 110, 136–138, 209
Leyte, 38, 45, 172
Locsin, Carlos, 109, 210, 216, 218, 248n. 33
Loney, Nicholas: credit facility, 107, 117; and *hacienda* labor, 92, 127–129, 138, 253n. 12; historic role, 5, 93, 99–100, 231n. 2; view of Chinese mestizos, 98, 100, 108–109; view of Spaniards, 105, 247n. 26
Lopez, Eugenio, 166, 251n. 76, 258n. 76
Lopez, Vicente, 271n. 94
luck, 11, 51, 62, 72–75, 240n. 19. See also bet; cockfighting; gambling
Luzuriaga, Eusebio Ruiz de, 100, 246n. 17
Luzuriaga, Jose Ruiz de, 104, 159, 172, 176
Lynch, Frank, 5, 29

Magos, Alice, 28, 35
Mainawa-on, 191, 225
Mallat, Jean, 106, 247n. 28
man of prowess: Big Man, 57, 67–68, 237n. 31; Big People, 5, 29, 64, 69; bravery, 29, 53, 55–56, 187; *maayo nga lalaki,* 53, 168, 173, 237n. 27; *mabuting tao,* 53, 237n. 27; *magaling na lalaki,* 53, 237n. 27; *malhechores* (evildoers), 53, 139; in Southeast Asia, 30. See also *datu; dungan*
Mapa, Dionisio, 264n. 6
Marx, Karl, 8
Masonic Capitalism, 11–12, 30, 61, 162–165, 185–186

Masonry, 17–20, 62, 159–165, 223–224, 260nn. 25–27
McCoy, Alfred, 4–5, 107, 164, 182, 231n. 2, 246n. 11
McIntyre, Frank, 204–205
Melanesia, 237n. 31
Mestizo. *See* Chinese mestizo; Spanish mestizo
Mestizo Power: and deception, 185–188; terror of, 183–185; under U.S. rule, 192, 223–228; victory over *babaylan*, 180–182, 188; victory over Friar Power, 174–176
Mexico, 17–19, 21, 45
Mindanao, 33, 44, 54, 171, 179, 234n. 30, 237n. 32
Mindoro Island, 198–199, 267n. 43
Ministerio de Fomento, 23
Ministerio de Gracia y Justicia, 23
Ministerio de Ultramar, 23, 91, 120, 124, 147, 158
Molto, Remigio, 118, 138–139, 200, 251nn. 73–74
Money: circulation of species, 258n. 74; and magical beliefs, 74, 161–164, 167, 223–225; precolonial, 241n. 25; vs. wages in kind, 134–135, 139. *See also* gambling
Montelibano, Alejandro, 250n. 64
Montilla, Agustin, 100, 246n. 16
Montilla, Bonifacio, 110

Nagano, Yoshiko, 231n. 1
Nash, June, 11
nationalism, 101, 105, 170, 198, 201–209, 211, 267n. 41
National Sugar Board, 148
Negros Island: civil government, 264n. 6; constitution, 189–190; as emporium of wealth, 117–118; Federal Cantonal government, 178–180, 190, 263n. 62; Negros Occidental provincial government, 190–191, 193, 195–196; Negros Oriental provincial government, 190; population, 127, 220; provincial subdivision, 92; Provisional Revolutionary Government, 173, 178–179, 184

Negros Occidental towns: Bacolod, 104, 115, 158, 172, 181, 193; Bago, 172, 174, 224; Binalbagan, 43, 204, 211, 236n. 11, 271n. 99; Cabancalan (Kabankalan), 139, 168–169, 250 n. 64, 267n. 44; Cadiz (Nuevo), 149, 251n. 76; Cauayan, 193; Escalante, 193, 267n. 44; Ginigaran, 104, 159; Ilog, 169, 193–194; Isabela, 169, 180, 193, 211–212, 258n. 76, 266n. 32; Jimamaylan, 102, 169, 172, 180; La Carlota, 91, 102, 110–112, 149, 173, 194, 224, 271n. 99; La Castellana in Isabela, 181, 184, 224–225, 248n. 39; lists containing, 103 table 1, 150–151 table 6, 152 table 7, 246n. 20, 247nn. 22, 24; Manapla, 149, 268n. 53; Minuluan (Talisay), 102, 128, 149, 152–153, 158, 160, 191, 255n. 41, 267n. 44, 271n. 99; Pontevedra, 111–112, 115, 172, 244n. 51, 251nn. 76–77; San Enrique, 251n. 77; Saravia (E.B. Magalona), 102, 104, 113, 138, 158, 250n. 64; Silay, 102, 148, 172, 201, 224, 255n. 41, 258n. 76, 268n. 55; Victorias, Nuestra Señora de las, 1–2, 113, 224, 268n. 53
Negros Oriental towns: Bais, 102; Dumaguete, 183–184; lists containing, 104 table 1, 246n. 20, 247nn. 22, 24; Manjuyod, 170; Nueva Valencia, 159; Siaton, 167; Tayasan, 159; Tolong, 168; Zamboanguita, 167
Nueva Ecija, 248n. 36

Ong, Aihwa, 11
opium, 247n. 28, 251n. 77
Ossorio, Miguel, 211, 268n. 53–54; 271nn. 91, 97
Otis, Elwell, 190

Palawan, 48
Pampanga: cockpits, 147; *encomienda*, 238n. 35; farm sizes, 250n. 66; *gobernadorcillo*, 238n. 36; model farm, 91; resident Spaniards, 101; share-

Index   311

cropping, 220–221; sugar mills, 202, 268n. 55; sugarcane production, 93, 109, 242n. 34, 248n. 32; wage rates, 220, 272n. 108
Panay Island: *babaylan* resistance on, 165–168, 170, 224; Guimaras Strait, 128, 146; revolution, 172; workers from, 222. *See also* Iloilo
Pangasinan, 93, 242n. 34
Papa Isio. *See* shaman
paternalism, 64–65, 76, 141–145, 175–176, 187, 219
Peele, Jonathan, 24
Perez, Miguel, 106, 248n. 37
Phelan, John, 42
Philippine Assembly, 190–191, 196
Philippine Commission, 190, 192, 198
Philippine Constabulary, 180, 182, 193, 222, 224
Philippine Legislature, 198–200, 205–206, 266n. 26, 271nn. 92, 98
Philippine National Bank (PNB), 197, 201–206, 209, 211–213, 270n. 84
Philippine Sugar Association, 109
Philippine Sugar Centrals Agency, 203
Pigafetta, 48
Postal Savings Bank, 204
power, 26, 49, 54–55, 57–59. *See also* power encounters
power encounters: in class relations, 89–90, 215; cosmic struggles, 26–28, 30, 36, 47–49; friars and capitalist merchants in, 26, 30–31, 61–62; friars and local spirits in, 40–44, 166–170; interpersonal relations as, 28–29; natives amid, 44–47, 52, 178, 228; status hierarchy and, 29–30. See also *dungan; engkanto;* Friar Power; Mestizo Power
Primo de Rivera, 157
Puerto Rico, 83, 120, 123–124, 198

Quezon, Manuel, 211

race: American views of natives, 190–191; and class-caste categories, 17,
99, 101; and *hacenderos,* 97–102, 105–106, 160–161, 214; native claim to equality, 105–106, 160–161, 168; Spanish views of *indio,* 90–92, 142, 168, 245n. 57, 248n. 38, 261n. 42
Rafael, Vicente, 56, 68, 234n. 29
Real Sociedad Economica de Amigos del Pais, 84–86, 113, 245n. 57
Real Tribunal de Comercio de Manila, 84–88
Recollects, 140, 156–159, 164–165, 171, 199
reducción, 45
reducción de infieles, 157–158
relations of production: definition, 7; friar dominance, 76–77; preconquest, 65–67, 70; sharecropping, 75, 79–82, 88–90, 135, 243n. 46; wage labor, 215–223. *See also* sharecropping; wage labor
Rizal, Jose, 49, 60, 164, 169, 224, 260nn. 25–26
rosary, Catholic, 145, 161
Rubin, Lucas, 111, 248n. 37, 250n. 64
rural settlement pattern, 45–46, 149–153
Russell and Sturgis, 25–26, 107–108, 187, 251n. 76

sacada, 128, 222–223. *See also* labor recruitment
Sahlins, Marshall, 68
Samar, 38, 237n. 24
San Juan del Monte, 43–44
Santayana, George, 25, 233n. 21
Sayer, Derek, 7
Schurman, Jacob Gould, 190–191
Scott, James, 76
Scott, William Henry, 68, 239nn. 7–8, 240n. 11
sentencia, 49–50, 108, 113. *See also* cockfighting
Severino, Melecio, 264n. 6
shaman: Bangotbanwa, Estrella, 166, 261n. 32; Buhawi, Dios (Ponciano Elopre), 139, 167–168; *catalona,* 27, 44; challenged colonial order,

44, 50–52, 59, 71–72, 165–170; and Friar Power, 6, 42, 45, 51, 71, 165–170, 261n. 32, 262n. 48; friars as, 35–41; gender of, 27, 30, 50; Intrencherado, Flor, 224–225; Juan Perfecto, 262n. 48; Kamartin, 168; and Mestizo Power, 224–225; Papa Isio (Dionisio Papa), 168–169, 180–184, 186; precolonial, 27, 30, 235n. 31, 237n. 23; *pulahanes*, 48; and U.S. colonialism, 181–184. *See also* man of prowess

sharecropping: to access native labor, 87–90, 241n. 26; *agsa* tenancy beginnings, 79–80, 89; on friar estates, 76, 78, 241n. 22; *kasamahan*, 78, 80–82, 89; mestizo and *indio* in, 78–82, 130; outlawed, 81–82; and peasant autonomy, 60–72, 75–77, 81–82, 87, 89–90, 145–146, 257n. 59; on sugar *haciendas*, 126–128, 134–135, 141–146, 154–155, 162, 215–222, 256n.53, 272n. 107; sugar tenancy law, 221; transition to, 67, 75–77. See also *buwis*

Shiraishi, Takashi, 11

Smith, James, 189–190

social control, 132–135, 143–146. *See also* geographic mobility

sorcery, 188, 234n. 30

Sota, 41, 173–174, 177–178, 185, 192, 195, 226, 262n. 56

Southeast Asia, 4, 30, 58, 88, 140, 148, 186, 189, 190, 236n. 19

Spaniards: capitulation in Negros, 172–173, 176–177, 183–184; and centrifugal sugar mills, 206–208, 267n. 44, 272n. 104; creole vs. peninsular, 101–102; effects of foreign trade on, 20, 118–119, 156–157, 251n. 74; *hacenderos*, 82–83, 100–106, 110–112, 115, 118, 251nn. 73, 77; piety, 36–37. *See also* Friar Power; friars; race; Spanish colonial state

Spanish colonial state: agricultural policies, 82–86, 90–92, 117 (*see also* sugarcane production); and banditry, 86–88 (*see also* geographic mobility); commercial policies, 16, 19, 22–24; Filipino elites act like, 178, 199; fragmentation of, 139–140, 148, 157, 165; neutrality vis-a-vis *hacenderos*, 90, 92, 135–138; subsidy to Spain, 243n. 36; U.S. defeat of, 171–172. *See also* friars; Friar Power

Spanish mestizo, 34, 43, 159, 183, 233n. 21

spirit beings: *anito*, 27, 40–41, 44; Bathala Meycapal, 234n. 22; *bibit*, 46; *diwata*, 27, 38, 41, 72; *dwende*, 33, 35, 73; *kama-kama*, 36; *kapre* (*Cafre*, kaffir), 33–36, 40, 235n. 4; Laon, 27, 41, 174, 234n. 22; Macaptan, 27; Maka-ako, 226–228; *maligno*, 33, 35; *multo, murto*, 33, 35–36; *nono* (crocodile), 43, 50; *papu*, 29, 36, 59, 166, 174; *santilmo*, 33, 35; Santo Niño, 38, 41, 74, 236n. 12; *sigben*, 36; *sirena*, 33, 35; *tag-lugar*, 33, 35; *tamawo*, 225; *tikbalang*, 36, 46; *yawa*, 161–163, 223. See also *engkanto;* Sota; spirit-world

spirit-world: analysis of, 10; *conquista espiritual* (spiritual conquest), 33–38, 40–43, 70, 169; and *hacenderos*, 161–164, 173–176, 186–188, 223–224, 227–228; indigenous beliefs in, 26–30, 235n. 32, 236n. 16; individual strategies to relate to, 71–75, 81, 145–146, 161; *orasyon*, 51; preconquest vis-a-vis Hispanic, 36–38, 169–170, 177–178; and resistance, 44–46, 48–49; structure of, 27, 35–36, 56, 225–226; and sugar factories, 1–2, 215, 225–226. See also *anting-anting;* shaman; spirit beings

Sturgis, George, 25–26, 30–31, 61–62, 159, 163, 187

sugarcane production: locusts, 27, 195, 266n. 26; mill-*hacienda* crop division, 199, 209–215, 271nn. 96, 99, 272n. 103; in Negros, 98–102, 103–104 table 1, 117, 194–195,

206–208, 210, 216; pre-1860s, 76, 82–86, 90–92, 242nn. 32–33; rinderpest, 146, 194; and spirit-world, 145, 225–226; U.S. limitation of, 212, 271nn. 93, 98. See also *hacendero*
Sugar Central Board, 200–201
sugar mills: Bank Centrals, 200–206, 209, 211, 213, 268n.46, 269n. 66; centrifugal, 197–199, 201, 210, 267nn. 43–45, 268nn. 53–57; foreign staff, 203–204, 269n. 71; muscovado, 85, 117, 142–144, 199, 206, 209–210, 250n. 71, 256n. 56; and spirit-world, 1–2, 215, 225–226; and sugar planters, 209–215, 271nn. 96, 99; wage labor in, 220, 222
Sugar News, 109, 204, 210, 215
sugar trade: British policy, 82; China market, 124, 197, 267n. 39; commercial agents, 118–120, 124, 160, 201; exports, 20, 24–25, 84, 100, 121–123 table 2; Iloilo port, 4, 9, 12, 93, 98, 100, 231n. 5, 266n. 28; and Philippine independence, 209, 214, 218, 271nn. 92–94, 272n. 106; prices, 100, 108, 201, 202, 205, 214–215, 243n. 38; Pulupandan port, 195, 266n. 28; Spanish policies, 83, 120–124; and spirit-world, 108, 225–226; U.S. market, 120–121, 123–124, 180, 197–199, 206, 209, 214, 271n. 98
Sultan Kudrat, 234n. 30

Taft, Howard, 198
Taussig, Michael, 11, 22
Tejido, Domingo, 106, 248n. 37
Thompson, E.P., 10, 68–69
Tondo Province, 16, 77, 84, 92, 242nn. 32, 34
Tovar, Antonio, 119, 126, 131, 140, 142–143
Turner, Terence, 10

United States: colonialism, 171, 178–182, 190; Congress, 198–199, 209, 214, 267n. 41, 272n. 106; gambling in, 248n. 34; sugar market, 120–121, 123–124, 180, 197–199, 206, 209, 214, 271n. 98. See also American colonial state

Varona, Francisco, 109, 113–115, 159–161

wage: in kind, 134–135, 139; rates, 135, 138–139, 220, 245n. 57, 255n. 37, 272n. 108; and spirit-world, 225–226
wage labor: *duma-an*, 155, 188, 219; intermediate forms, 218; problems of plantation, 4, 82, 87–89, 127–135; unemployment, 220; "unfree" form, 219–221, 226
Wickberg, Edgar, 21, 59–69, 258n. 71
Williams, Raymond, 11
Wolf, Diane, 11
Wolters, Oliver, 30
women: and friar concubinage, 37, 42–43, 50, 159; and gambling, 53, 236n. 21; and marriage, 59–60, 63, 69, 238n. 38–39, 239n. 8; precolonial gender system, 30, 50, 236n. 13; shamans, 27, 30, 237n. 23; spiritual arbiters, 42–43, 50, 73; wage rates, 220
Wood, Leonard, 204–206, 270n. 84
Worcester, Dean, 130–131, 146, 186
World War II, 214, 222, 270n. 90

Ynchausti y Compañia, 119, 184

Zambales, 40, 45
Zamora, Carlos, 171

About the Author

Filomeno V. Aguilar, Jr., teaches in the School of History and Politics at James Cook University in Queensland, Australia. He was educated at Ateneo de Manila University (B.S.), the University of Wales (MsC), and Cornell University (Ph.D), where he was awarded the 1992 Lauriston Sharp Prize. This is his first book.